OFFSHORE ENGINEERING

ELECTRICAL

VOLUME 2

Michael J. Dennis

Published by New Generation Publishing in 2012

Copyright © Michael J. Dennis 2012

First Edition

www.newgeneration-publishing.com

New Generation Publishing

About This Book

This book, in two volumes, has been produced for engineers and technicians in the electrical discipline who work, or wish to work, in the offshore oil and gas industry.

The book aims to provide all of the knowledge elements required by electrical engineers and technicians who will be assessed as authorised electrical persons, to work on offshore rigs and production platforms. Naturally the specific systems and company procedures for such assessment can only be learnt on the job.

Knowledge is provided from elementary electrical theory, power generation and distribution, electrical systems and units, safety, protection and commissioning; covering all aspects of electrical engineering offshore. Where it is considered helpful, reference is made to the differences or similarities to onshore systems. All parts of the book are fully illustrated.

Self assessment questions and answers are provided at the end of each section, to ensure knowledge gain and retention.

A comprehensive contents list is provided at the start of the book as well as an index at the end. Thereby the two volumes become an ongoing reference tool.

The author spent 24 years in the offshore oil and gas industry. He has worked offshore as an electrical engineer on production platforms, and has been responsible for the electrical training of engineers and technicians from many drilling and production companies operating in the UK and Norwegian sectors of the North Sea, the Middle East, Africa and the Far East. He therefore has extensive knowledge of electrical engineering gained on the job, as well as from writing training programmes, lecturing and training and assessment of engineers and technicians throughout the offshore production and electrical disciplines.

Contents
Volume 2

PART 5 ELECTRIC MOTORS

CHAPTER 1 TYPES OF A.C. ELECTRIC MOTORS

1.1   GENERAL

By far the greatest consumers of electric power on an offshore or onshore installation are electric motors. In providing mechanical power to their driven loads they draw electric power from the system - and this means active power (kilowatts). However, in addition to the mechanical power converted from the electrical, the motor, like any other machine, suffers 'losses': that is to say, some power is consumed within the motor by friction, windage and internal heating. Such losses are not passed on to the mechanical drive, but the energy for them must nevertheless be drawn from the electrical system. Therefore a motor will always give out less mechanical power than the electrical power which it draws; its output must then be less than its input. The ratio $\dfrac{\text{output}}{\text{input}}$ is called the 'efficiency' (symbol η) and is therefore always less than one. It is usually expressed as a percentage.

Mechanical output was formerly (and often still is) expressed in horsepower, but with SI units it should be expressed in kilowatts (mechanical). One horsepower equals 0.746kW, or approximately ¾kW (a useful rule-of-thumb). Many nameplates and tags, however, will still be found to be marked in horsepower.

Electrical power is always expressed in watts or kilowatts (electrical). If it is necessary to distinguish between mechanical and electrical kilowatts, a suffix 'm' or 'e' should be used. Thus:

$$\text{efficiency } (\eta) = \frac{kW_m}{kW_e}$$

Example: A 100 hp motor draws 88kW$_e$ of electric power.

$$100 \text{ hp} = 75kW_m \qquad \therefore \ \eta = 75/88 = 0.85$$

$$\text{Efficiency} = 85\%$$

Efficiency is not constant but varies with the mechanical loading on the motor. It is normally highest at full load, falling off rapidly as loading is reduced.

Apart from the active power drawn by the motor to convert to mechanical power, most motors also draw reactive power (kilovars) to magnetise themselves. This results in a mixture of active and reactive power entering the motor, showing as a power factor less than unity. The matter of reactive power and power factor of a motor is dealt with in Chapter 4.

1.2   SYNCHRONOUS MOTOR

The synchronous motor is an important type because of its unique performance. It is relatively costly, and its design is complex. For that reason it is not used on offshore installations, and even onshore only in special applications.

In construction it is identical with the synchronous generator which is described in Part 2 'Electrical Power Generation' and used to generate power on all platforms. A synchronous **generator** is driven by a prime mover (steam or gas-turbine or diesel). It converts the mechanical power received into electrical power which it pumps into the network. But that same machine, if uncoupled from its prime mover and coupled to a mechanical load (such as a brake, pump or compressor as shown in Figure 1.1) will, without switching, continue to run in synchronism with the a.c. electrical supply. It will then behave as a motor however, drawing power from the mains and converting it to mechanical power which it delivers to the load. Moreover since it runs in synchronism with the a.c. mains, it must run at constant speed, whatever the loading. It is this constant speed facility which singles out the synchronous motor from all other types and makes it suitable, despite the cost, for drives which demand exact constant speed.

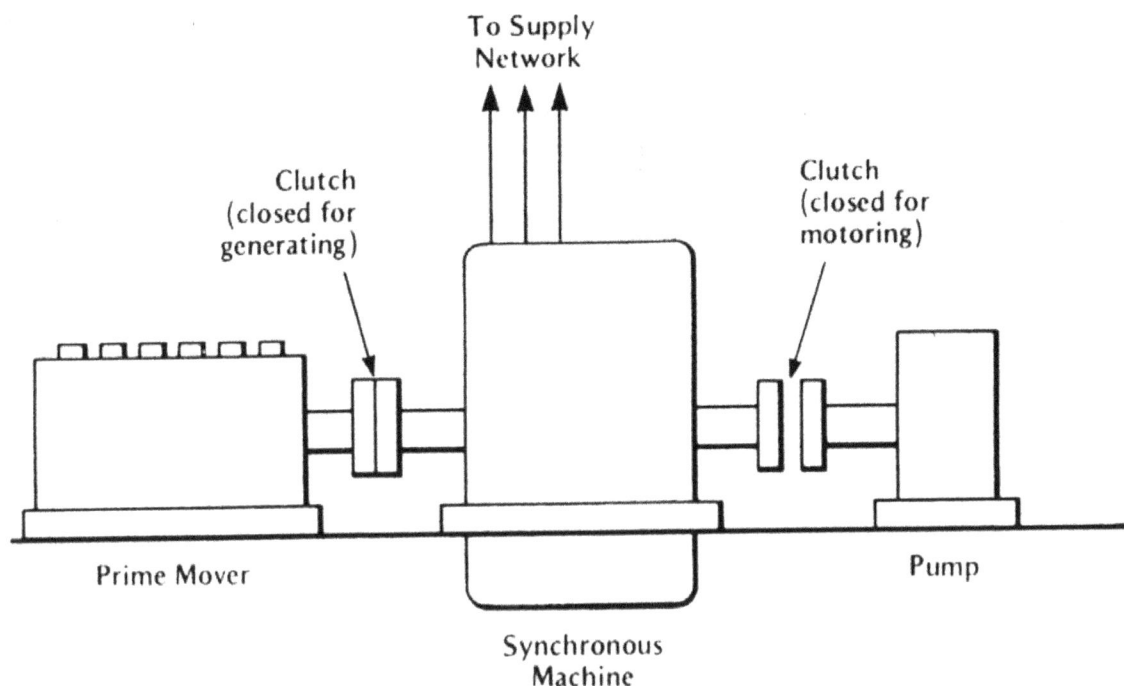

FIGURE 1.1
SYNCHRONOUS MACHINE AS GENERATOR OR MOTOR

Another advantage is that, since the motor draws its excitation from a separate exciter and not from the mains, it is possible, by controlling the excitation, to run a synchronous motor at unity power factor so that it draws no reactive power, and all its current contributes to useful work. This can be a great advantage to a network system where there are other large induction motor drives with their heavy reactive power demands. Indeed, it is not uncommon to run a large synchronous motor at a small leading power factor by over-exciting it, in order to correct for other lagging loads and to maintain system voltage.

It is also possible to use such a motor as a 'synchronous condenser'. In this application it drives no mechanical load, but it is over-excited so that it generates only reactive power. Its use in this mode for system power factor correction and maintaining system voltage is explained in Part 6 'System Control'.

A synchronous motor must, like the generator, have independent excitation, which may be brushless. It must also be provided with starting arrangements which will run it up to speed so that it may be synchronised to the system before driving its load.

Since there is no difference in construction between a synchronous generator, a synchronous motor and a synchronous condenser, all three are often referred to simply as a 'synchronous machine'. Indeed in some installations a single synchronous machine may be used in any of the three modes as desired.

The chief disadvantages of a synchronous motor, as compared with an induction motor of the same power, are increased size and complexities, and more particularly the difficulty of starting it. Without special additional features it is not self-starting, and these add materially to the cost.

1.3   INDUCTION MOTOR (SQUIRREL CAGE)

There is another type of motor which works on an entirely different principle: it is the 'induction motor'. It is very widely - one might say almost exclusively - used throughout offshore and onshore installations for industrial drives. In this application it is always used on 3-phase supply systems.

The principle of operation is described in detail in the next chapter.

1.4   INDUCTION MOTOR (SPLIT PHASE)

This is a modification of the normal 3-phase induction motor to enable it to be run on a single-phase supply. Its principal use is with small domestic equipment where normally only single-phase supplies would be available. This too is described in the next chapter.

1.5   OTHERS

There are many other types of a.c. motors which have been developed for special applications. They include:

- Commutator motor
- Repulsion motor
- Reluctance motor
- Synchronous-induction motor
- Schrage motor.

These will not be met with on offshore installations and are unlikely to be found in onshore installations. No further descriptions are therefore given.

CHAPTER 2 PRINCIPLES OF OPERATION OF THE INDUCTION MOTOR

2.1 THE SQUIRREL CAGE INDUCTION MOTOR

2.1.1 The 3-phase Motor

The normal induction motor operates only on a 3-phase supply. It consists of a stator wound with a 3-phase winding in slots on the inside of the yoke. The windings are distributed around the stator so that in a so-called '2-pole' motor their axes are spaced 120° apart.

Each winding end is brought out to a terminal. The other ends are usually star-connected within the machine. In some instances the machine may be delta-connected, in which case all six ends may be brought out to six or three terminals.

If a 3-phase supply is connected to the three stator terminals, such an arrangement gives rise to a rotating magnetic field. This completes one revolution in one cycle of supply; that is to say, with a 60Hz system frequency the field rotates at 60 rev/s, and with a 50Hz supply at 50 rev/s. This is called the 'synchronous speed'.

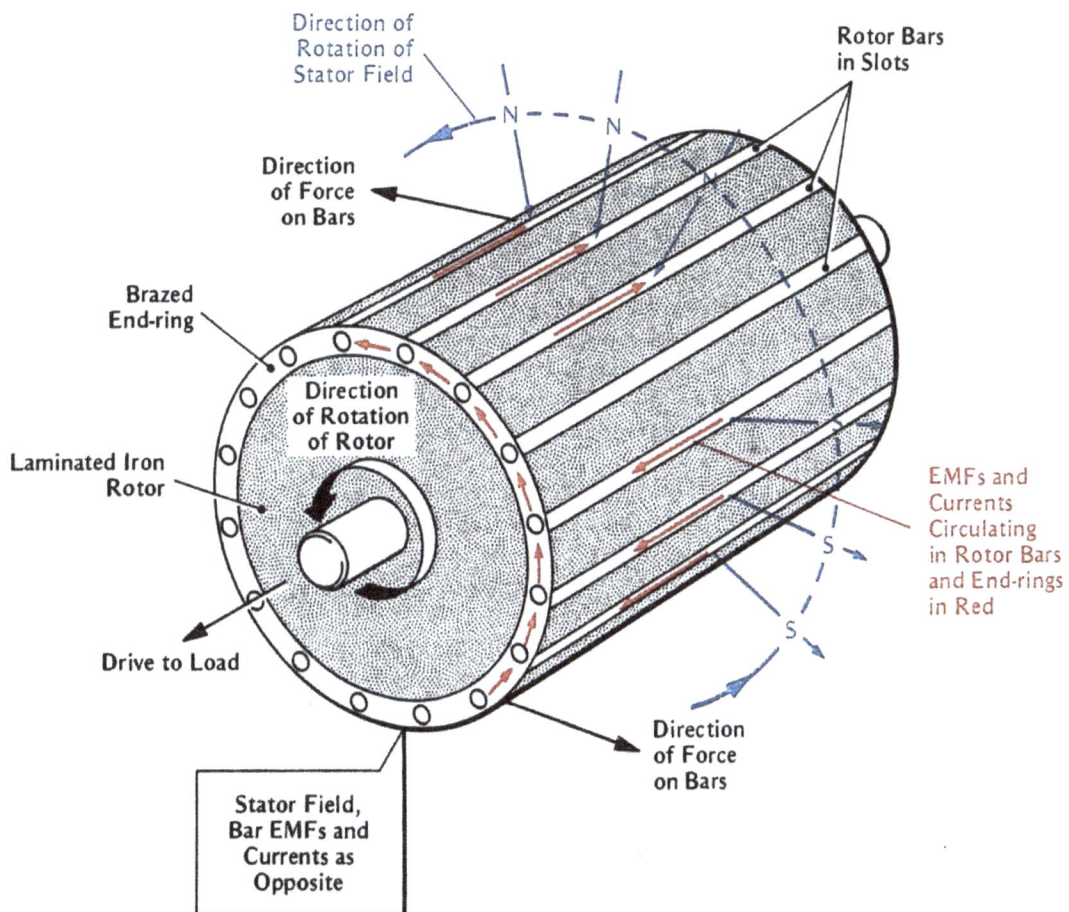

FIGURE 2.1
MAGNETIC FIELD, EMFs AND CURRENTS IN ROTOR

Inside the stator is a rotor which is free to rotate and has no electrical connections to it - that is to say, no brushgear. The rotor could be of solid iron, but it is more usually laminated and fitted with uninsulated copper or aluminium bars in slots round its outer edge, the bars being short-circuited at both ends by brazed metal rings, as shown in Figure 2.1.

The bars and rings together form a cage-like assembly, and this form of motor, by far the most numerous, is called 'squirrel cage'. There are variations on this arrangement, but they are not found offshore, and not often onshore.

Figure 2.1 shows only the rotor. The stator and its windings already described in the earlier part are not shown, but the rotating field which the stator produces is indicated in blue. This is the only manifestation of the stator that the rotor 'sees'.

While the rotor is at rest and therefore stationary, the rotating field from the stator passes through all the rotor conductors in turn, generating emfs in them by Faraday's Law of Induction. Because all the rotor bars are short-circuited, these emfs cause currents to flow through the bars and end-rings as indicated in red in the figure. The direction of the currents will change as the passing field changes from N to S and back again.

The rotor bar currents react with the magnetic field surrounding them to produce a sideways mechanical force on each bar, as explained in Chapter 6 of Part 1. By Fleming's Left-hand Rule these forces act on all bars in the same direction and produce a combined torque on the rotor in the direction of the field rotation. This causes the rotor to accelerate from rest. In more simple terms the rotor can be considered as being 'dragged around' by the rotating field.

As the rotor accelerates, the relative speed of the field passing it becomes less, the frequency and amount of the emfs and the rotor currents get less, and the mechanical force on the conductors and the torque on the rotor, after increasing for a time, also become less. Finally the rotor settles down to a speed slightly less than that of the field (the synchronous speed), when the emfs are at very low frequency and are just sufficient to maintain the rotor current and torque against the load. This settled speed is typically 1 to 2% below synchronous, and the difference is called the 'slip'. For example, a 2-pole, 60Hz motor would have a synchronous speed of 3 600 rev/mm; the motor nameplate may show the motor's rated speed as 3 540 rev/mm, which would indicate a slip of 1.67% at full-load. At less than full-load the slip is reduced and the speed approaches nearer to synchronous. Conversely, as mechanical load is applied, the slip increases, and the rotor's induced emf and hence its current rises, which gives an increase in driving torque to meet the load.

Thus induction motors have nearly, but not quite, a fixed speed just below synchronous, and it cannot be controlled (as in a d.c. motor) except by complicated and expensive equipment which is not installed in offshore or onshore installations. For this reason it is often called an 'asynchronous motor'.

As the rotor gathers speed, it generates increasing back-emf in the stator, causing the stator current to fall. At full speed, as load is applied, the slip and the rotor currents increase, causing increasing stator currents (and therefore motor 'load') by transformer action.

When the creation of rotating fields was discussed in the earlier part it was explained that, with one stator winding per phase, the field made one revolution per cycle. If, however, there were two parallel windings per phase, making a total of six windings disposed at 60° (instead of 120°) intervals, the field would make only half a revolution per cycle. A machine so wound is called a '4-pole' motor; at 60Hz its field rotation would be only at 1 800 rev/mm, and this therefore would be its synchronous speed.

As already explained, the full-load speed is just below synchronous, differing from it by the small slip, so that the nameplate rated speed of 4-pole motors at 60Hz will be around 1 750 rev/mm. These in fact are far more numerous than the faster 2-pole motors. Similarly 6-pole motors (synchronous speed at 60Hz 1 200 rev/mm) will have rated speeds of about

1 160 rev/mm. With 50Hz supply systems the rated speeds for 2-, 4- and 6-pole motors will be about 2 950, 1 475 and 980 rev/mm respectively.

It should be particularly noted that the induction motor is not strictly a constant-speed motor, as its speed varies over a small range of up to about 2% between no-load and full-load. Neither is it a variable-speed motor in the sense that it can be used for variable-speed drives, as can be done with d.c. motors. For most practical purposes however, such as for driving pumps, compressors and similar plant, it can be regarded as a constant-speed machine. Only where **exact** constant speed is needed would it be necessary to go to a synchronous motor.

2.1.2 The Single Phase Motor

The 3-phase motor described above depends on the creation of a rotating magnetic field from static windings in a 3-phase stator. However, it is possible to create a rotating field from a single-phase supply provided that it is split into two parts whose space-displacement and time-displacement are equal - in this case 90° instead of the 3-phase 120°. Such a motor is called a 'split-phase' motor. The principle is described in Part 1, Chapter 16.

Figure 2.2 shows how the split is achieved. It has four pole windings spaced at equal 90° intervals round the stator. Opposite pairs of poles have opposite polarities, so that, when one is producing an N field, the other is producing an 5, and vice versa.

The other pair of windings, 90° displaced, are fed through a capacitor of sufficient size to cause the current through the second pair of windings to lead by nearly 90° in time on that in the other pair.

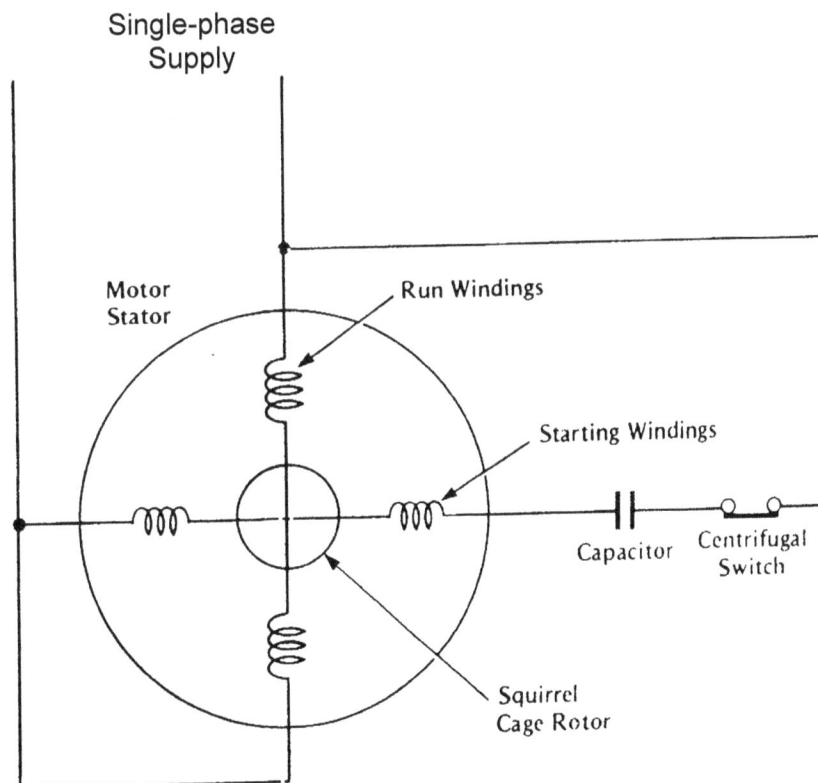

FIGURE 2.2
SPLIT PHASE, SINGLE PHASE (CAPACITOR) INDUCTION MOTOR

$t = 0°$ $t = 90°$ $t = 180°$ $t = 270°$ $t = 360°$

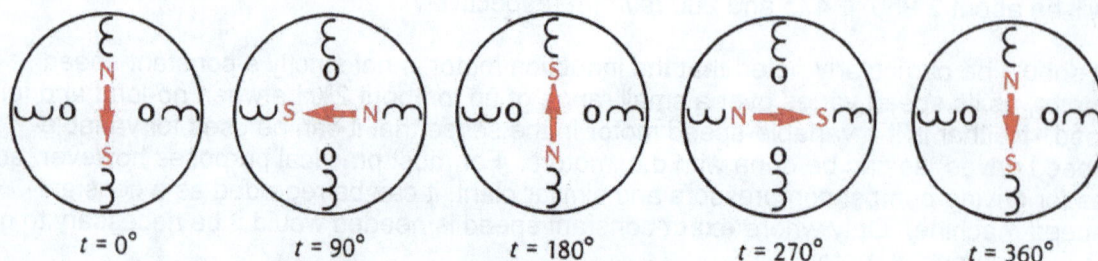

FIGURE 2.3
SPLIT PHASE MOTOR STATOR FIELD

Figure 2.3 shows the polarities at 90° (quarter-cycle) time intervals. At time $t = 360°$ (on the extreme right) the 12 o'clock pole is N maximum and the 6 o'clock S maximum. Both the 3 o'clock and 9 o'clock poles are zero because their currents, leading 90° on those of the other two, are passing through zero at that instant.

One-quarter of a cycle (90°) earlier, moving from right to left, at $t = 270°$, the 9 o'clock pole, whose current leads 90° on that of the 12 o'clock, was at maximum N, and the 3 o'clock at maximum S. The poles at 12 and 6 o'clock were zero because their currents were passing through zero at that instant.

Similarly one-quarter of a cycle still earlier ($t = 180°$) 6 o'clock was N and 12 o'clock was S, with 3 and 9 o'clock both zero. And at $t = 90°$, 3 o'clock was N and 9 o'clock S, with 12 and 6 o'clock both zero. And so back to $t = 0°$, which is the same as $t = 360°$.

It can be seen from Figure 2.3, moving now from left to right - that is, from $t = 0°$ to $t = 360°$ - that the N-pole progresses steadily from 12 o'clock at $t = 0°$, through 3 o'clock, 6 o'clock and 9 o'clock back to 12 o'clock. The S-pole progresses in the same direction but 180° behind.

Therefore a clockwise rotating field, completing one revolution in one cycle of time, has been produced from a single-phase supply. It should be added that the above explanation has been idealised. The current in the 3 and 9 o'clock poles will not lead the full 90° on that in the other poles because of resistance and inductance, but the split is quite sufficient to produce an adequate rotating field. This arrangement is not so efficient as a proper 3-phase motor, but it meets the requirements for smaller motors.

Once the rotating field has been obtained in this manner from a single-phase supply, the squirrel-cage rotor will run up to speed and follow it, with the usual slip, exactly as for a 3-phase motor.

When a split-phase single-phase motor has been run up to speed, it will continue to run and to deliver power even if the capacitor winding is disconnected. Therefore all but the smallest of such motors have their 3 o'clock and 9 o'clock windings, and the capacitor itself, short-rated. Once the motor has run up, a centrifugal switch opens the capacitor circuit. When the motor stops it reconnects it. In such a motor the 3 and 9 o'clock windings are referred to as the 'starting winding' and the other as the 'running winding'. On very small motors the centrifugal switch may be omitted and both windings left in circuit.

The split-phase motor is widely used for domestic equipment, where only single-phase supplies are usually available. Like the 3-phase equivalent, it is direct-on-line started and exhibits similar slip, torque and current characteristics. It is sometimes called a 'capacitor motor'; the capacitor is usually mounted on the outside of the motor case.

Other methods are sometimes used to achieve the phase difference between the starting and running windings, but the capacitor method is by far the most common.

2.2 THE WOUND ROTOR INDUCTION MOTOR

The induction motors so far described use a rotor of the squirrel-cage type, in which the conductors are uninsulated and where the rotor-bar currents, induced by the field passing at slip speed, are limited only by the resistance of the bars themselves and of the end rings and by the reactance of the rotor. There is no external control of these currents. It will be shown later that, when a motor is started by switching it direct onto line, the initial starting currents are very high, and there is no means of controlling them at the motor.

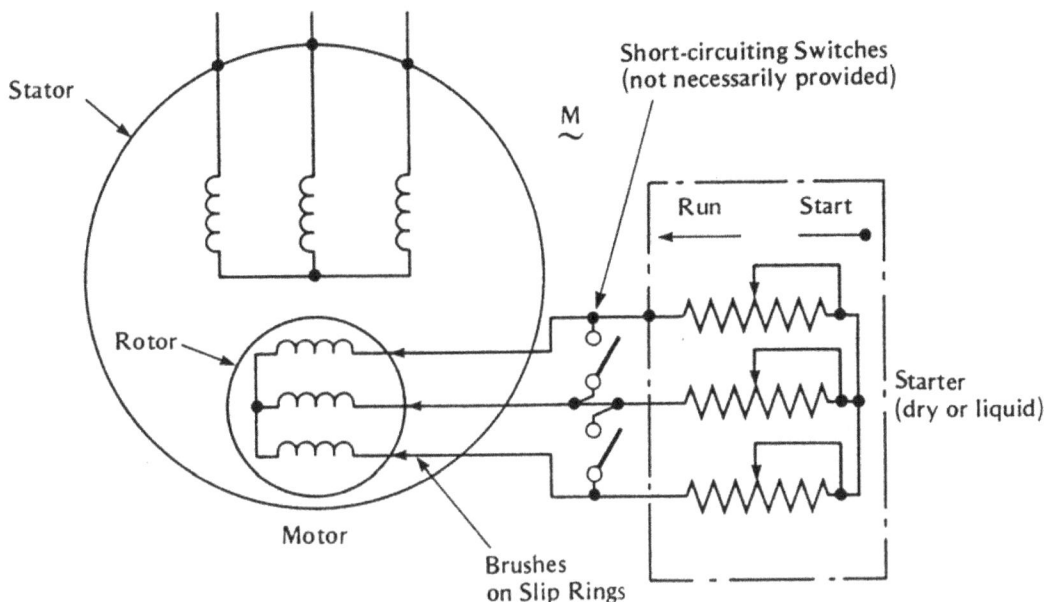

FIGURE 2.4
WOUND ROTOR INDUCTION MOTOR

Although these high starting currents are normally allowed for in the design of the system, it is sometimes necessary to reduce them. One way of achieving this is by using a wound rotor. Such a machine has, instead of a squirrel-cage rotor, one that is wound like a normal machine with insulated conductors in slots. Three separate phase windings are provided on the rotor, and they are brought out to three sliprings, as shown in Figure 2.4.

Under normal running conditions the three sliprings are short-circuited by an external switch or contactor, but during starting they are connected to a set of variable resistances. They introduce into the rotor circuit extra resistance which limits the rotor current and hence the current drawn by the stator. These 'starting resistances' are progressively reduced as the motor gathers speed (much like the old d.c. starters) until at full speed they are shorted out altogether, and the machine behaves as a squirrel-cage motor with short-circuited rotor bars.

It should be noted that the resistances are purely for starting and are therefore only short-rated. They are not for speed control, since speed is virtually constant and is determined by the system frequency. Inserting the rotor resistances has only a slight effect on the slip and so very little effect on the speed.

Starting resistances for wound-rotor machines may be of any variable-adjustment type, or they may be fixed with tappings selected in turn by contactors where starting control is automatic. In some cases a liquid resistor is used where three plates are gradually lowered into a tank containing an electrolyte; this gives a continuously variable control.

Wound-rotor motors and their control gear are far more expensive than the simple squirrel-cage motor. None are installed in any offshore and few, if any, in onshore installations.

CHAPTER 3 ENCLOSURES AND COOLING

3.1   ENCLOSURES

3.1.1 General

All motors are made with a stator yoke which acts as a case for enclosing the motor. Enclosures are of many different types, depending on the service to which the motor will be put. The minimum requirement is that it will protect the motor from direct damage caused by articles being dropped on it, and that it will protect the operator or any passer-by from danger either by contact with live conductors or with rotating parts.

Enclosures progress from this minimum through various stages of increasing protection to a form which allows the motor to be completely submersed in water and subjected to a specified pressure. There are additional requirements for motors to be used in hazardous areas; they are discussed in Part 9 'Electrical Safety'.

3.1.2 Classification of Enclosures

The types of motor enclosure are indicated by a system of coding. There are two such coding systems: one has been in general use for many years, but it has for some time been superseded by a more complicated code based on international agreement. It is listed in BS 4999, Part 20, and is indicated below.

It consists of the letters 'IP' followed by two digits whose meanings are as follows:

First Digit		**Second Digit**	
0	Non-protected	0	No special protection
1	Protected against solid bodies not less than 50mm	1	Protected against dripping water
2	Protected against solid bodies greater than 12mm	2	Protected against water drops up to 15° from vertical
4	Protected against solid bodies greater than 1 mm	3	Protected against spraying water
5	Protected against dust	4	Protected against splashing water
		5	Protected against water jets
		6	Protected against conditions on ship's deck
		7	Protected against immersion
		8	Submersible to specified pressure

Thus a motor classified 'IP44' indicates a machine protected against ingress of particles greater than 1mm and of splash-proof design

3.2   COOLING

3.2.1 Temperature Rise

The maximum temperature to which the stator windings may be allowed to rise by the cooling system depends on the type of insulation material round the conductors. Motors are classified according to the insulating material used, and to each class is allotted a maximum ultimate temperature. The classification is as follows (according to BS 2757 /IEC 60085).

Class	Typical Insulating Material	Ultimate Temperature
Y	Cotton, silk, paper, etc., unimpregnated	90°C
A	Impregnated cotton, silk, etc.; paper; enamel	105°C
E	Paper laminates; epoxies	120°C
B	Glass fibre, asbestos (unimpregnated); mica	130°C
F	Glass fibre, asbestos, epoxy impregnated	155°C
H	Glass fibre, asbestos, silicone impregnated	180°C
C	Mica, ceramics, glass, with inorganic binders	>180°C

It should be noted that the classification letters do not follow an alphabetical sequence. This is because there were originally only three classes - 'A', 'B' and 'C'. Later intermediate classes were added, and it was decided not to disturb the original three which are well understood. Most motors on offshore installations are Class 'B' or 'F'. This classification applies not only to motors but also to generators, transformers and similar electrical equipment.

Certain of the higher-temperature materials may be hygroscopic and therefore not always suitable in any particular environment, particularly where dampness is severe.

It should be particularly noted that the classification depends on the ultimate temperature to which the insulating material may be subjected, for it is this which determines whether or not it will suffer damage when heated. It does not depend on temperature rise alone. If, for instance, the ambient temperature is 40°C, a Class 'B' material may be used if the designed temperature rise will not exceed 90°C, so making the ultimate maximum temperature 130°C. Designed temperature rises must therefore take into account the greatest expected ambient temperature in which the machine will operate.

3.2.2 Classification of Cooling Methods

The various methods of cooling are indicated by a system of coding based on international agreement. It is listed in detail in BS 4999, Part 106, and consists of the letters 'IC' followed by two digits signifying the cooling circuit arrangement and the method of supplying power to circulate the coolant. The meanings of the digits are as follows:

First Digit

0 Free circulation
1 Inlet duct ventilated
2 Outlet duct ventilated
3 Inlet and outlet duct ventilated
4 Frame surface cooled

5 Integral heat exchanger (using surrounding medium)
6 Machine-mounted heat exchanger (using surrounding medium)
7 Integral heat exchanger (not using surrounding medium)
8 Machine-mounted heat exchanger (not using surrounding medium)
9 Separately mounted heat exchanger

Second Digit

0 Free convection
1 Self-circulation
2 Integral component mounted on separate shaft
3 Dependent component mounted on the machine
5 Integral independent component

6 Independent component mounted on the machine
7 Independent and separate device or coolant system pressure
8 Relative displacement

This classification applies not only to motors but also to all rotating machines such as generators.

Where it is desired to specify the nature of a coolant, the following letter-code is used in conjunction with the cooling code:

Gases	{	air	A
		hydrogen	H
		nitrogen	N
		carbon dioxide	C
		helium	L
Liquids	{	water	W
		oil	U

When nothing but air is used, the letter 'A' may be omitted.

A motor formerly referred to as 'TEFC' (totally enclosed, fan-cooled) would have an enclosure code 'IP55' or 'IP56' and a cooling code 'IC41'.

CHAPTER 4 POWER FOR INDUCTION MOTORS

4.1   OUTPUT AND LOSSES

A motor is installed for the purpose of driving some item of plant such as a pump, compressor or fan, and it imparts mechanical power to its driven load. This is true power, and the motor takes it in electrically in the form of active power (i.e. kilowatts) from the main electrical system.

Most of the input electrical power is passed on to the driven load, but some of it is lost within the motor itself. These losses can be listed as follows:

- Friction
- Windage
- Iron losses
- Copper losses due to load
- Copper losses independent of load.

Since the speed of the motor is virtually constant, the first two (friction and windage) may be considered constant at all loads. 'Iron losses', which are due to the imperfect magnetisation of the iron, depend on the magnetising flux and so on the applied voltage; as this is constant, the iron losses may also be regarded as constant at all loads.

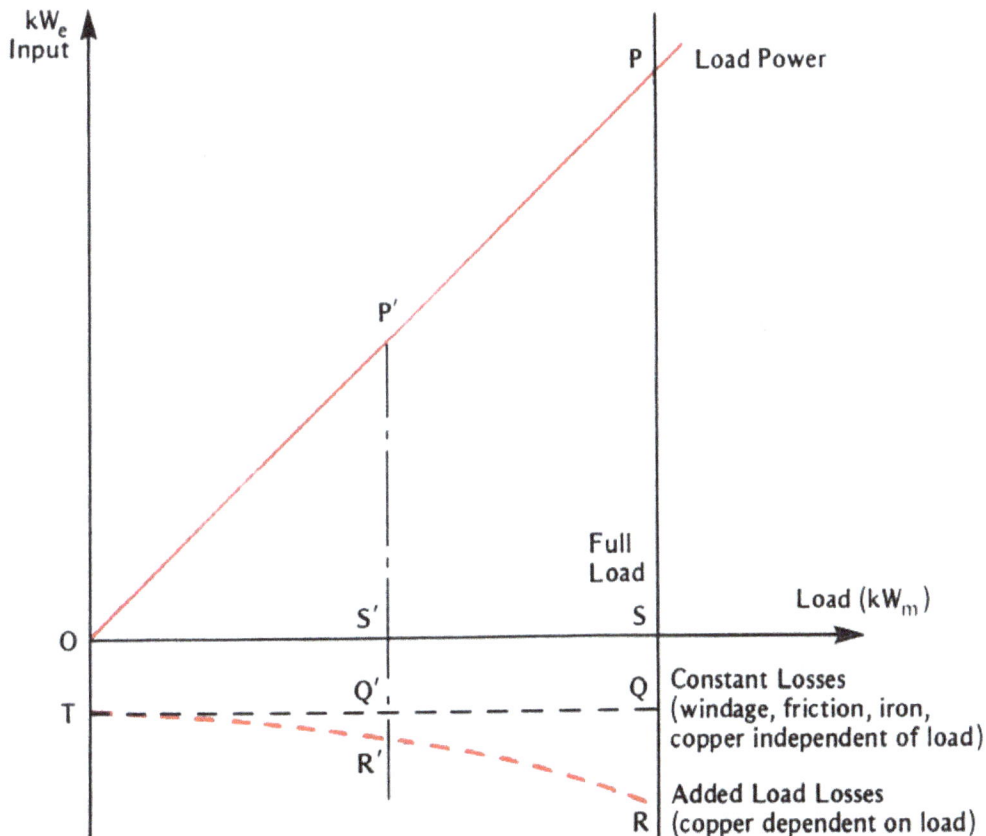

FIGURE 4.1
MOTOR POWER AND LOSSES

The principal current drawn by the machine is the active current which powers the load and which therefore varies as the load varies. The heating loss due to any current is I^2R, therefore the 'copper losses' due to the load current are variable and are roughly proportional to the square of the load. These variable losses occur in both the stator windings and the rotor bars.

Finally, even when there is no mechanical load on the motor, it is still drawing some active power to supply the losses; it is also drawing magnetising (reactive) power. Both these result in an appreciable no-load current, which itself gives rise to some I^2R losses. These are the 'copper losses independent of load' and are also almost constant.

The losses for which a motor must draw active power, in addition to the power required to supply the load and how they vary with load, are shown in Figure 4.1. There is a constant element which is always present, even at no-load, and a variable element which is roughly proportional to the square of the load.

Suppose the input power and losses (both in kW_e) are plotted to the same scale against the mechanical output load (in kW_m). Draw the line representing the input power actually needed for the load above the zero line; it will be a straight line OP at 45° through O. Draw the loss curves below the zero line. They will consist of two parts: a constant part representing friction, windage, iron losses and copper losses independent of load; this is the horizontal line TQ. The other part representing copper losses depending on load will be a 'square law' curve which, when added to the constant loss line TQ, gives the total loss curve TR. The point S represents full-load mechanical output.

Then at full-load the part PS is the mechanical output power, SQ the constant losses, and QR the variable copper losses. The line PR is then the total power required to meet the load and to supply all losses - that is, the total input power in kW_e.

$$\text{The efficiency } (\eta) = \frac{\text{output}}{\text{total input}}$$

$$= \frac{PS}{PR} \text{ at full-load}$$

and this is clearly less than unity.

At some other load S' the efficiency is $\dfrac{P'S'}{P'R'}$, and it can be seen that this is less than the full-load efficiency $\dfrac{PS}{PR}$. This shows that the pattern of losses causes the efficiency to fall with reduction of load until, at no-load where the output is nil, the efficiency becomes zero.

All losses result ultimately in heating, and the motor must be designed to dissipate heat at the maximum rate at which it can be produced without overheating the machine. For small motors their cases can usually do this, but larger motors normally need a fan. The largest may require cooling through an air/water heat exchanger. Cooling is further discussed in Chapter 3.

4.2   MAGNETISATION

As explained in Chapter 2, the operation of an induction motor requires a rotating magnetic field - that is to say, it needs to be magnetised. In Part 1 it was stated that magnetisation causes inductance in a circuit, and it was shown that inductance gives rise to 'inductive reactance' in an a.c. circuit. This causes the magnetising current to lag 90° on the applied voltage (neglecting resistance).

Whereas active current creates a demand for active power (watts), reactive current calls for reactive power, or 'vars'.

An induction motor therefore demands active power to meet its mechanical output load and to supply its losses, but it also demands reactive power to supply its magnetisation. Whereas active power will vary with the loading, as shown in Figure 4.1, the magnetisation remains virtually constant and independent of the loading, since the rotating field is present whether the motor is working or running light. Therefore the reactive (var) loading can be regarded as constant while the active (watt) loading varies with loading.

Figure 4.2 shows pictorially how the active and reactive power to a motor is generated, distributed and consumed. Active, or true, power originates from the generator's prime mover as mechanical output from the engine. On the other hand reactive power emanates from the generator's excitation system through its main field.

In Figure 4.2 the active power route is indicated in red, and the reactive power route in blue. Both powers come from the generator itself through a common cable. At the switchboard they give a common current indication on the generator ammeter, and both combine to give a power-factor indication. Then they separate to give independent wattmeter and varmeter indications. They recombine to feed the motor through a common cable. At the motor the reactive power is used to magnetise the machine, and the active power supplies the (variable) mechanical load and also the losses.

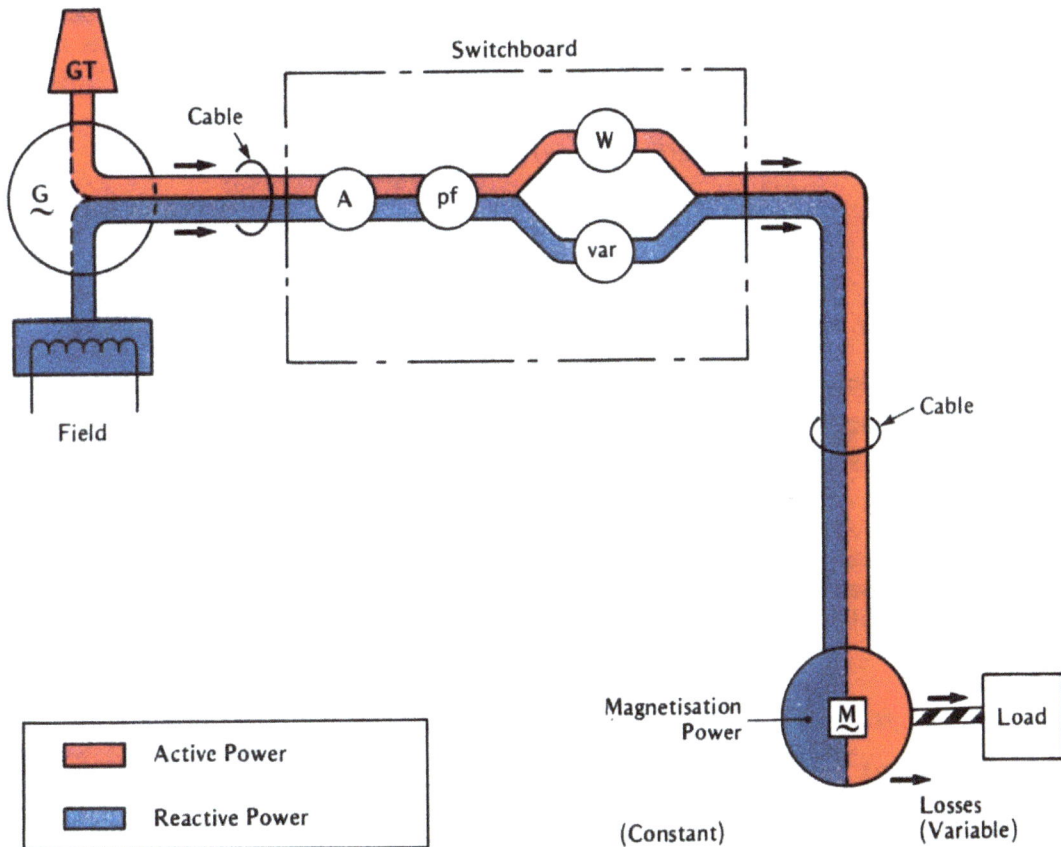

FIGURE 4.2
ACTIVE AND REACTIVE MOTOR POWER

The following example shows how these two types of power combine at various loadings.

Example   Q   A 6.6kV, 400 hp motor has a full-load power factor of 0.8. What are its load current and power factor at full-load, half-load, quarter-load and no-load? Assume a full-load efficiency of 90%. Neglect no-load losses.

A   400 hp is equivalent to ¾ × 400 = 300kW$_m$

If efficiency is 90%, full-load input power is $\frac{300}{0.9}$ = 333kW$_e$.

Prepare a table as follows ready for insertion of values as the calculation proceeds:

	KW$_e$	kvar	kVA	I (amps)	cos φ
Full	333	250	416	36	0.8
½	167	250	301	26	0.55
¼	83	250	263	23	0.31
No-load	0	250	250	22	0

(The two figures in 'boxes' were given in the question or deduced as above. The remainder are calculated as follows.)

At full-load:

$$kVA = kW_e \div cos\ φ$$
$$= 333 \div 0.8$$
$$= 416$$

$$kvar = kVA\ sin\ φ. \text{ As } cos\ φ = 0.8, sin\ φ = 0.6$$
$$= 416 × 0.6$$
$$= 250$$

kW$_e$

At ½-load   kW$_e$ = ½ × 333 = 167
At ¼-load   kW$_e$ = ¼ × 333 = 83
At no-load   kW$_e$ = 0

Enter in the table.

kvar

As the reactive load is constant and so does not change with loading, the calculated figure of 250kvar for full-load applies also at all other loads, including no-load.

kVA

At ½-load $kVA = \sqrt{kW_c^2 + kvar^2}$ $= \sqrt{167^2 + 250^2}$ $= 301$

At ¼-load $= \sqrt{83^2 + 250^2}$ $= 263$

At no-load $= \sqrt{0^2 + 250^2}$ $= 250$

I

At all load $I = \frac{kVA}{\sqrt{3} \times kV}$

At full-load $= \frac{416}{\sqrt{3} \times 6.6}$ $= 36A$

At ½-load $= \frac{301}{\sqrt{3} \times 6.6}$ $= 26A$

At ¼-load I $= \dfrac{263}{\sqrt{3} \times 6.6}$ $= 23A$

At no-load $= \dfrac{250}{\sqrt{3} \times 6.6}$ $= 22A$

cos φ   At all loads   cos φ $= \dfrac{kW_e}{kVA}$

At ½-load $= \dfrac{167}{301}$ $= 0.55$

At ¼-load $= \dfrac{83}{263}$ $= 0.31$

At no-load $= \dfrac{0}{250}$ $= 0$

(Note: The zero power factor at no-load is theoretical, as no-load losses have been neglected. In practice the losses at no-load are appreciable, and a typical no-load power factor is 0.2 to 0.3.)

All the above calculated figures have been inserted in the table.

Deductions from the Table

If the table is examined carefully, certain conclusions, some perhaps surprising, can be drawn.

(i)   As loading is reduced, both motor current and the kVA are also reduced, but by no means in proportion. Due to the continued presence of the magnetising current, the current and kVA at no-load are still more than half their values at full-load (in the example taken). This means that the reading of a local motor ammeter must be taken with reserve as an indication of motor loading.

(2)   The power factor also falls with reduction of load; theoretically to zero at no-load but in practice to a figure about 0.2 to 0.3 because of the continuing losses. On an offshore or onshore installation where many motors may be running at the same time, it is unlikely that all the motors will be running at full-load simultaneously, and some may even be running light. Therefore the overall power factor of the combined motor load is likely to be well below that given on the motors' individual rating plates. It is of course improved overall by the addition of high power factor loads such as lighting and heating.

CHAPTER 5 STARTING OF INDUCTION MOTORS

5.1 GENERAL

The a.c. motors throughout all offshore and onshore installations are almost exclusively of the induction type, with squirrel-cage rotors. They may be of high voltage supplied from HV busbars, or they may be of low voltage supplied from one or other of the LV busbars; the latter form the great majority of offshore and industrial motors.

The range of powers, extends from about 1 2000 hp (9 000kW mechanical) down to 1 hp or even less. Motors below about 300 hp (220kW$_m$) run from the 440V or 415V system, but those of higher power are usually of the high-voltage type.

When a motor is started direct-on-line ('DOL' starting), at the instant of switch-on the rotor is stationary, and the motor behaves as a short-circuited transformer. The initial starting current at standstill is therefore very large, though at low power factor. Typically it may be four to six times the normal full-load current, with a power factor of the order of 0.3 to 0.2.

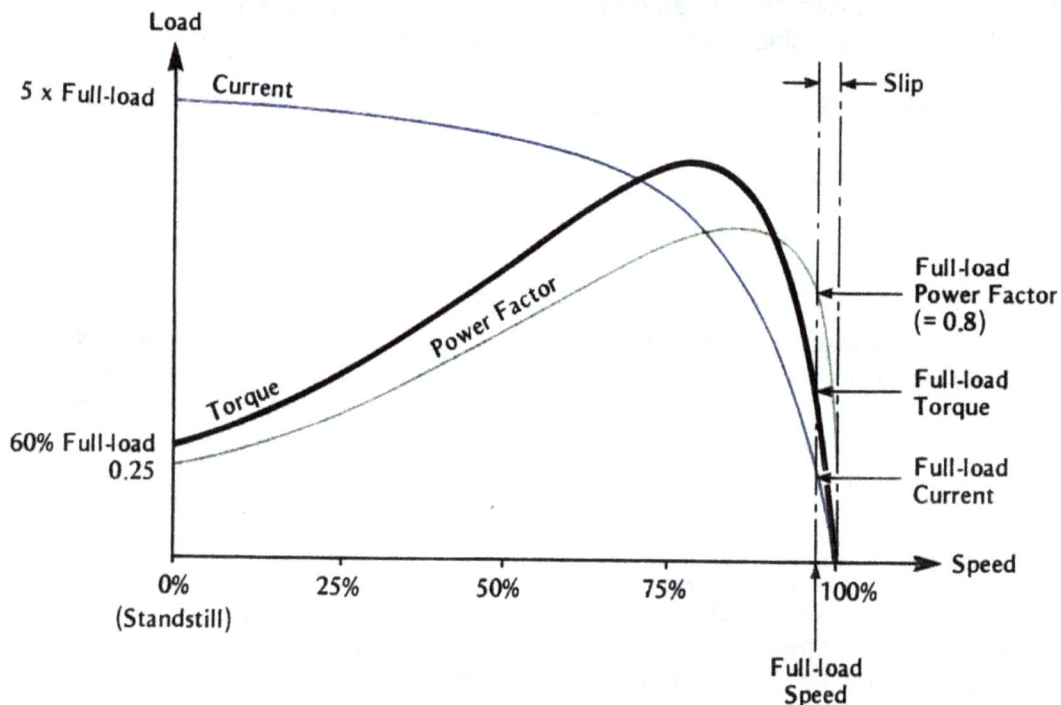

FIGURE 5.1
INDUCTION MOTOR CURRENT, TORQUE AND POWER FACTOR/SPEED CURVES

Figure 5.1 shows the varying conditions as a motor runs up from standstill to full speed. As the motor begins to move and to generate back-emf, the low power factor reactive current steadily falls off, but the motor begins to draw more active power as it accelerates. The net effect is that the initial starting current persists fairly constant through most of the run-up period, but the power factor steadily improves. Finally the speed settles down to just below synchronous, that is with a small 'slip', and the current falls sharply to whatever the load demands. This is indicated as current/speed (blue) and power factor/speed (green) curves in Figure 5.1, which also shows the characteristic 'Rock of Gibraltar' torque/speed (black) curve.

With very few exceptions all offshore motors are direct-on-line started, whether of the high-voltage or low-voltage type. The 440V or 415V motors, which form the great majority, are usually remotely started through contactors in separate panels of LV switchboards (also called 'Motor Control Centres' or 'MCCs'), which are described in greater detail in Part 3 'Electrical Power Distribution'. HV motors are remotely started through circuit-breakers or vacuum or air-break contactors backed by High Rupturing Capacity (HRC) fuses in HV switchboards.

The run-up time of a direct-on-line started motor depends on:

- The starting torque of the motor
- The inertia of the driven load
- Whether or not the driven end is loaded.

Motors are designed for different starting torques. The 60% of Figure 5.1 is only typical, but the range is much wider, depending on the application. In general, a high starting torque gives a lower efficiency.

Clearly the inertia of, say, a centrifugal fan is far less than that of a motor generator set, and the run-up time would be much shorter. Loading the driven end (as distinct from starting light) will clearly extend the run-up time.

Torque depends on the square of the applied voltage. Therefore an induction motor is very sensitive to voltage drop at its terminals, however caused. For example, a 20% drop in applied voltage (i.e. to 0.8 nominal) will result in only 64% (0.8^2) of the designed torque. In extreme cases this may prevent a motor from even starting at all, especially if there is much breakaway friction ('stiction'), as with reciprocating drives. This is further discussed in Chapter 6.

5.2 REDUCED VOLTAGE STARTING

If it is desired to reduce the heavy starting current to a level well below that which would be experienced with direct-on-line starting, various methods are available which reduce the voltage applied to the motor at the instant of starting, and restore it to full value when the motor has run up to speed. Two of the principal methods are described below. It should be noted however that reduced voltage starting also means reduced starting torque, so that, where high starting torques are needed, these methods may not always be suitable.

5.2.1 Star/Delta Starting

Figure 5.2(a) shows the conventional method of starting a motor direct-on-line.

Figure 5.2(b) is known as 'star/delta' starting. All six ends of the motor windings are brought out, and a 3-pole changeover hand-switch or contactor connects them alternatively in star or delta. (The changeover switch is not, of course, inside the motor as indicated in Figure 5.2(b), but it is shown there for electrical clearness.) It is drawn in the 'star' position; delta is the running condition, where full line voltage is placed across each phase winding. When starting, the motor is temporarily connected in star, so that the voltage appearing across any one phase winding is line voltage divided by √3 (see the inserts). Thus a 440V machine would have 254V applied to each phase - this is the 'reduced voltage' - and the starting line current taken at that voltage is reduced to one-third of what it would have been with a delta connection. When the motor is up to speed and the starting current has fallen to its running value, the motor is reconnected into delta and takes its full voltage across each phase. As it is then running at nearly synchronous speed, the changeover will cause only a small increase of current.

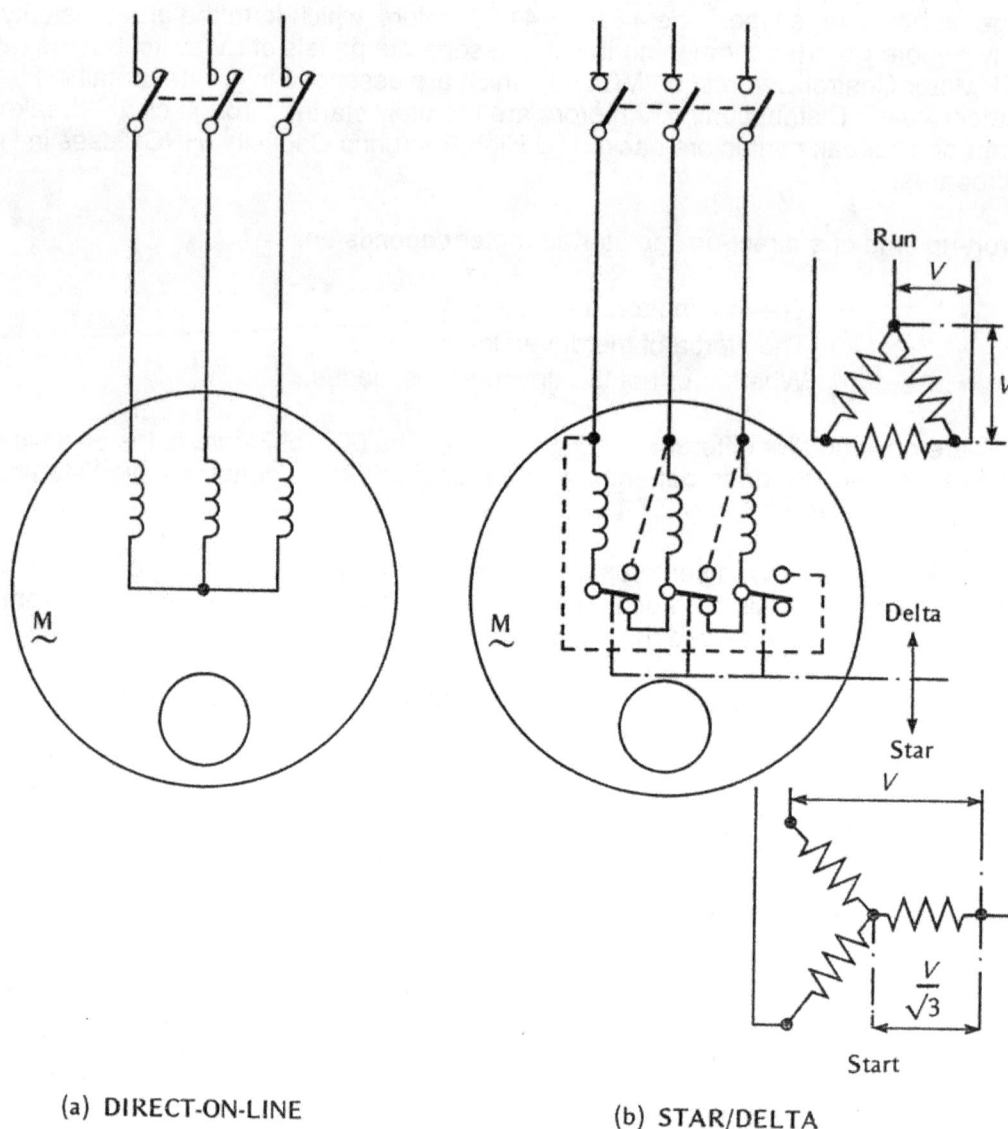

(a) DIRECT-ON-LINE (b) STAR/DELTA

FIGURE 5.2
STAR/DELTA STARTING

The changeover from star to delta may be by hand changeover switch; it must be provided with an interlock to prevent it operating until it has been returned to the star position. The star and delta positions are usually tallied START and RUN respectively. Alternatively the reconnection may be by contactor, especially with an auto-started motor. Interlocks will ensure the correct sequence of operating.

Delta-connected motors are less easy to protect than star-connected ones and are little used on offshore installations. Consequently star/delta starting is seldom found offshore but is very common in industrial installations, in which case the machines run as delta-connected motors.

5.2.2 Auto-transformer Starting

Figure 5.3 is a different form of reduced voltage starting which uses an auto-transformer to provide the lower starting voltages. (An ordinary double-wound transformer could be used, but, as the ratio is close, an auto-transformer is just as effective and is smaller and cheaper.)

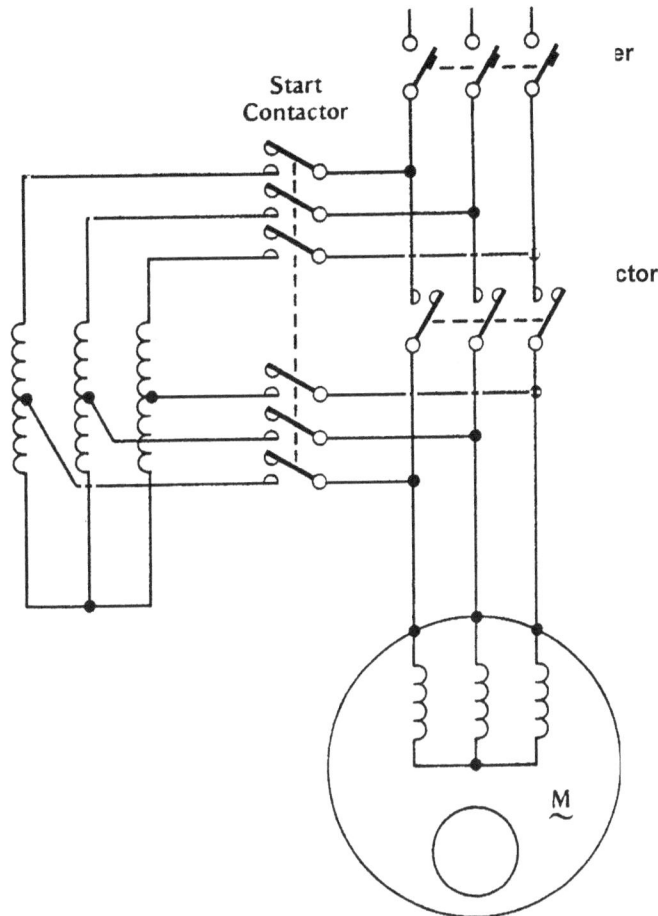

FIGURE 5.3
AUTO-TRANSFORMER STARTING

Here the motor is started on the auto-transformer tap (typically 50 to 70%). It runs up on the reduced voltage, taking an appreciably lower starting current. When it is up to speed the auto-transformer and its tap are disconnected by a hand-switch or contactors and the motor is connected directly across the full-voltage mains. As in the case of star/delta starting, the motor will be running at nearly synchronous speed at the end of the first auto-transformer stage, so that the changeover to full voltage will cause only a small increase of current. This method, unlike star/delta, allows a star-connected motor to be used.

As with direct-on-line starting, the run-up time depends on the motor design, on the inertia of the driven end and on whether the machine starts loaded or light. In the case of auto-transformer starting it will also depend on the voltage ratio of the auto-transformer and hence on the voltage applied during the first stage.

Certain features of auto-transformer starting should, however, be noticed:

(a)   The 'start' and 'run' switches, if of the contactor type, must be positively interlocked so that both cannot be closed at the same time, otherwise part of the auto-transformer primary will be shorted out.

(b)   The changeover from 'start' to 'run' must be time-delayed. When power is taken off a large running motor, it remains magnetised for a time by its rotor current until it has been damped out. While it continues, the motor will be generating back-emf. If the full voltage is applied while this is happening, it might come in anti-phase, with consequent currents of the order of a short-circuit. Time delays of about two seconds are usual with large motors.

21

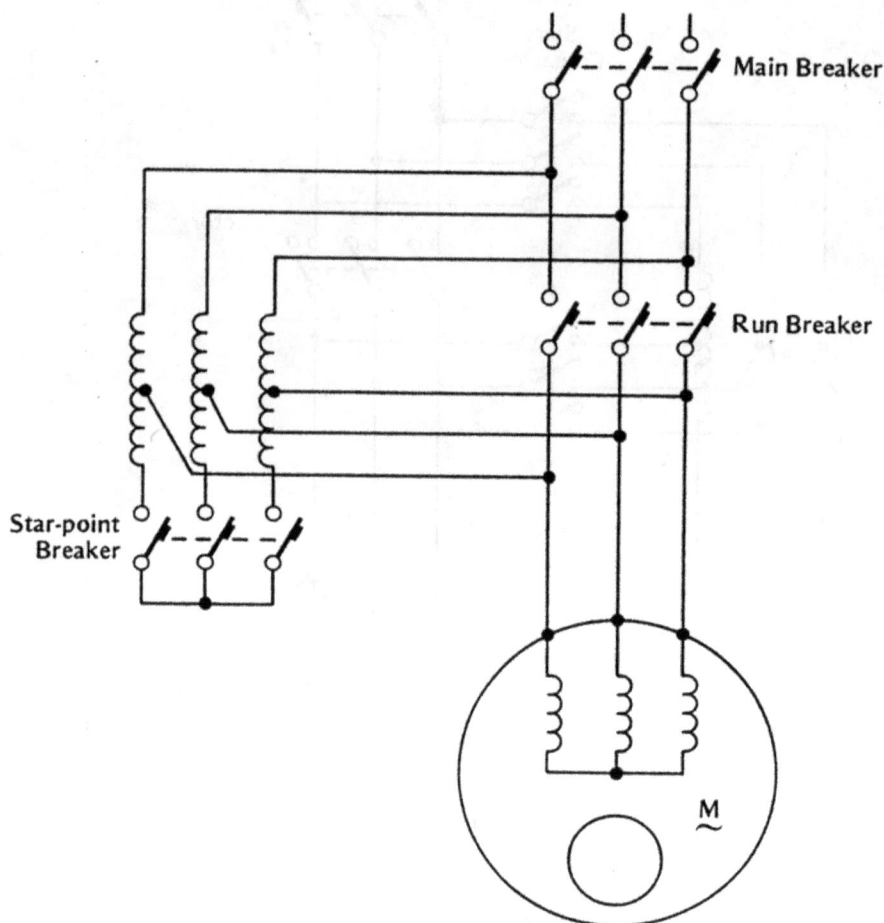

FIGURE 5.4
KORNDORFFER METHOD

5.2.3 Korndorffer Method

The disadvantages are overcome by a modification to the basic auto-transformer method, known as 'Korndorffer' starting, shown in Figure 5.4.

Here the auto-transformer remains constantly connected, but it is not actually energised - that is, it does not carry any current - until its star-point is completed by a circuit-breaker or contactor. As soon as the main breaker and star-point breakers are closed, the motor, which is permanently connected to the auto-transformer taps, receives its reduced voltage and starts its first stage run-up.

When it is up to near synchronous speed, the star-point breaker opens, leaving the motor still connected to the mains through the auto-transformer primary section acting as a choke. Immediately after the star-point breaker has opened, and without special time delay, the 'run' breaker closes, connecting the motor direct to the full-voltage mains to complete its second stage and shunting out the choke. At no time during transition from first to second stage has the motor been totally disconnected from the mains; therefore its back-emf can never become out-of-phase with the mains voltage.

In large motors the sequence of switch closing and opening is automatic and is tied in with other mechanical sequences such as pre-start purging of the motor casing, running up of lubricating pumps and correct opening of valves on the driven load side. The completion of each stage is monitored by timing relays, and, unless it is completed within a certain preset time, the main breaker is tripped and the start aborted, with appropriate alarm lamps lit on the main control panel.

On most installations the only motors large enough to require auto-transformer starting are typically those driving the Gas Compression, Pipeline Boosters and Re-injection Compressors. Typically these motors are rated between 7 160kW$_m$ and 9 250kW$_m$ (9 600 hp and 12 400 hp) and are started using 18MVA generators as supply source and in each case the Korndorffer method is used

5.3   EFFECT ON SYSTEM OF STARTING A LARGE MOTOR

One very important consequence of starting these very large motors must be realised. Even with reduced voltage starting and the consequent limiting of the starting current, the call on generator capacity is considerable. Because the starting power factor is low, most of the starting current is reactive, at least at first, and it is shown in Chapter 3 of Part 2 that it is the reactive load current which causes voltage drop in generators. In the case of the 9 250kW$_m$ re-injection compressor motor the reactive starting load is about 23Mvar, which is more than one generator, even of size 18MVA, can reasonably cope with. At least two generators should be on line to enable such a motor to be started, and even then their AVRs will be extended to the limit to maintain voltage. The situation will be aggravated if the generators already have an appreciable standing load, and load-shedding may be found necessary to enable a start to be made.

In the extreme case the voltage dip even with full AVR action may be so great, and the starting torque so heavily reduced, that the motor may not be able to break away against its 'stiction' and the motor may fail to start at all.

In any case a severe voltage dip on the whole offshore installation system may be expected until the motor is up to speed, and that very dip may itself cause the motor to be sluggish in running up even if it does not prevent it. If the dip is large or prolonged, other motor contactors may trip on undervoltage, and operators must be prepared to find this. Those installations with only two generators will be more susceptible than those with four.

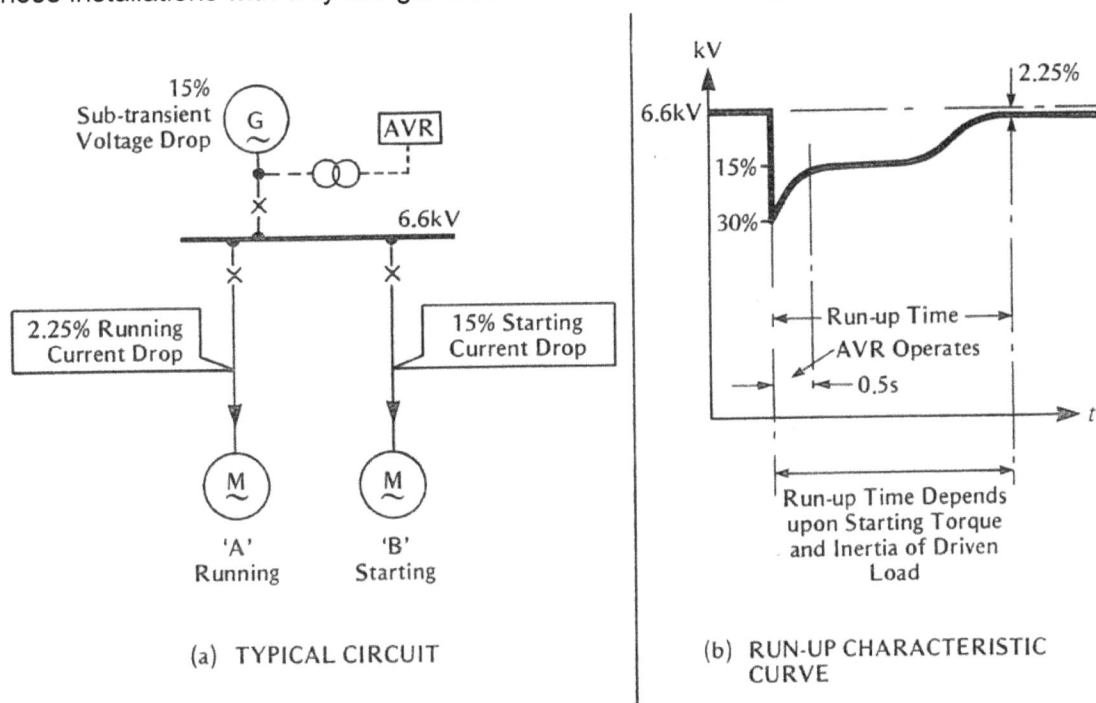

(a) TYPICAL CIRCUIT

(b) RUN-UP CHARACTERISTIC CURVE

FIGURE 5.5
STARTING OF LARGE MOTOR

Figure 5.5(a) is an example of such a situation, and Figure 5.5(b) shows the voltage situation at the motor terminals when a large motor is direct-on-line (i.e. single-stage) started while another is running. It is assumed that the AVR is already compensating for all other running loads, including that of the large motor which is already running.

In the example shown, at the instant of start and before the AVR can react, there is a sudden voltage drop of 30% due partly to the reactive starting current flowing through the generator reactance and partly to voltage drop in the feeder cable. This is followed by a short period of partial recovery of about half a second while the AVR reacts and increases excitation to the limit; the voltage rises somewhat but stays at 15% dip during most of the run-up period because of continuing voltage drop in the cable. When the motor is up to speed and its current falls to its 'run' value, the voltage rises further as the cable drop is reduced until it settles at its controlled value near 6.6kV. The voltage at the motor terminals, however, will still be a little down (typically 2¼%) because of the voltage drop in the feeder cable due to the motor's running current.

Note that the intermediate drop level (15% in Figure 5.5(b)) to which the voltage partially recovers under AVR action will depend on the standing load and the extent to which the AVR is already committed. The amount and speed of recovery will depend on the margin left in the AVR.

It must be stressed that it is the reactive loading on the generators due to the motor starting which is the critical factor. The active (MW) load demanded by the gas compression motors during starting is well within the capacity of one generator unless it is already heavily loaded. When shedding loads, therefore, those with large reactive components (e.g. motors) should be selected rather than, say, heating. The Mvar meter on the switchboard is the one to watch.

The whole question of how far a generator may be loaded by running motors, or by motors which are about to be started, is discussed more fully in Chapter 2 of Part 6 'System Control' under the heading 'Capability Diagram'.

Starting a large motor in an extensive onshore installation, which is backed up by a big supply network, is less of a problem; the principles set out above remain, although their effect may be less noticeable.

5.4   REPEATED STARTING

If, when a start signal is given, a motor does not start, there is an understandable urge to 'try again'.

Each time an attempt is made to start a motor and the contactor actually closes but the motor does not move, a severe starting current of the order of five to six times full-load current flows continuously through the contactor contacts and the windings of the stationary motor. This gives rise to heating (I^2R) at a rate 25 to 36 times normal for the motor, and without ventilation. When in due course the overcurrent protection operates and trips the contactor, not only has a large amount of heat been generated inside the motor, but also the contactor has to break some 500% of the normal current, and at a low power factor - a most severe condition.

If another attempt is made immediately, the overcurrent protection resets. It has the same time delay as before, and therefore an equal amount of heat is released to be added to the first. Each starting attempt will therefore raise the motor winding temperature still further. Too many attempts will raise it to such a temperature that there is severe risk of damage to the insulation and of an immediate or early breakdown, even though the protection is working properly. Cables and motor cable boxes may also be thermally overstressed, with similar risk of breakdown.

The repeated opening of the contactor under such severe conditions is likely not only to overheat it but to burn, and possibly weld, the contacts. Many examples of this have been reported. If this happens and the contactor cannot open because of welding, all protection of the motor is lost, and a burnout is almost inevitable, and possibly fire.

Some motors are now provided with thermistor or resistance thermal detector elements embedded in their windings. They detect the temperature rise from whatever cause and, if it exceeds a certain preset level, they trip **and lock out the motor** until it has cooled (always assuming, of course, that the contactor has not welded-in). The operator can seldom know whether any particular motor is so protected, and he should observe a general rule that if, after two attempts, a motor does not start, he should make no further try but call in Maintenance.

Indeed some operators have laid down a general instruction that motors over 50kW are only suitable for two successive starts **when cold**. Another one start attempt is allowable after a cooling period of 30 minutes at standstill. No more than three starts may be attempted in any one hour.

For certain very large motors the manufacturers may make special stipulations not only on the number of starting attempts, but also on the maximum permitted number of successful starts over a given time. For example, for the large gas compression motors (7.1MW to 9.2MW) they stipulate not more than two attempted starts in succession, not more than one successful start in any 30 minutes and not more than 300 starts in one year.

5.5 EXAMPLE OF A MOTOR STARTING CALCULATION

At the end of Chapter 4 a calculated example was given on how to estimate the current and power taken by a motor when **running** at various loads. Now an example is given of the calculation needed to estimate the current and power (active and reactive) of a motor when starting. The same motor as was assumed for the Chapter 4 example is used, but certain additional starting information is given, such as would be found on the motor's rating plate.

Example:

Q   A 6.6kV, 400 hp motor has a full-load power factor of 0.8 and efficiency of 90%. Its standstill current at starting is 5 times its full-load current and at 0.2 pf. What are its starting current, and its active and reactive starting loads?

A   As in Chapter 4, 400 hp is equivalent to ¾ x 400 = 300kW$_m$.

If efficiency is 90%, full-load input power is $\dfrac{300}{0.9}$ = 333kW$_e$.

If full-load power factor is 0.8, full-load input = $\dfrac{333}{0.8}$ = 416kVA.

At full-load, kVA = $\sqrt{3}kV \times I$, where kV and I are line voltage and current.

In this case kV = 6.6 and kVA has been calculated 416.

$\therefore$ $416 = \sqrt{3} \times 6.6 \times I$

Or $I = \dfrac{416}{\sqrt{3} \times 6.6} = 36.4$A (at full-load)

Starting current is 5 x full-load current

$= 5 \times 36.4 = 182$A (at standstill)   ....(i)

Starting power factor is given as 0.2 (= $\cos \varphi$) ....(ii)

$\therefore$ $\sin \varphi = 0.98$ (from tables)

$\therefore$ Active power
$$= \sqrt{3}kV \times I \times \cos \varphi$$
$$= \sqrt{3} \times 6.6 \times 182 \times 0.2$$
$$= 416kW \qquad\qquad ....(iii)$$

$\therefore$ Reactive power
$$= \sqrt{3}kV \times I \times \sin \varphi$$
$$= \sqrt{3} \times 6.6 \times 182 \times 0.98$$
$$= 2\,040kW \qquad\qquad ....(iv)$$

Summarising the figures (i), (ii), (iii) and (iv):

Starting current	= 182A	(36A)
Starting power factor	= 0.2	(0.8)
Starting active power	= 416kW	(333kW)
Starting reactive power	= 2 040kvar	(250kvar)

If these four figures are compared with the corresponding quantities in the top line of the table of Chapter 4 for the motor running on full-load (repeated above in brackets alongside each), it will be seen that, when starting:

(a)   The starting current is very much greater (it was given as 5 times).

(b)   The starting power factor is very much lower (it was given as 0.2).

(c)   The starting active power is of the same order as the full-load running active power (in this case a little higher).

(d)   The starting reactive power is vastly greater than the full-load running reactive power (in this case eight times).

As the motor runs up and approaches synchronous speed, these figures gradually change until, at whatever load the motor is to run, they settle at the figures in the table of Chapter 4.

Since the voltage drop in a generation and distribution system is due principally to the reactive loading, it becomes clear that a voltage dip will be experienced whenever a motor, particularly a large motor, is started. The sudden demand for considerable reactive power will cause the voltage to drop before the AVR is able to correct it. This appears on an installation as a momentary dip of the lights.

CHAPTER 6 MOTOR OPERATION AT REDUCED VOLTAGE

6.1   GENERAL CONSIDERATION

Due to system disturbances, especially when starting large drives on limited generating capacity, motors may at any time find themselves operating on a voltage lower than the nominal value.

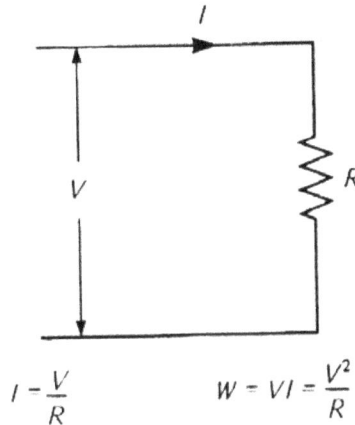

$$I = \frac{V}{R} \qquad\qquad W = VI = \frac{V^2}{R}$$

FIGURE 6.1
CURRENT AND POWER IN A SIMPLE RESISTOR

If such a thing happened when voltage were applied to a simple resistor, as shown in Figure 6.1, the current, as determined by Ohm's Law, would always remain proportional to the applied voltage. For example, a 10% reduction of voltage (i.e. to 0.9 times nominal) would result also in a 10% reduction of current (i.e. to 0.9 times its previous value), and the **power**, which is the product of voltage and current, would be reduced to 0.9^2, or 0.81, times its previous value - that is, a power reduction of 19%. For small voltage reductions the percentage power drop is approximately double the percentage voltage drop.

This is quite simple with a fixed resistance, but induction motors do not behave in this way, as will be explained below. It is because they do not present a fixed impedance to the voltage, equivalent to the R of Figure 6.1. In fact Ohm's Law does not appear to apply. The behaviour of motors when the applied voltage varies must be examined in more careful detail.

6.2   EFFECT ON TORQUE

It was explained in Chapter 2 that an induction motor operates by the interaction of the rotating stator field and the currents in the rotor bars. The magnitude of the field depends directly on the applied voltage; also the emf induced in the rotor, and so the rotor bar current, depends on the strength of that field. So the rotor current too depends on the voltage, and, since the developed torque is the product of both, that torque is proportional to the square of the applied voltage. Expressed in symbols:

$$T \propto V^2$$

It will be seen therefore that torque is very sensitive to voltage variation. Taking the figures from the previous example, a 10% reduction of voltage (i.e. to 0.9 times nominal) reduces the torque to 0.9^2, or $0.81\,T$. For small voltage variations the percentage drop of **torque**, like that of the power, is also approximately double the percentage voltage drop.

For small voltage variations this might be acceptable, although it would increase the motor's run-up time from starting and could so cause some overheating. It can, however, pose a serious problem if the variation is large. For example if, when starting a very large motor, the system voltage fell by 30% (i.e. to 0.7 times nominal - a situation not unknown), the torque would fall to 0.7^2, or $0.49\,T$ - that is, to about half its nominal value. If this were the starting torque of a large motor with a high-friction drive, it might not be enough to overcome the initial break-away friction, or 'stiction', and the motor would then not start at all.

Consider now the effect of a running motor suffering a sudden reduction of applied voltage of 10% - that is, to 0.9 times nominal. It has already been shown that the torque will be reduced by 19%, that is to $0.81\,T$.

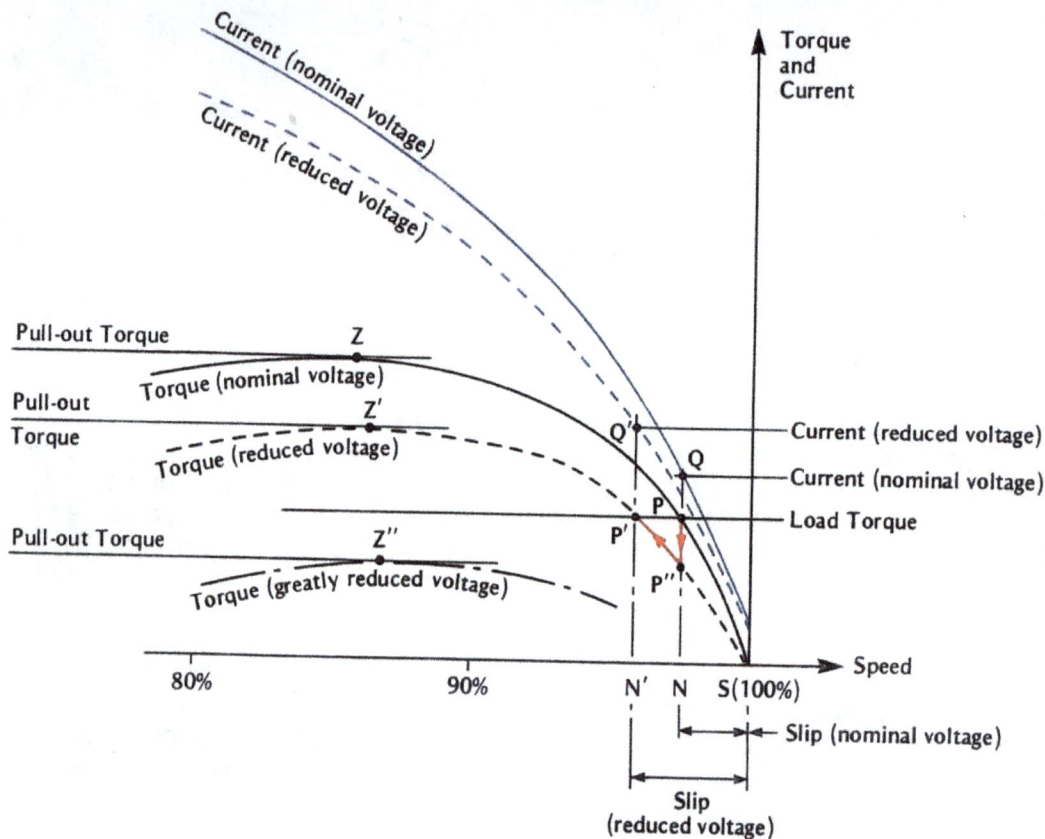

FIGURE 6.2
INDUCTION MOTOR OPERATING ON REDUCED VOLTAGE

The full lines of Figure 6.2 are reproductions, on a wider scale, of the right-hand parts of the torque/speed (black) and current/speed (blue) characteristics of Figure 5.1 in Chapter 5, for a motor operating at its rated, or nominal, voltage. The dotted curves are the corresponding characteristics for the same motor operating at a 10% reduced voltage. As already explained, the reduced-voltage torque/speed curve will lie at all points 19% below the nominal curve.

Point S is synchronous speed and point N the actual speed when a stated load torque is applied with the motor working at nominal voltage. SN is then the slip, point P the working point for the given load torque, and point Q the working current point.

If now, while the motor is still loaded, the voltage is reduced by 10%, momentarily the external load torque and the speed remain unaltered. All that happens is that the characteristics change suddenly from the full-line to the dotted-line curves, and the torque working point drops from P to P".

The torque being delivered by the motor (NP") is now less than the mechanical torque being imposed on it by the external load. Since the latter has not changed, the motor and its driven load will begin to slow down, and the speed point will move to the left from position N. The slip will increase, and, as it does so, the working point will move to the left from P" up the dotted characteristic. As it does so the motor torque will increase. When it reaches a new point P', it is once again equal to the load torque; the deceleration ceases and the motor settles down to a new speed N' slightly less than its previous full-voltage speed N, and with a slightly increased slip. The progressive change in torque is indicated in Figure 6.2 in red.

6.3   EFFECT ON CURRENT

On the same Figure 6.2 the point Q represents the current with full voltage applied and with the given load. After voltage reduction the new speed is at point N', and the new current is given by point Q' on the intersection of the vertical through N' and the reduced-voltage dotted current curve. This is **higher** than the current (Q) at full voltage - that is to say, the current rises as voltage is reduced, in direct contrast to Ohm's Law.

This result should not be surprising if considered from the power point of view. The mechanical power extracted from the motor is the product of torque and speed. The speed decreases only slightly since there has been a slight increase in a small slip, and the load torque can be regarded as almost constant. Consequently the mechanical power output is also nearly constant, demanding a near-constant electrical power input. But electrical power ($VI \cos \varphi$) is proportional to the product of V and I, so, if V goes down, I must go up more or less in inverse proportion.

6.4   PULL OUT AND STALL

In Figure 6.2 the dotted torque curve was drawn for only a 10% voltage reduction, and the maximum available, or 'pull-out', torque shown there (point Z') is greater than the load torque. Therefore a point P' will be established where the motor torque can rise to meet the demanded load torque, even with the 10% voltage reduction, and the motor settles down to a new steady but slightly lower speed.

But if the voltage reduction had been greater, the dotted reduced-voltage torque curve would be far lower (being proportional to V^2), and its maximum, or pull-out, value at point Z" might then be lower than the demanded load torque, as indicated in part by the chain-dotted curve in Figure 6.2. The point P' could then never be found where the motor torque can rise to meet the load torque, and deceleration would continue until the motor stalled to a stop. Its current would rise to its starting or 'locked-rotor' value at that voltage and would flow continuously, causing rapid overheating (see Chapter 10, Paragraph 10.3, 'Overcurrent / Overloading'). The motor would have to rely on its automatic protection to clear it off the line and prevent burnout.

Even without a voltage reduction, if the load torque should rise excessively - for example due to a malfunction of the driven machine, or to a bearing seizure - to a level exceeding the pull-out peak torque curve of Figure 5.1, the torque delivered by the motor cannot then rise to meet it, and the motor will decelerate into a stall. The lower the voltage, the lower the pull-out torque.

CHAPTER 7 POWER FACTOR CORRECTION

7.1 THE INCENTIVE

The larger consumers of electric power onshore pay not only for the energy they actually use (in kWh units) but also a contribution towards the capital cost of the supply system. The amount paid for this element is usually based on the maximum kVA demand (note, not kW) over an accounting period. This system of tariffs is further discussed in Part 4 Electrical Systems.

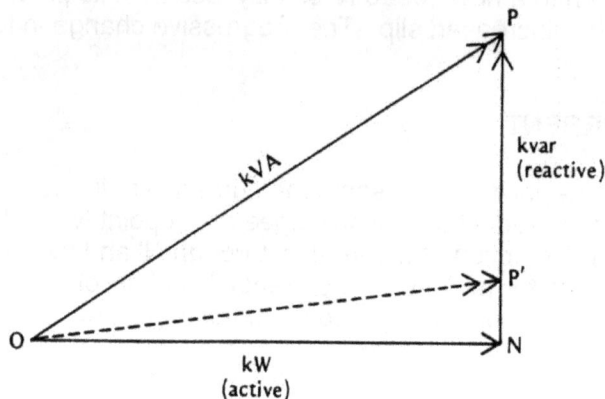

FIGURE 7.1
POWER TRIANGLE FOR INDUCTIVE LOAD

The kVA being demanded at any time is the vector combination OP of active power (kW) and reactive power (kvar), as shown in Figure 7.1. The active power reflects the mechanical output of the various motors of an installation, together with lighting and heating, whereas the reactive power is required for their magnetisation. It depends mainly on the total number and size of motors running at any instant and is independent of their loading. From Figure 7.1 it can be seen that, the smaller the reactive element NP can be made, the nearer the kVA demand (OP') can be made to approach the active load element (ON) and so achieve its least value for any given active load.

Ideally therefore the minimum running cost of an installation is achieved if the reactive loading (vars) can be eliminated altogether, so that the plant draws **only active** load from the system - that is to say, it runs at unity power factor so far as the supply system is concerned. This is known as 'power factor correction'.

It can be of great importance to onshore installations that take their power from an Area or Supply Authority and have to pay them for it under a tariff such as described above. It is of less importance to an offshore installation which does not obtain its energy under a tariff, but a good power factor is nevertheless desirable as it uses plant more efficiently. The methods described below will be found onshore but not offshore.

7.2 THE PRINCIPLE

It is shown in Chapters 22 and 23 of Part 1 that the current which flows in a purely inductive circuit when an a.c. voltage is applied lags 90° in phase on that voltage, whereas in a purely capacitive circuit the current leads 90°.

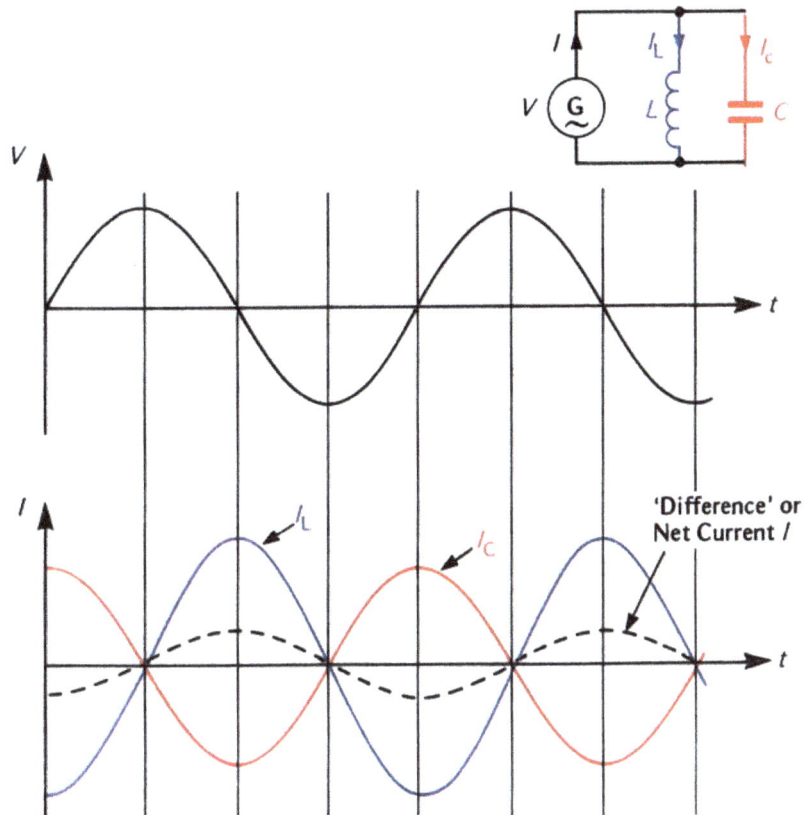

FIGURE 7.2
LAGGING, LEADING AND NET CURRENT

Figure 7.2 shows an a.c. voltage applied to both a purely inductive circuit *(L)* and a purely capacitive circuit *(C)* together.

The inductive current wave is I_L (blue) the capacitive I_C (red). Since the former lags, and the latter leads, 90° on the voltage wave, they are 180° apart with respect to each other - that is to say, they are 'anti-phase', the positive parts of the one coinciding with the negative parts of the other. The magnitudes of the two current waves depend on the impedance of the inductor (Z_L) and that of the capacitor (Z_C) according to the a.c. version of Ohm's Law, namely:

$$I_L = \frac{V}{Z_L} \text{ and } I_C = \frac{V}{Z_C}$$

In Figure 7.2 the dotted curve is the difference between the I_L and I_C curves - or more strictly the algebraic sum, since I_C is negative with respect to I_L at all points along it. If there is little numerical difference between I_L and I_C (as shown in the figure), the difference curve will be very small indeed. If $I_L > I_C$ it will be in phase with I_L (as shown), but if $I_L < I_C$ it will be the other way up, in phase with I_C.

In the special case where $I_L = I_C$ numerically, there is no difference at all. Between them the two circuits then draw no **net** current whatever from the mains, even though current passes through each and circulates between them, passing from one to the other and back again. The circuits are said to be 'in resonance'.

31

This suggests that, if we have a circuit containing inductance, such as a motor, which draws lagging reactive power (vars) from the mains, it can be completely offset by placing in parallel with it another circuit containing only capacitance. The value of that capacitance is chosen such that the leading vars drawn by it just counterbalance the lagging vars drawn by the motor. If this is done, the pair will between them draw no net vars from the mains, but reactive power will circulate back and forth between the two. The capacitor can in fact be regarded as supplying all the magnetising vars to the motor, instead of the mains being called upon to do so.

Of course any **active** power needed by the motor for its driven load and losses will continue to be drawn from the mains. It is only the demand for reactive power that has been completely removed from the mains and is now met from the capacitor. Since only active power then comes from the mains, it is supplied at unity power factor, and the kVA demanded is reduced to the lowest value possible - namely equal to the kW demand.

7.3   THE PRACTICE

This description suggests a practical means of so correcting a motor that it draws only active power from the mains.

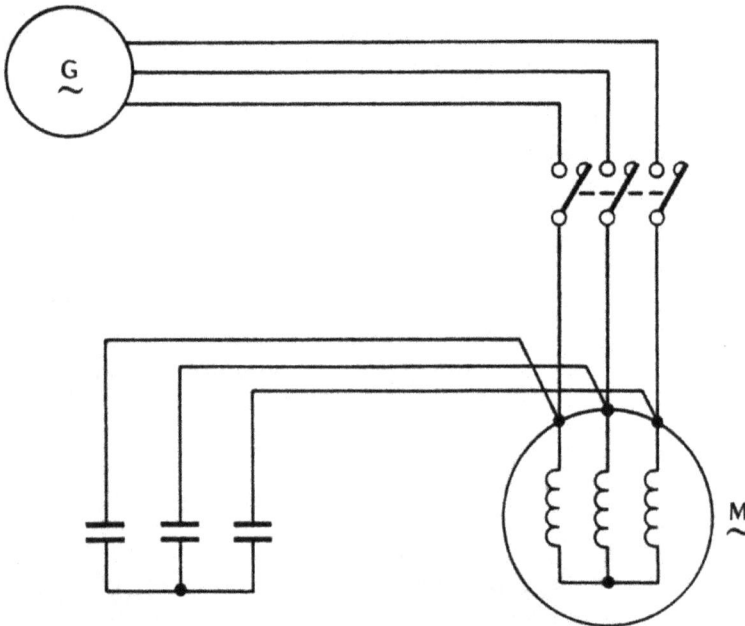

FIGURE 7.3
CAPACITOR CONNECTIONS TO A MOTOR

A 3-phase set of capacitors is connected in parallel with the motor terminals as shown in Figure 7.3. They will be switched by the same contactor as is used to start the motor; this ensures that the capacitors are only in circuit when the motor itself is. The capacitance value is chosen so that the reactive power in kvar (leading) drawn by the capacitors is as nearly as possible equal to the reactive power in kvar (lagging) drawn by the motor to magnetise itself. And since this magnetising power is constant and does not vary with the motor loading, the chosen capacitors will compensate at all motor loads. An example of the calculation for choice of capacitor size is given overleaf.

Example:

Q   A 240 hp motor has a power factor of 0.8 at full-load and an efficiency of 85%. What size of capacitors is required to provide full correction?

A   240 hp = 180kW$_m$. If efficiency is 85%, total input power is $180 \div 0.85 = 212kW_e$.

As power factor is 0.8, the total input kVA is 212 ÷ 0.8, or 265kVA.

Since power factor (cos φ)	=	0.8, sin φ = 0.6
Total reactive power	=	kVA × sin φ
	=	265 × 0.6
	=	159kvar

This is the total leading reactive power to be supplied by the bank of three capacitors. Each capacitor therefore should have a rating of 53kvar. (Note that power capacitors, unlike those used in electronic circuits, are usually rated in 'kvar' at a stated voltage and frequency (or sometimes 'kVA'), which is the same thing, as resistance is negligible. Their capacitance in µF or mF can be calculated if desired, but it is not of any use in this calculation.)

Since the capacitors are connected directly to the motor terminals and therefore down-stream of the starting contactor, any charge left in them on switching off will be dissipated in the motor windings.

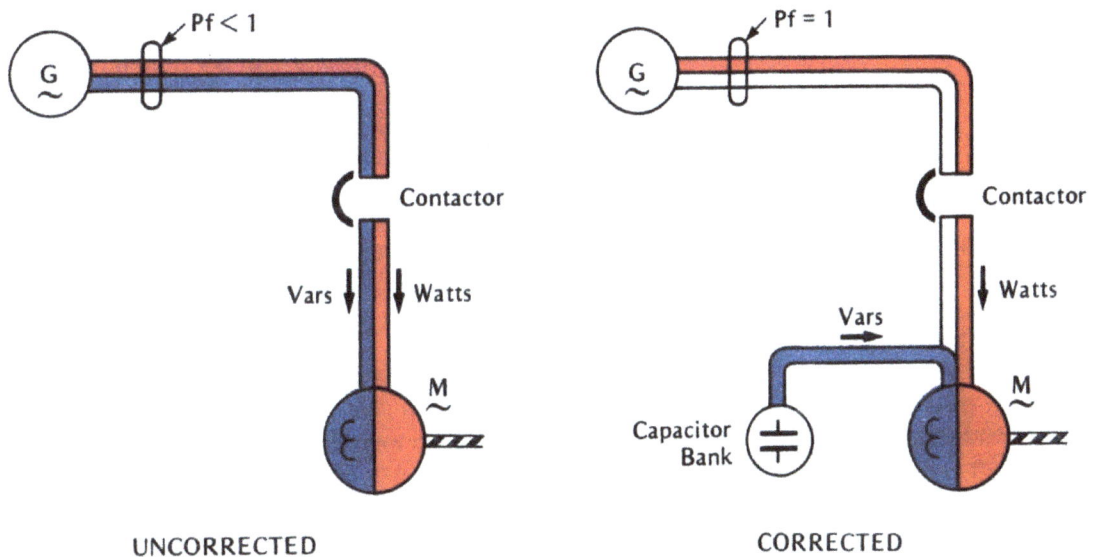

FIGURE 7.4
EFFECT OF POWER FACTOR CORRECTION BY CAPACITOR

It has already been explained in Figure 7.2 that the leading current drawn by a capacitor is completely opposite in sign to the lagging current drawn by an inductor. In fact a leading current can be regarded as a **negative** lagging current. Following this line of argument, the leading vars drawn by the capacitors can be regarded as negative lagging vars going in, or positive lagging vars coming **out**. In the right-hand picture of Figure 7.4 the blue parts represent lagging vars, which now come **from** the capacitor to the motor, so cutting out the need to draw them from the mains. This diagram explains perhaps more clearly how a capacitor bank corrects the power factor of a motor by providing its lagging vars for magnetising instead of the mains doing so. As far as the mains are concerned, they see a motor which only requires active power and therefore operates at unity power factor.

In large industrial installations some of the bigger motors may be provided with individual capacitors, but with smaller machines a single capacitor bank might be installed to correct a group of motors. In that case they would be sized to correct for an average number of motors running. If more than the average number were on-line there would be some under-correction; if less, there would be over-correction.

Power factor correction is not confined to motors, though this is its main application. Of interest may be the application to an induction furnace in a steelworks. Here a crucible containing steel pieces for melting is heated by induction from an alternating current flowing in a coil round the crucible. When cold, the steel is highly inductive and so causes a heavy demand for reactive power in the heating coil, with a consequent low power factor. As the steel heats, its magnetic properties change; the inductance drops, and with it the demand for reactive power.

If sufficient capacitors were installed to correct for the initial cold state (usually banks of several in parallel), there would be progressive over-correction as the steel became hotter. This would require capacitors in the banks to be switched out in sequence.

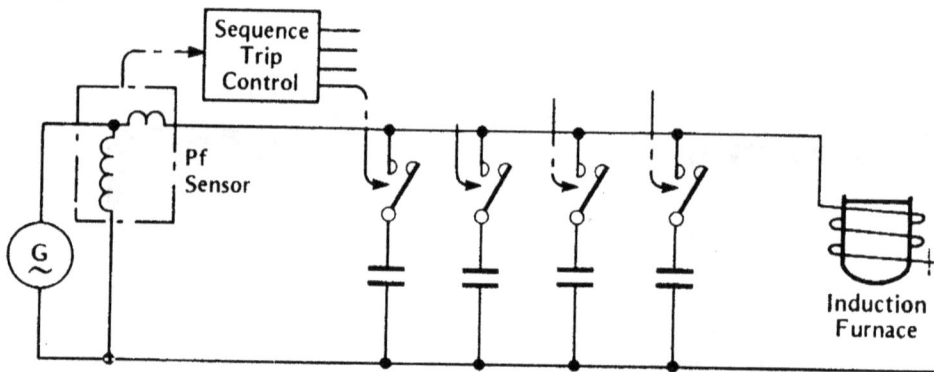

FIGURE 7.5
INDUCTION FURNACE AUTOMATIC POWER FACTOR CONTROL

Figure 7.5 shows a typical automatic control system. It monitors the reactive power being drawn and controls the number of capacitors needed as the melt progresses, so keeping the power demand from the mains as near unity power factor as possible all through the melting process.

7.4   CLOSENESS OF CORRECTION

Ideally a motor's reactive power demand should be exactly nullified by the correcting capacitors, but this is not always practical, nor is it necessary.

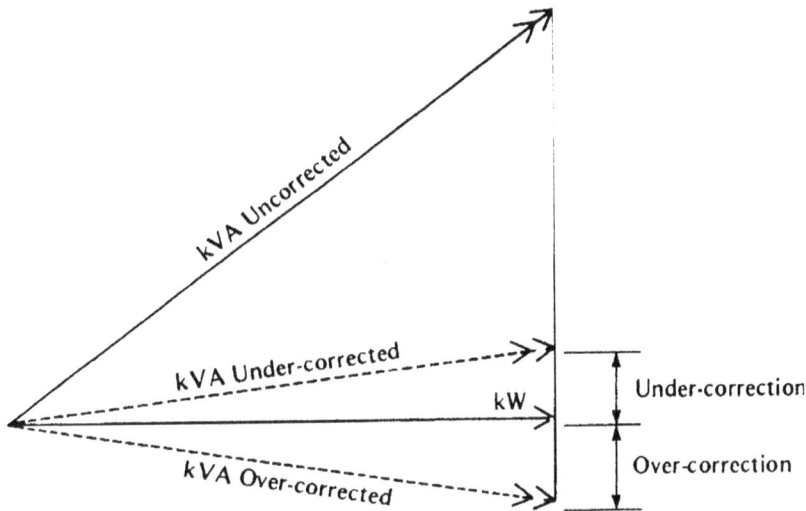

FIGURE 7.6
UNDER-CORRECTION AND OVER-CORRECTION

Figure 7.6, which is a development of Figure 7.1, shows the kW, kvar and kVA of a motor uncorrected, slightly under-corrected and slightly over-corrected. It is the purpose of power factor correction to reduce the kVA to the lowest possible level - ideally to equal the kW. It can be seen from the figure that, so long as the under- or over-correction is not too large, the kVA does not differ much from the kW; indeed, the difference is a 'second order' effect.

Capacitors, although simple pieces of equipment, are nevertheless costly to provide and install, especially for high-voltage plant. Therefore no more should be spent on them than will show an overall gain compared with the cost of a high maximum demand kVA charge.

This calls for a nice calculation which takes account of the tariff, the cost of the capacitor equipment and its installation. A careful balance is required, and it will probably result in slight under-correction. If circumstances subsequently alter, such as increased tariffs, further capacitors can always be added.

7.5   POWER FACTOR CORRECTION OF NETWORKS

So far this chapter has considered only local power factor correction of consumer equipment, mainly motors. This may be required to achieve the minimum running cost for the consumer's plant.

Supply authorities' networks onshore should also be operated at as high a power factor as possible, both for economic reasons and more particularly to maintain voltage levels on the system. How this is done is explained in Part 6 'System Control'.

In particular the use of the 'synchronous condenser' in this role is discussed in Chapter 1, para. 1.9, of that part.

CHAPTER 8 MOTOR TESTING

8.1   GENERAL

The general requirements for the testing of all rotating electrical machines, including motors, are laid down in BS 4999 :1976, Part 60.

As far as induction motors are concerned, the principal requirements are:

- Manufacturer's Tests
- On-site Test
- Insulation Resistance Testing.

8.2   MANUFACTURER'S TESTS

Manufacturer's tests are given three classifications: 'Basic' (formerly called 'Type Tests'), 'Duplicate' and 'Routine Checks'. Basic tests are mainly to prove a new design. They include exhaustive tests to ensure that the design meets the specification and all other performance requirements. They are normally carried out only on the 'first of class' motor, and a test certificate is usually provided to confirm the tests. Basic tests may, on special request, be repeated on the first machine of a new, large order, but this is not usual.

Duplicate tests are for performance. They are applied to a motor that is of the same design and construction as one previously made (and in no way altered) and which has already undergone basic tests. The duplicate tests are to ensure that the motor is still in accordance with the original design.

Routine checks are tests to show that each motor has been assembled correctly, is able to withstand the appropriate high-voltage tests and is in sound working order both electrically and mechanically.

The three classes of test are listed in Table 1.

TABLE 1 - MANUFACTURER'S TESTS

Test	Basic	Duplicate	Routine
Resistance of windings (cold)	X	X	-
No-load losses and current	X	X	X
Locked rotor - current	X	X	-
- torque	X	-	-
Secondary induced voltage (wound-rotor type only)	X	X	X
Temperature rise	X	-	-
Power factor and efficiency	X	-	-
Momentary overload	X	-	-
High voltage	X	X	X
Vibration	X	-	-

Most of these tests are self-evident, but two are further explained overleaf.

8.2.1 Locked Rotor Test

This test is carried out with the rotor locked and at a voltage reduced so that the stator and rotor currents do not exceed their normal full-load values. It simulates the standstill situation at the moment of starting without motion actually taking place. By measuring the current and torque at this reduced voltage, the actual standstill current and torque when starting at full rated voltage may be calculated.

8.2.2 High Voltage Test (also called a 'Withstand' test)

At this test a high voltage is applied between the frame and all the motor windings (the stator windings only in the case of squirrel-cage motors), with all other conductors, metal and auxiliary (i.e. heater) circuits bonded to the frame. The actual voltage applied is in accordance with Table 2 and is sustained for one minute. It may be at any frequency between 25Hz and 100Hz. It is primarily an insulation test for the motor's windings and is included also in the routine checks to ensure that there has been no fault during assembly of any individual motor.

TABLE 2 - HIGH VOLTAGE TESTS

Windings	Test Voltage (rms)
Motor stator windings: <100V, <1kVA >1 000V, <1kVA 1 – 10 000kVA >10 000 kVA and <2 000V 2 000 — 6 000V 6 000 — 17 000V >17 000V	500V + twice rated voltage 1 000V + twice rated voltage 1 000V + twice rated voltage, min 1 500V 1 000V + twice rated voltage 2.5 times rated voltage 3 000V + twice rated voltage Special agreement
Rotors of wound-rotor machines: (non-reversing motors)	1 000V + twice open-circuit standstill voltage with normal stator voltage applied. Min 1 500V

8.3   ON SITE TESTS

Any motor installed on an offshore or onshore installation may be assumed to have undergone its full routine check tests, and its prototype a full basic or duplicate test. On-site tests are therefore only needed to check the original installation and commissioning, and thereafter to ensure that no deterioration has taken place. The remainder of this chapter deals only with tests for deterioration.

Deterioration can occur for many reasons: among them are entry of dampness or water leakage in the motor or cable-entry box, overheating of the windings due to overloading or prolonged stalling, or mechanical faults such as vibration or bearing failure.

Both dampness and overheated windings can cause reduced insulation resistance of the windings. After drying out, the motor should be insulation-tested to ensure that insulation resistance has been restored. Deterioration can be progressive, especially when a motor is little used, and a regular programme of insulation resistance testing every motor should be drawn up and the results logged. After temperature correction (see para 8.4) the resistance levels should be plotted, and, if there is progressive deterioration, this will be immediately apparent.

After repairs to a motor set an insulation resistance test should always be carried out before reconnection if the motor or its connections have in any way been interfered with.

High voltage withstand tests should never be needed on site unless a major overhaul has been carried out, in which case it would be an engineering or manufacturer's concern.

8.4 INSULATION RESISTANCE TESTING

Instruments such as a 'Megger' are provided to operate at 250V, 500V or 2 500V, and the correct one must be used depending on the rated voltage of the motor to be tested. Normally motors over 415V and high-voltage motors would require a 2 500V megger. Motors in the 100V range would need the 500V instrument; a higher-voltage instrument might itself damage the insulation.

When the tester is connected and the handle wound up, the voltage should continue to be applied until the needle settles down to a steady value; this might take one minute or more.

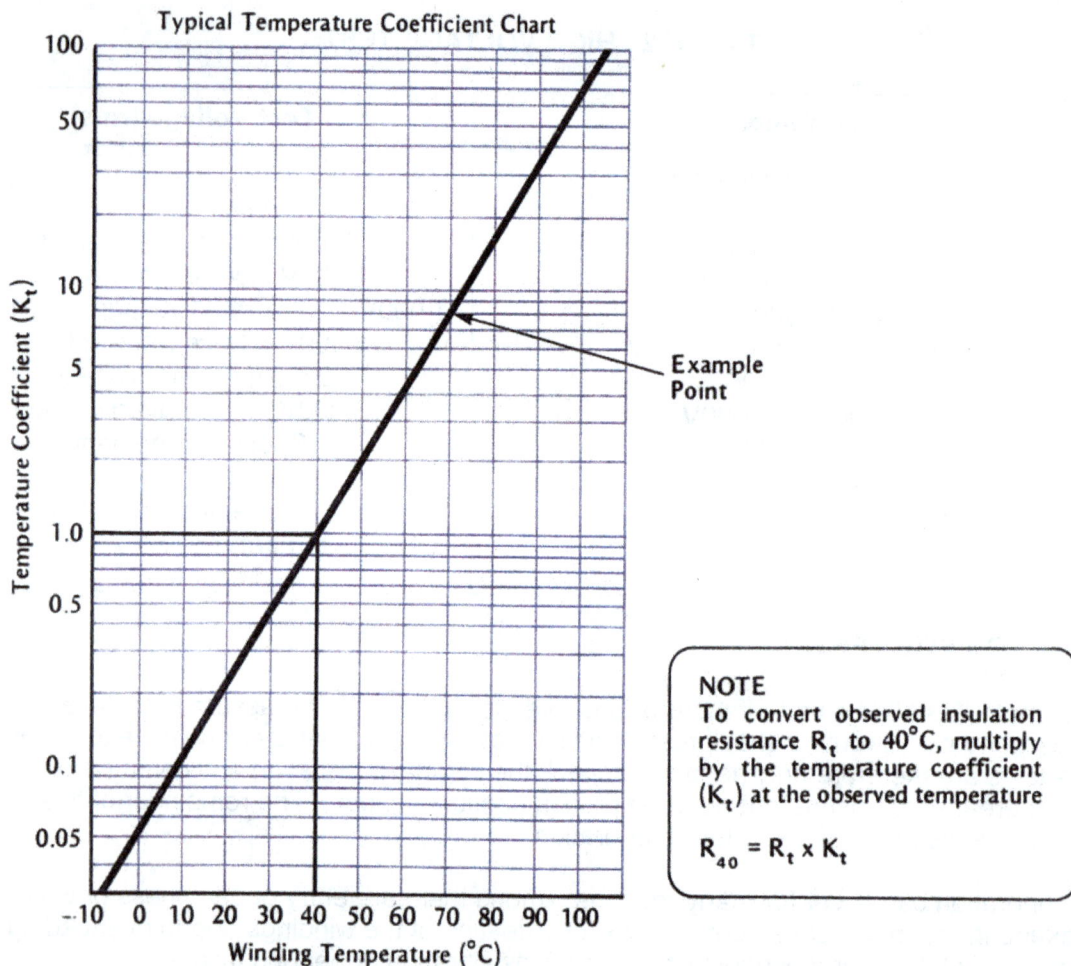

Typical Temperature Coefficient Chart

NOTE
To convert observed insulation resistance R_t to 40°C, multiply by the temperature coefficient (K_t) at the observed temperature

$$R_{40} = R_t \times K_t$$

FIGURE 8.1
INSULATION RESISTANCE TEMPERATURE COEFFICIENT

When testing the insulation resistance of a winding, all other conductors, metalwork (stator and rotor) and auxiliary circuits such as for temperature protection and heaters should be connected to the frame with light wire (fuse wire will do), and the test voltage applied between winding and frame. Where the three phase windings are independent and brought out to six terminals (fairly rare), it is advisable to make a test also between pairs of windings. However, most motors are star-connected with their star-point internal and permanently made, so inter-phase tests are not possible.

Insulation resistance is very dependent on temperature, and, in order to compare one reading with another, it is necessary to reduce the value to a common temperature. This is usually 40°C. Unlike the resistance of a conductor, which rises with temperature, the resistance of insulation falls rapidly with increase of temperature.

The graph of Figure 8.1 is used to make this correction by means of a 'temperature coefficient'.

For example, if the observed reading (R_t) is 10 megohms when taken at 70°C, then, using the graph, the temperature coefficient (K_t) is 8.0, and the corrected reading (R_{40}) at 40°C is then 10×8.0 = 80 megohms. It can be seen from this example that the correction is considerable when the winding is hot at normal working temperatures.

The recommended minimum value of insulation resistance for items of plant is given in manufacturers' literature. As a guide where precise information is not available, the minimum acceptable value (R_m) for a motor stator winding is given by

$$R_m = (kV + 1) \text{ megohms when corrected to } 40°C,$$

where kV is the motor's rated voltage in kilovolts. Thus:

for 415 or 440V motors	R_m =1.4 MΩ
for 6.6kV motors	R_m =7.6 MΩ
for 11kV motors	R_m =12 MΩ

CHAPTER 9 INTRODUCTION TO MOTOR PROTECTION

Motors, like any other item of electrical equipment, lend themselves readily to automatic protection against abnormal operation and possible damage. This invariably takes the form of quick disconnection from the supply to prevent - or at least to limit - any damage. The protection can vary from a simple set of fuses to a sophisticated protection system consisting of sensors or relays, or both.

Protection in general, and motor protection in particular, is dealt with in detail in Part 10 'Electrical Protection'. It will be sufficient here to enumerate the possible types of protection which may be found with motors on offshore or onshore installations.

Type of Protection	Remarks
Fuses	Against overcurrent from any cause such as mechanical overload, short-circuit or prolonged stalling. Also to back-up contactors.
Overcurrent	Instantaneous or inverse-time, electromagnetic or thermal relay. Against overcurrent from any cause (as for fuses).
Earth Fault	Against an earth fault in the motor supply circuit or the motor itself.
Stalling	Against stalling while running still connected.
Single-phasing	Against loss of one phase, by means of a 'Negative Phase Sequence' device or relay.
Undervoltage	Against loss of supply when running, preventing automatic restart when voltage is restored.
Motor Protection	A combination of thermal overcurrent (inverse-time) and single-phasing protection. Sometimes with a stall relay included.
Winding Overtemperature	Against excess temperature in the motor windings, by embedded thermocouple, thermistor or resistance temperature device elements in the windings. These sensors are described in Part 7 'Control Devices'.

CHAPTER 10 MOTOR FAILURES

10.1   GENERAL

The squirrel-cage induction motor, as used offshore and onshore, is a very robust machine and should give little trouble in service unless it is abused.

Nevertheless, like any machine, it can be subject to failure, and this chapter deals with those failures that are most likely to be met. Excluded from the list are such obvious things as physical damage or flooding, which can hardly be blamed on the motor itself.

Failures in a motor may be electrical or mechanical. They are discussed below.

10.2   INSULATION FAILURE

Failure of insulation of the stator windings, which carry the full system voltage, can cause a current leakage to earth or between phases. If this is not detected and the motor taken out of service, it can quickly lead to a complete breakdown and a flashover to earth or between phases - that is, a short-circuit. Short-circuits, even though starting between two phases, spread rapidly to envelop the third phase and produce a full 3-phase short-circuit.

Insulation breakdown can occur if the original insulating material was faulty, but it is much more likely to be due to damp or overheated insulation.

If a motor has become damp through a long period of disuse, it is always advisable to dry it out with its own heaters or a temporary heater before voltage is applied. An insulation resistance test will indicate when a satisfactory state has been reached.

Deterioration of insulation may not at first be enough to cause breakdown, but if repeated or prolonged it is progressive. The surge of current when starting a motor in this vulnerable state may just be enough to cause the final breakdown. Again, periodic insulation resistance tests should indicate any progressive deterioration that may be taking place.

10.3   OVERCURRENT/OVERLOADING

During the course of its service life a motor may have become overloaded and so subject to overcurrent in its windings. Such overcurrents may perhaps not have been enough to operate the motor's overcurrent protection but sufficient to overheat it if prolonged, remembering that heating is proportional to the **square** of the current. Thus a 20% overcurrent will cause over 40% heat generation, leading to possible and eventual insulation breakdown. Continuous running at reduced voltage can produce the same effect (see Chapter 6).

Stalling of a motor while still connected and running can lead to a similar overcurrent (but not 'overload') situation, where the current rises to its starting value and is sustained without ventilation until the motor protection operates. With a typical starting ('locked-rotor') current five times full-load current, the heating rate is 25 times that at full-load, aggravated by lack of ventilation. Such a situation, if sustained even for a few seconds, can quickly lead to insulation failure and breakdown.

Repeated attempts to start, or even repeated successful starting over a short period, can also cause overheating of the windings, as explained in Chapter 5. For this reason some operators have made a strict rule against repeated starts, which is set down in that chapter.

10.4 CABLE BOX FAILURE

One area where flashover is not uncommon is the cable entry box to the motor. This is usually a 'trifurcating box', where the 3-core power cable enters through a gland, and inside it the outer cable sheath is removed and the three cores are led separately to the motor terminals.

It has been found sometimes that during installation individual cores may have been bent too sharply, so cracking their insulation and presenting a weak spot for eventual breakdown.

A cable box failure can be very dangerous, as the release of energy caused by a flashover in the confined space of the cable box can lead to what amounts to an 'explosion', with danger of fire and to personnel. In high-voltage motors where the fault level is also high (see Part 10 'Electrical Protection') the release of energy can be enormous.

10.5 ROTOR BURNOUT

Rotors of squirrel-cage motors have uninsulated conductors lying in slots in the laminated iron; there is therefore no problem of insulation failure as with the stator, and furthermore the iron rotor in direct contact with the conductor bars provides a good 'heat sink'. Rotors can have other problems however.

The rotor conductors carry very considerable currents when running, and even more when starting. A rotor is susceptible to similar overcurrents as occur in the stator, and for the same reasons as given in para. 10.3. The usual construction of the squirrel-cage itself is for the conductors to be brazed into the end-rings. If the quality of the brazing is not good, the rotor currents may cause excessive heating at the joint due to poor contact. This can become progressive, leading to final breakdown of the joint, severe arcing and eventually total burnout, which could well involve the stator too.

When motors are dismantled for overhaul, special attention should be given to the brazed end-ring joints for signs of overheating.

For motors with wound rotors and sliprings (very rare), the rotor would have the same insulation problems as described for the stator, to which may be added possible trouble with the sliprings and brushes.

10.6 BEARINGS

Most motors have rolling bearings, and sometimes one bearing is shared with the driven load. Larger motors usually have journal bearings.

The calculation of the 'life' of a rolling bearing is a complicated business involving the maker's data for that bearing (based on a stated speed and total running hours) modified by the actual operating speed and the total running hours required. The designer calculates the correct bearing for the job, and it should give the life he expects of it.

However, this does not always happen, especially if the bearing has been subjected to loads in excess of those calculated, or if it has undergone particularly severe conditions such as excessive temperature or vibration.

Bearings will consequently fail in service from time to time and will need to be replaced. Failure is usually indicated by increasing noise and can often be confirmed by putting a rod in contact with the bearing case to the ear.

When a bearing is replaced it **must** be replaced by the correct one recommended by the manufacturer. Not all bearings, even of the same dimensions, are interchangeable, since bearings with differing degrees of internal clearance are manufactured.

Particular care is required with greasing. A new bearing is usually provided in its box packed with a preservative grease. This is **not** the running grease and must be completely removed and replaced by the correct quantity of the correct running grease.

Bearings must be greased regularly as part of the planned maintenance routine. **They must never be overpacked with grease and never overgreased when running.** This causes viscous friction and consequent overheating, leading to melting of the grease. One of the major causes of bearing failure is due to overgreasing.

Journal bearings, which are likely to be found only on the largest motors, should give little trouble so long as they continue to have proper lubrication. In many large motors part of the starting sequence is to pressurise the bearings to 'float off' the journal before movement takes place.

The quite large stray magnetic fields in the area of the bearings induce voltage in the journal itself, and, as the whole rotor is floating on a bed of oil, these charges cannot escape. If they build up sufficiently, spark-over can occur between the journal and the metal shell, breaking down the oil film and possibly damaging the metalled bearing surface.

Some large machines are provided with insulated bearings and with a brush running on the shaft near the bearing and connecting it to the earthed frame. This discharges any build-up of voltage on the journal and should prevent the sparking problem. Periodic checking of the earth continuity of this brush is desirable while the motor is running.

10.7   VIBRATION

Vibration is one of the major causes of failure in rotating electrical machines. It may arise from a number of causes:

> Mechanical unbalance
> Electromagnetic unbalance
> Thermal unbalance
> Unbalance induced by starting and restarting.

Mechanical unbalance is due to the centre of gravity of the rotating mass not being exactly on the centreline of rotation. It is checked by an overspeed test carried out by the manufacturer and, if small, is compensated for by adding small balance weights at the proper points on the rotor rim. This, however, is a 'basic' test and is not normally repeated in production machines.

Provided that the production machines are correctly assembled with good quality control, there is no reason why their rotors should be out of balance. If they were it should have become apparent - and been corrected - during the normal routine check tests.

A squirrel-cage rotor is very robust, and there is little scope for movement of the conductor-bars to cause unbalance and therefore vibration. It is just possible that one of the small balance weights may work loose or even be shed; this would reveal itself by noticeable vibration. In such a case the motor should at once be stopped and the cause found and corrected.

Electromagnetic unbalance can be due to an unequal air gap around the rotor, causing unequal magnetic pull as it rotates. It can also be due to an electrical fault in the rotor windings, causing distortion of the magnetic field system in the air gap.

Thermal unbalance may be due to uneven thermal strains distorting the material of the rotating mass.

During the starting of a motor the currents in the three phases are asymmetrical at first and cause a directional pull on the rotor due to the d.c. components of the currents. This is short-lived but may have a cumulative effect over a period of use.

Some of the largest motors are provided with 'Vibration Monitors'. There are sensors at each bearing which, through an electronic circuit, give an alarm if the vibration reaches a certain preset lower level and which trips the motor if it reaches a preset higher level.

Vibration is measured by accelerometers mounted at suitable points on or near the motor. They measure acceleration directly and, by suitable integrating circuits, can also be made to indicate vibration velocity or vibration amplitude. All these quantities are periodic, and it is customary to express them in 'root-mean-square' terms: thus acceleration is expressed as 'mm/s^2 (rms)', velocity as 'mm/s (rms)' and amplitude as 'mm (rms)'.

Some operators set acceptable vibration velocity limits in their specifications for electric motors. Typically it should not exceed 3mm/s (rms) measured on the bearing on a horizontal plane (for horizontally mounted motors) through the shaft centreline and with the completely assembled motor running at no-load. A vibration sensed by the monitor exceeding 5mm/s is regarded as an alarm level and above 11mm/s as critical. The vibration measured on the shaft relative to the bearing must not exceed 12mm/s (rms).

In practice it has been found that, although the vibration limits on test in the maker's works have been satisfactorily met, they are often exceeded when the motor is erected on site.

CHAPTER 11 USEFUL FORMULAE

Power:

$$1 \text{ hp} = 0.746 \text{kW}_m \qquad (= \tfrac{3}{4} \text{ kW}_m \text{ approx.})$$

$$1 \text{kWm} = 1.34 \text{ hp} \qquad (= \tfrac{4}{3} \text{ hp approx.})$$

Efficiency:

$$\eta = \frac{kW_m \ (output)}{kW_e \ (input)} \quad \text{for a motor}$$

$$kW_e = \frac{kW_m}{\eta}$$

Power Factor:

$$\cos \varphi = \frac{kW_e}{kVA}$$

$$kVA = \frac{kW_e}{\cos \varphi}$$

Voltage, Current and Power:
(3-phase)

$$kVA = \sqrt{3}kV \times I$$
$$kW_e = \sqrt{3}kV \times I \times \cos \varphi$$
$$kvar = \sqrt{3}kV \times I \times \sin \varphi$$

where these are rms line voltage and rms line current

(Note, when using these formulae on 415V or 440V motors, kV = 0.415 or 0.44 respectively.)

CHAPTER 12 QUESTIONS AND ANSWERS

12.1 QUESTIONS

1. Name three types of a.c. electric motor. What are the fundamental differences between a squirrel-cage induction motor and a synchronous motor?

2. How may a synchronous motor be used to reduce the power factor of the system?

3. Describe briefly the principal elements of a squirrel-cage induction motor, and how they react to cause the motor to run.

4. When would a wound-rotor induction motor be used?

5. If a motor is rated at 400 hp and has an efficiency of 85%, how many kilowatts does it draw from the supply at full load?

6. A design of induction motor can run on a single-phase supply. How is this done? Make a sketch.

7. A '2-pole' motor (with one winding per phase) is supplied from a 60Hz main. At what speed would you expect it to run? How does this speed vary with loading?

8. A '4-pole' motor (with two windings per phase) is supplied from a 50Hz main. At what speed would it run?

9. How is the enclosure of a motor classified against entry of (a) solid particles; (b) water? What is the range of protection against each?

10. What do you understand by the 'class of insulation' of a motor? Give an example.

11. What class of insulation would you specify for a motor to run in an ambient of 40°C and with a designed temperature rise at full load of 110°C?

12. What type of enclosure would you expect to see on a motor sited on the weather deck of a platform?

13. Describe two ways by which a motor is cooled.

14. What goes to make up the 'losses' in a running motor? Indicate those which are dependent on the loading and those which are not.

15. What kind of power does an induction motor draw in order to:

    (a) drive its load
    (b) magnetise itself
    (c) provide its losses?

16. How does the reactive power (in vars) drawn by an induction motor vary with motor loading?

17. If an induction motor draws 200A at full load, what sort of current would you expect to see at no-load?

18.   A 440V, 160 hp induction motor has a rated full-load power factor of 0.8 and a full-load efficiency of 85%.

   Calculate: (a) the active power
              (b) the reactive power
              (c) the total kVA
              (d) the current
              (e) the power factor

at full, half, quarter and no-load. (Disregard no-load losses.)

19.   How are squirrel-cage induction motors usually started? When started in this way, what momentary current and power factor would you expect to see?

20.   On what factors does the run-up time of a direct-on-line started motor depend?

21.   Name two alternative methods of starting so as to reduce the starting current. Describe briefly the principle of each.

22.   If an induction motor is started in either of the ways suggested by Q21, what will be the effect on the starting torque as compared with DOL starting?

23.   How does Korndorffer starting differ from simple auto-transformer starting? Name some applications of this method on offshore installations.

24.   What is the effect on the supply network of an offshore installation when starting a large motor? Explain why it occurs. Would you expect a similar effect with an onshore installation?

25.   The motor of Q18 has a starting current 5 times full-load current at a power factor of 0.25.

   Calculate: (a) the starting current
              (b) the starting kvar
              (c) the starting kW_e

26.   What limits should be observed when restarting motors?

27.   An induction motor is running and driving a steady load. The voltage suddenly drops by 20%. What then occurs, and what effect will be noticed on the switchboard instruments?

28.   How does torque vary with voltage?

29.   Under what conditions will an induction motor stall on drop of voltage?

30.   Why is power factor correction often employed in shore installations?

31.   Describe briefly, with sketch, how power factor correction may be applied to an induction motor.

32.   An induction motor draws at full load $200kW_e$ at 0.85 pf. How much capacitance is needed (in kvar) to correct power factor completely? If correction is complete at full-load, how will it be at half-load?

33. What types of test does a motor undergo? What tests would be carried out on a motor of a production run?

34. What on-site tests would you expect to carry out (a) as a routine; (b) after a repair; (c) after a major overhaul?

35. Why is it necessary to correct an insulation resistance test reading for temperature?

36. Name any three types of malfunction against which a motor may be protected?

37. What situations can cause deterioration of a motor's insulation? How can they arise, and what checks would you make if deterioration is suspected?

38. What are the most common failures in motors?

39. What is the special danger of a flashover occurring in a motor's cable box?

40. What special care should be taken when replacing a bearing?

12.2   ANSWERS

(Figures in brackets after each answer refer to the relevant chapter and paragraph in the text.)

1.   Synchronous, induction (squirrel-cage), induction (wound-rotor), induction (split-phase), commutator, repulsion, reluctance, synchronous-induction and Schrage. The principal differences between the synchronous and squirrel-cage induction motor are:

Synchronous	**Induction**	
Constant speed	Not quite constant speed	
Requires excitation	Self-excited	
D.C. field system	Rotating field from stator	(1.2 - 1.5)

2.   The power factor of a synchronous motor may be raised or even made leading by increasing the excitation. Thus a synchronous motor need not impose a low power factor on the system (as would an induction motor), or, by running it over-excited, it could be made to compensate for other low power-factor machines.   (1.2)

3.   Stator with a fixed 3-phase winding, giving rotating field. Rotor with uninsulated conductor bars short-circuited through end-rings. Rotating field, in passing the rotor bars, induces emfs in them, causing currents to circulate through the bars and end-rings. These currents react with the field, causing a sideways thrust in the same direction in each bar and so a torque to drive the rotor.   (2.1)

4.   Where the starting current of a direct-on-line started induction motor may be unacceptably high, or where a slow, controlled start is needed, a wound-rotor induction motor could be used, with resistance (liquid or otherwise) gradually cut out of the rotor circuit until the sliprings are left short-circuited.   (2.2)

5. $353kWe$ $(= 300 \div 0.85)$. (NB. 400 hp 300kW$_m$)   (1.1)

6.   One winding connected direct to the single-phase supply; a second winding displaced 90° connected through a capacitor. This gives a rotating field in which a squirrel-cage rotor turns. Sketch as Figure 2.2.   (2.1.2)

7.   Approximately 3 550 rev/mm at full-load. Increases towards 3 600 rev/mm as load is decreased.   (2.1.1)

8.   Approximately 1 470 rev/mm at full-load.   (2.1.1)

9.   Enclosures are classified by the letters 'IP' followed by two digits. The first (0 to 5) indicates protection against entry of solid bodies from zero (unprotected) to 4 (down to 1mm) and 5 (against dust).

   The second digit indicates protection against entry of water from zero (unprotected), through drip- and splash-proof to 8 (submersible).   (3.1 .2)

10.   Insulation is classified by its ability to withstand a maximum operating temperature, which is the sum of the maximum ambient temperature and the designed maximum temperature rise. The class of insulation depends on the material used.   (3.2.1)

11.   Class F. (Ultimate temperature is 40°C ambient + 110°C rise 150°C).   (3.2.1)

12.   IP56 (formerly 'totally-enclosed, fan-cooled' or TEFC).   (3.2.2)

13. Small motors: natural ventilation without or with an internal fan.
    Large motors: internal or external fan.
    Very large motors: air cooled with fan and air/water heat exchanger.　　　(3.2.2)

14. Losses include:　　*　friction
    　　　　　　　　　*　windage
    　　　　　　　　　*　iron losses
    　　　　　　　　　　copper losses due to load current
    　　　　　　　　　*　copper losses due to magnetising current.

    Those marked * are virtually independent of loading.　　　(4.1)

15. 　　(a)　Load drive - active power, kW_e.
    　　(b)　Magnetising - reactive power, kvar.
    　　(c)　Losses- active power, kW_e.　　　(4.1)

16. Reactive power does not vary with loading but remains constant at all loads.　(4.2)

17. Just over half full-load current - say 120A.　　　(4.2 Example)

18.

Load	kW	kvar	kVA	I (amps)	pf
full	141	106	176	231	0.80
half	71	106	128	168	0.55
quarter	35	106	112	147	0.31
no-load	0	106	106	139	0

19. Normal method is 'direct-on-line' (DOL). Very large current (4 to 6 times full-load current at standstill) and very low power factor (typically 0.2 to 0.3).　　(5.1)

20. Run-up time depends on:

    • starting torque
    • inertia of driven load
    • whether driven end is loaded.　　　(5.1)

21. Alternative ways are 'reduced voltage' starting. These are:

    • star/delta
    • auto-transformer
    • Korndorffer (auto-transformer modified).

    With star/delta the motor is temporarily reconnected into star, so that applied voltage per phase is reduced by a factor $\frac{1}{\sqrt{3}}(= 0.58)$ and the starting current reduced to one-third. After the motor has run up, it is reconnected into delta and runs as a delta machine.

    With auto-transformer starting the applied voltage is reduced by an auto-transformer to whatever level is desired. After the motor has run up, the auto-transformer is disconnected and full line voltage is applied to the motor, which is star-connected.

    With the Korndorffer modification, during the disconnection of the auto-transformer and reconnection to the line, the motor is at no time free of the line supply.　(5.2)

22. Starting torque will be reduced. Since torque is proportional to the square of voltage, in the star/delta case it will be reduced to a fraction $0.33 (= 0.58^2)$. In the auto-transformer case the reduction of torque will depend on the transformer tap used. (5.2)

23. Korndorffer is a modified form of auto-transformer starting, but it differs from the latter in that the motor is at no time disconnected from the mains and can therefore never get out of synchronism. Immediate reconnection to full mains voltage is therefore possible. Used on offshore installations for very large gas reinjection and pipeline boost motors. (5.2.3)

24. Because a large motor takes a very heavy current at low power factor when starting, its reactive power demand (kvar) is considerable and may constitute a high proportion of the generators' rating. This is just the condition to cause a severe voltage drop in the system, even if the AVR is eventually able to rectify it. The heavy voltage drop will also reduce the starting torque of the motor itself and may even prevent its starting altogether. It will certainly increase run-up time.

    On onshore systems the large, low power-factor starting current would in general have far less effect, since the generating capacity of the shore network is so much greater and can absorb large reactive power demands without noticeable voltage drop. (5.3)

25.   (a)   1155A.
        (b)   852kvar.
        (c)   220kW. (5.5)

26. If a motor **fails** to start, not more than two attempts (from cold) are to be made. If a motor of 50kW or more successfully starts but is at once stopped again, only one more start is allowed immediately. A third start may not be made for a further period of 30 minutes, nor may there be more than three starts in any one hour.(5.4)

27. Voltage drops by 20% - i.e. to 0.8 times nominal.
    Torque drops to $0.8^2 = 0.64$
    ∴Torque falls momentarily to 64% of what it was before voltage drop, rising again to its previous value as the motor increases its slip and settles at its new speed.

    If the loading remains constant, as the voltmeter reading falls by 20% the ammeter reading will rise by approximately the same amount. (6.2 and 6.3)

28. Torque varies as the square of the applied voltage. (6.2)

29. If, on the characteristic reduced by the square of the new voltage, the new pull-out torque is less than the load torque, the motor will not stabilise at the new reduced voltage but will continue to lose speed until it stalls. (6.4)

30. Large consumers are charged both on the units consumed and on their kVA maximum demand. Since kVA is the vector sum of kW and kvar, it will be at its minimum if the kvar is at or near zero. Lagging reactive kvar can be balanced-out by capacitors which take an equal amount of leading kvar, thereby keeping the power cost to the minimum. (7.1)

31. Power factor correction consists of connecting the right size of capacitors in parallel with the motor, so that the mains provide only active power at unity power factor. The kVA is then reduced to its minimum value, equal to the kW. Sketch as Figure 7.3. (7.2)

32. 200kW$_e$ at 0.85 pf is equivalent to $\dfrac{200}{0.85}$ = 235kVA.

Since cos φ = 0.85, sin φ = 0.53.

∴ Reactive load = 235 sin φ = 125kvar lagging.

Hence capacitance needed for full correction 125kvar leading, i.e. approximately 42 kvar per phase.   Since the kvar of a motor does not vary with load, this same capacitance will continue to give full correction also at half-load.   (7.3)

33. - Basic tests
     - Duplicate tests
     - Routine tests.

Basic tests would be carried out on the first motor of a series. On a motor of a production run only routine tests would be made.   (8.2)

34. Normally only periodic insulation resistance tests as routine or after a repair. After a major overhaul a HV withstand test may be needed (carried out by the Contractor).   (8.3)

35. Insulation resistance is very sensitive to temperature variation, falling rapidly with increase of temperature. In order to compare readings taken at different temperatures, each reading must be corrected to its corresponding value at 40°C so that successive readings may be compared and any trend noted.   (8.4)

36. Any three of: Overcurrent
                     Earth fault
                     Stalling
                     Single-phasing
                     Undervoltage
                     Winding overtemperature.   (9)

37. Deterioration of insulation can occur through damp or through overheating of windings. Damp can occur if a motor is left unused and unheated for long periods.

Overheating can result from overloading, from repeated attempts to start, from too-frequent starting or from operating for long periods at reduced voltage.

Insulation state can be monitored by insulation resistance testing. Regular testing, with plotted results, can show a trend, even if the insulation is at that time satisfactory.   (10.2)

38. - Insulation failure or breakdown
     - Cable-box failure
     - Rotor burnout
     - Bearing failure
     - Excessive vibration.   (10.2 - 10.7)

39. Because a cable box has only a small volume, the release of energy in it due to a flashover can cause pressure of explosive proportions.   (10.4)

50. When a rolling type of bearing is replaced (ball, roller or thrust), care must be taken to ensure that not only the right size is used but also the one with the correct degree of clearance is selected. In other words, the replacement must be the exact duplicate of the one removed.
Also, when unpacking a new bearing, all the packing grease must be removed, the bearing flushed out and refilled (but not overfilled) with the correct running grease.   (10.6)

PART 6 SYSTEM CONTROL

CHAPTER 1 GENERAL PRINCIPLES

1.1   INTRODUCTION

There are certain obvious and fundamental differences between onshore and offshore installations. Chief of these, from the point of view of Electrical System Control, is that an offshore installation - a drilling or production platform - has to have its own generators. The system operator, therefore, has control of his source of supply. The safe and correct operation of large gas-turbines and the generators they drive introduces a complication into the control of an offshore installation that is not present onshore, where the supply of electricity is taken from the Electricity Supply Authority for the area.

Other differences between the two kinds of installation concern the special loads that must be fed. An offshore platform, because it is isolated, is completely dependent upon its own resources and therefore has to have living quarters, freshwater production plant, sewage treatment plant, lifeboats, navigation lights, foghorns, emergency radio and special escape-route emergency lighting. This list is not exhaustive.

For these reasons system control of an offshore installation involves considerations that do not apply onshore, and this book is primarily aimed at a typical offshore installation. Where a chapter, or part of a chapter, refers exclusively to one type of installation, a statement to that effect is made. Elsewhere, the reader is left to draw his own conclusions as to what is relevant.

1.2   PURPOSE

It is the prime purpose of any electrical supply network to provide a continuous supply of power to meet the requirements of the load as it varies from time to time. It is the duty of the person in charge to control the network to meet this load requirement without overloading any part of the system and, so far as is possible, always to have sufficient margin of plant available to meet an emergency situation without risking a shutdown.

To control a network successfully, the Electrical Authorised Person must have a thorough knowledge of the system and be aware of the loading limitations of each item of plant. There must be the closest co-operation with other departments so that the authorised person is made aware in advance of any likely changes in loading, such as the starting or stopping of large motors, so that he can start up or shut down generating sets or make such other arrangements as may be necessary to ensure integrity of the supply.

1.3   SECURITY OF SUPPLIES

Because of plant failure or emergency conditions on an offshore installation it may not be possible to maintain supplies to all items of load all the time. Consequently loads are classified in order of priority. If a progressive failure of generation or an emergency shutdown occurs, then load is shed and the network is sectionalised automatically so that only those priority loads which can be met by the available generation are left connected to the system.

Starting with the loads requiring the highest security of supply, load priorities are:

(a) **Battery Supported Loads**

These cannot withstand any interruption of supply or are required to operate after a total loss of all generation - for example, emergency radio, public address system, fire and gas detection system, emergency lighting in the living quarters and escape routes and navigating lights and foghorns.

(b) **Minimum Life-support Loads**

These include the battery systems as in (a) above and additionally those loads necessary to sustain life on the installation. These include the living quarters, other essential lighting, freshwater supplies, sewage treatment and lifeboat davits.

(c) **Normal Life-support and General Services Loads**

These sustain minimum life-support loads as in (b) above and in addition include certain common utility services such as sea water, ventilation and compressed air. These loads, together with very limited production loads, can be supplied from the secondary (sometimes called 'Essential Services') generation.

(d) **Main Loads**

These are concerned with the full production of crude oil and gas, the injection of sea water and the export or re-injection of gas. The main generation is required to supply this load.

An onshore installation also needs supply security, but the emphasis is different, being concerned mainly with production integrity.

1.4 PLANT RATINGS

A network uses generators, cables, switchgear and transformers to produce power and to convey it to each individual load. Each item of plant has a 'Continuous Maximum Rating' (CMR) and may suffer damage if loaded beyond this.

The electrical output of an offshore generating set is limited by the maximum mechanical power developed by its gas-turbine or diesel engine as well as by the heating effect of the current in the generator windings. It is convenient, therefore, to consider its rating in megawatts (MW) rather than in amperes. Because of this mechanical power limitation a generating set has no sustained overload capacity.

Some large gas-turbines have two mechanical power ratings - base load and peak load. The base load rating is usually about three-quarters of the peak load rating and can be sustained continuously. When the turbine is producing an output above the base load rating, the time interval between overhauls is reduced; thus peak load rating, although it can be sustained, should be used only in an emergency. No operator action is required to obtain the peak load rating. Base load and peak load running hours are recorded on separate running-hours indicators, the changeover from one to the other being automatic depending on the power output.

Cables and switchgear suffer quick damage from overheating if loaded beyond their CMR. The overheating occurs very quickly after the onset of an overload, and this equipment too has virtually no sustained overload capacity. Because the loading of the gear is limited by the heating of the conductors, the rating is expressed in amperes or kilovolt-amperes (kVA).

Liquid-filled transformers are different in that a circulating cooling system is provided to remove the heat from the windings and transfer it to the air surrounding the unit. Because the cooling liquid takes time to heat up, overloads are permissible for short periods - for example, 130% of full load for two hours - provided that the unit was not fully loaded immediately before the onset of the overload. Longer periods of overload are permitted if the ambient temperature is low. A transformer is rated in kVA, but the important thing is that the temperature rise of the windings must not be allowed to exceed 65°C, and that the temperature rise at the top of the liquid coolant must not exceed 60°C with a sealed or conservator-fitted transformer.

Thus, when controlling a network, the loading of generators is checked by observing the associated wattmeters, whereas ammeters show the loadings on feeders, switchgear and transformers. These indicating instruments, together with others, are provided for all important circuits on the Electrical Control Panel.

1.5   FIRM CAPACITY

1.5.1 Offshore

A network is said to be loaded within its firm capacity if any one item of generation or distribution plant becomes disconnected because of a fault without affecting the capability of the system to meet its load.

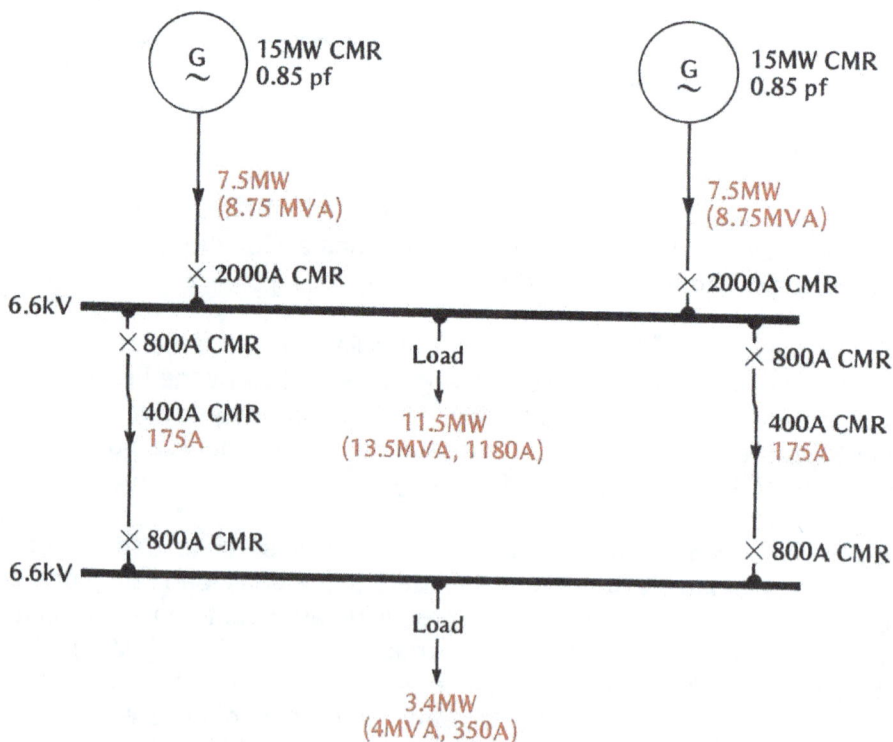

FIGURE 1.1
FIRM CAPACITY - TWO GENERATORS (OFFSHORE)

The hypothetical network shown in Figure 1.1 complies with this definition. It shows two 6.6kV generating sets, each rated at 15MW (CMR), 0.85 pf, feeding a 6.6kV switchboard through generator circuit-breakers each rated 2 000A (CMR). This switchboard is interconnected with another 6.6kV board by two cables, each rated at 400A (CMR), and with switchgear rated 800A (CMR). With the loadings shown (in red), one generating set can sustain the total load if the other becomes faulty. To maintain firm capacity, therefore, with two generating sets, each can only be loaded to half its CMR. Similarly, if one of the interconnecting cables trips out, the other can carry the load on its own, as can also the switchgear.

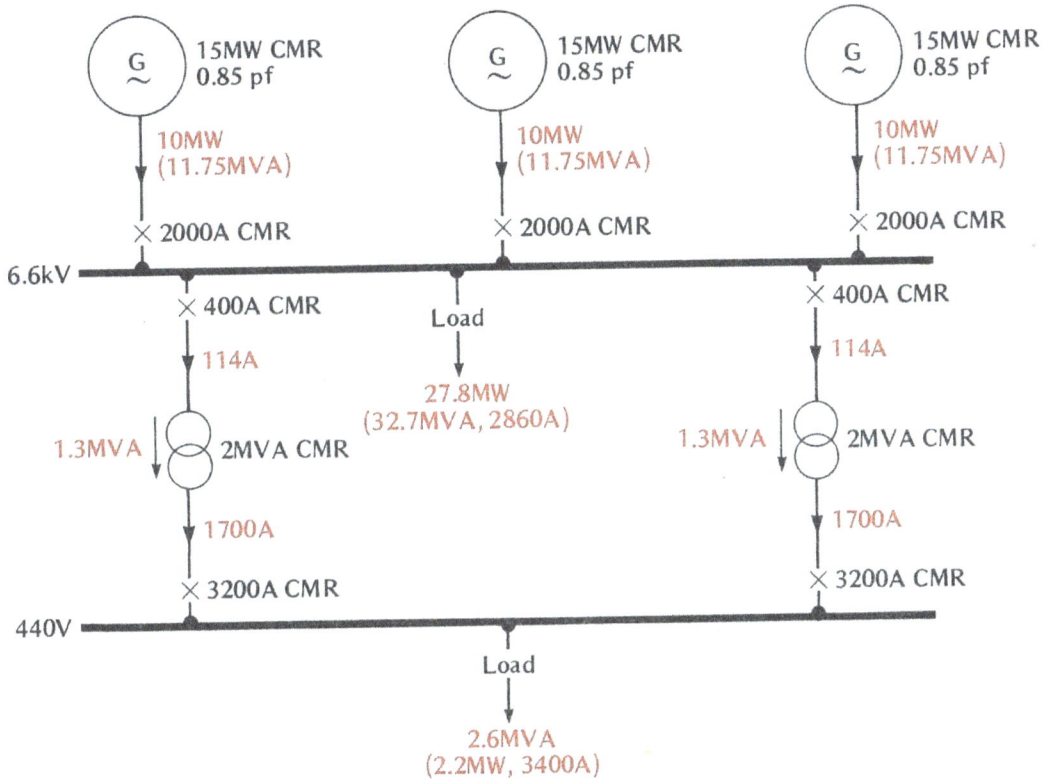

FIGURE 1.2
FIRM CAPACITY - THREE GENERATORS (OFFSHORE)

The network shown in Figure 1.2 employs three generating sets, with all CMRs shown in black and the actual loadings shown in red. It shows that in this case each machine may be loaded to two-thirds of its CMR while still preserving the firm capacity. The 130% overload capability of the transformers can be used to give them a firm capacity of 2.6MVA, instead of the 2.0MVA suggested by their individual CMRs, on the assumption that the overload will not last too long. One of the two 2MVA (CMR) transformers can then be lost without excessively overloading the other. However in this instance the rating of the low-voltage circuit-breaker (3 200A) will be slightly exceeded if one transformer is lost, and very rapid steps will be necessary to reduce the load on the 440V switchboard to prevent the breaker's overheating.

It is always preferable to operate the system with firm capacity in hand, but on the occasions when this is not possible, the risk of loss of supply or the need for load-shedding must be accepted.

The authorised person must always be aware of the state of the network, ensuring so far as possible that supplies are firm. There must be close co-operation with the Production departments to anticipate major load changes and to allow for adjustment of generating capacity. Safety must come first, and then security of supply.

1.5.2 Onshore

An onshore installation usually has two separate inputs from different parts of the Supply Authority's distribution system, only one being used at any one time. Interlocking between the two main supply circuit-breakers and the main bus-section breaker prevents the two inputs from being connected in parallel, and there is usually an automatic changeover facility.

Onshore, therefore, the term 'firm capacity' is mainly a matter for the Supply Authority. Within the most installations its significance is confined to distribution equipment.

As in offshore installations, a simultaneous failure of all normal supply sources in an onshore installation is catered for by the provision of a diesel-driven emergency, or 'basic services', generator.

1.6 NETWORK CONTROL

The typical network shown in Figure 1.3 represents most of the features to be found in an offshore installation.

The general service and normal life-support loads are fed from 6.6kV and 440V switchboards 'A' and the heavy main loads from 6.6kV and 440V switchboards 'B'. Two interconnectors provide links between 6.6kV switchboards 'A' and 'B'. All HV boards and elements are shown in blue, and all LV in red.

Three levels of generation are provided:

(a) **Main Generation**

This is represented by the two high-voltage 15MW gas-turbine sets. Together they are capable of carrying the full load of the offshore installation, including gas or water injection, and are the normal source of electrical power.

(b) **Secondary (or 'Sub-main') Generation**

The two 2.5MW high-voltage gas-turbine sets are arranged to start automatically on the failure of main generation to supply the general services and normal life-support loads. They can sustain normal production processes but not oil export or gas or water well injection. They are also used when the installation is not in production and there are no heavy 'main' loads.

(c) **Basic Services Generation**

A diesel-driven low-voltage generating set of relatively small power (350 to 1 000kW) provides power mainly for certain vital services, battery-supported supplies, minimum life-support services, essential lighting and main generation auxiliaries. The set may start automatically on loss of main supply, but on some offshore installations only hand-starting is provided.

Although the voltage of the high-voltage system is nominally 6.6kV on most offshore systems, it is normal practice to operate the generators at a slight excess voltage of 6.8kV to allow for voltage drop in the distribution system. This affects also the operating voltage level of the HV switchgear, HV motors and transformer primaries. Nevertheless **nominal** voltages will be used throughout these books - namely 6.6kV for high-voltage systems (in those offshore installations so fitted) and 440V for the low-voltage systems. Similarly 33kV, 11kV and 3.3kV are used for onshore systems, although the operating voltages are slightly higher.

The 6.6kV system would normally be fed from the main sets, the bus-section breakers being closed on 6.6kV switchboards 'A' and 'B' and both interconnectors in circuit; this arrangement gives the greatest security of supply. To meet peak loads, one or both secondary sets can be synchronised and run in parallel with the main sets.

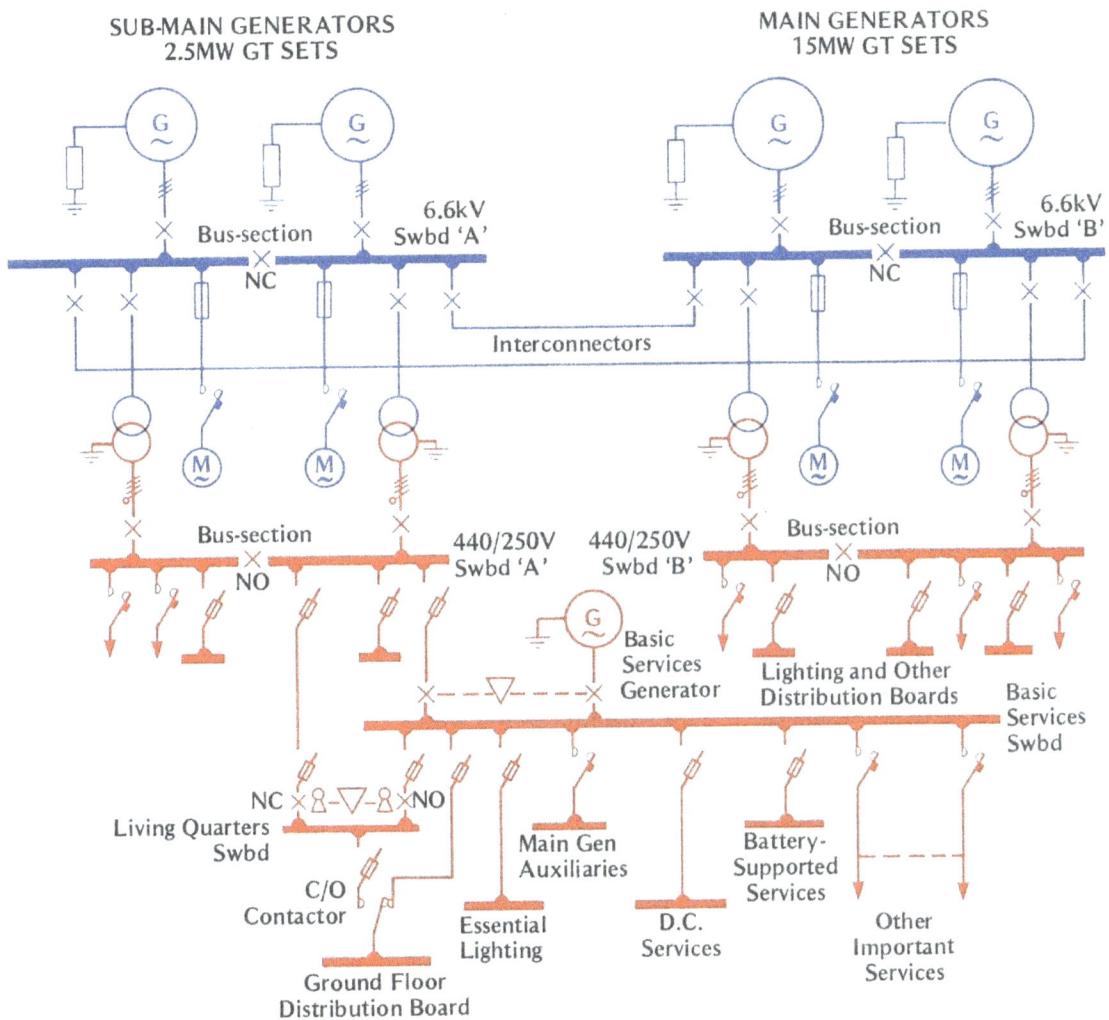

FIGURE 1.3
TYPICAL NETWORK

Each of the 440V switchboards 'A' and 'B' is energised through **both** of the associated 6 600/440V transformers but with the 440V bus-section breakers open. This is because the fault level at the 440V switchboard would be too high if the two transformers were feeding in parallel. Interlocks are provided to ensure that, except for the short time during switching, only two out of the three circuit-breakers are closed at one time - that is, either two transformers with the bus-section open or one transformer with the bus-section closed.

The living quarters switchboard is normally fed from 440V switchboard 'A'. However, should this feed not be available, an alternative is provided from the basic services switchboard. A key interlock prevents both incomers from being closed together.

Consider the situation where the whole platform load is being supplied by the two main sets in parallel. If one fails, the load will still be carried by the other without interruption of supply. Depending on how many are running, some large motors may be shed either by the automatic load-shedding equipment or by direct switching initiated by the generator circuit-breaker tripping (see Chapter 6). Should the second main set shut down, the supply would fail completely, causing all motors, both HV and LV, and the two interconnectors to trip on undervoltage, so separating system 'A' from 'B'. Sensing the loss of supply, the secondary sets start automatically, synchronise with each other and restore the supply to the 'A' system only. Should these secondary sets fail to start, the basic services diesel set must be started manually from its local panel except where, in some installations, it starts automatically. When this set is running and ready for load, the switchgear operates either automatically or by hand control, first to disconnect the basic services board from 440V switchboard 'A' and then to connect the generator. When the normal supply is restored, this procedure is reversed, and, after switching, the basic services generator is shut down.

The effect on the system when very large machines such as gas compressor drive motors are started is discussed in Part 5 'Electric Motors' and also later in this chapter. These motors, even when special arrangements such as Korndorffer starting are used, draw heavy **reactive** power (Mvar) from the system during the starting period even though the active power (MW) may be relatively small. Although this reactive power (some 23 Mvar for the largest machines) does not impose mechanical load on the gas-turbines, it must nevertheless be supplied by the generators and in some circumstances can cause severe system voltage drop.

1.7 REACTIVE POWER

A generating set is normally rated in MW at a certain power factor, usually 0.8 or 0.85 lagging. This describes its ability to supply active power and, by implication, reactive power at the same time. For example, the reactive power rating of a 15MW, 0.85 pf generator can be deduced from the impedance triangle of Figure 1.4.

Here the impedance triangle sides are each multiplied by the square of the full-load current, so that the resistance side represents I^2R watts, the reactance side I^2X vars, and the impedance side I^2Z volt-amperes. Thus, for the case quoted, the active power is 15MW and the phase angle φ is given by $cos\ \varphi = 0.85$, whence $\varphi = 31.8°$. Therefore the reactive power I^2X is $15 \times tan\ \varphi = 9.2$Mvar, and the apparent power I^2Z is $15 \div cos\ \varphi = 17.6$MVA.

This generator thus has a rating of 15MW and 9.2Mvar, and these outputs can be given **simultaneously**. But whereas the active power output of 15MW is strictly limited by the mechanical output of the prime mover, the reactive power output of 9.2Mvar is not so limited. It depends on the excitation available, and this normally provides for somewhat higher reactive output so long as the simultaneous active output is reduced.

How this and other limitations are taken into account is explained in the following chapter, which makes use of a device called the 'Capability Diagram'.

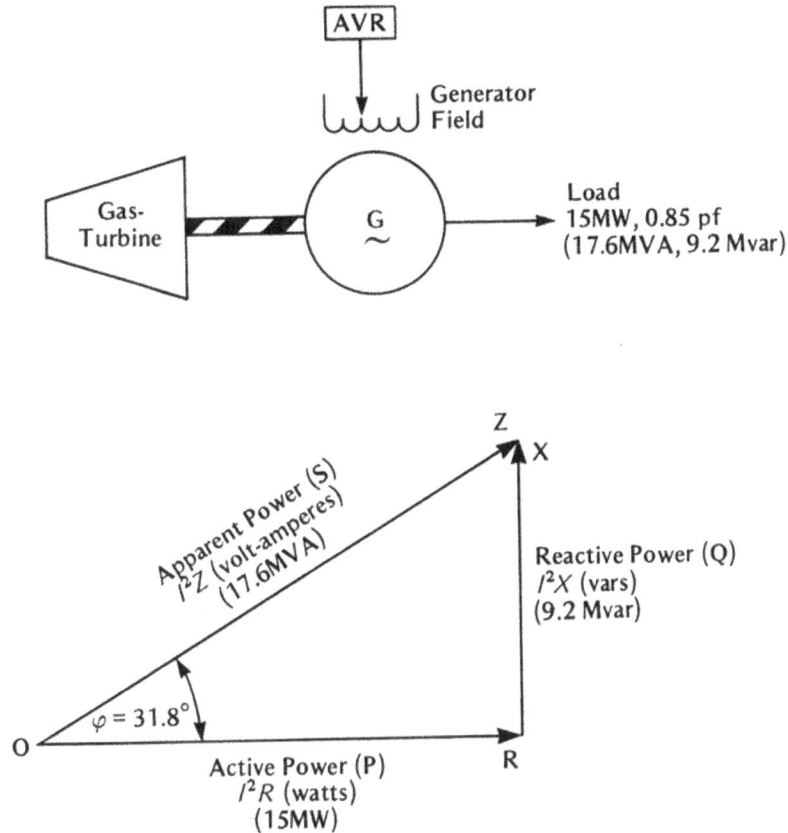

FIGURE 1.4
CALCULATION OF REACTIVE POWER

1.8 SYSTEM VOLTAGE DROP

It was shown in Part 2 'Electrical Power Generation' that the worst internal volt-drop conditions occur when a large reactive current flows through the internal reactance of the generator.

This principle does not apply only to generators. It applies also to transformers, cables and transmission lines - indeed to any element of plant which has significant inductance and therefore reactance. For example, a large reactive current passing through a transformer will give the worst internal voltage drop condition.

Therefore a power network system - as distinct from the individual motors that it feeds - is subject to appreciable voltage drop when reactive power is drawn through it. Not only is system voltage drop undesirable in itself but, as explained in Part 4 'Electrical Systems', the Area Supply Authorities have a statutory duty to hold the consumer's voltage between certain defined limits.

1.9 THE SYNCHRONOUS CONDENSER

System power factor could, in theory, be corrected by shunt capacitance just as is done with a motor, but static capacitors of a size needed for compensating a large power system would be far too large to be practical, and their switching to meet the varying load conditions would be an immense complication.

Fortunately the same result can be achieved by a single rotating machine - and one, moreover, capable of infinitely variable control. It is identical with the standard a.c. generator with a similar excitation system, which is usually brushless. A 'synchronous' machine of this type may be used as a generator if it is driven by a prime mover, or as a motor if it is started and connected to a mechanical load, as shown in Part 5 'Electric Motors'. Under the conditions explained in Chapter 2, para. 2.9, this same machine when over-excited can also behave as a bank of static capacitors. Hence its name 'Synchronous Condenser', using the old word 'condenser' and underlining the fact that it is not static but is merely a standard synchronous machine of the same type as an ordinary generator or synchronous motor. It has the additional advantage that the Mvar level, unlike that of a static capacitor, can be varied at will simply by altering the excitation. When used for system power-factor compensation it can be varied as required to meet changing load conditions on the system.

In practice any generator can be operated as a synchronous condenser simply by removing its prime-mover drive once it has been run up to speed and synchronised. This, however, is not often done, but special synchronous condensers may be provided at points on the network where compensation is best applied. In such cases there is no prime mover, and special steps are needed to start the machine. Sometimes it can be arranged for self-starting; in other cases a pony motor is used. In either event it is run up to speed, synchronised and then its excitation adjusted until it is exporting the quantity of megavars demanded by Central Control.

1.10 SYSTEM POWER FACTOR CORRECTION

To avoid voltage drop in the various elements of the system network it is desirable that the current in each element should have as high a power factor as possible.

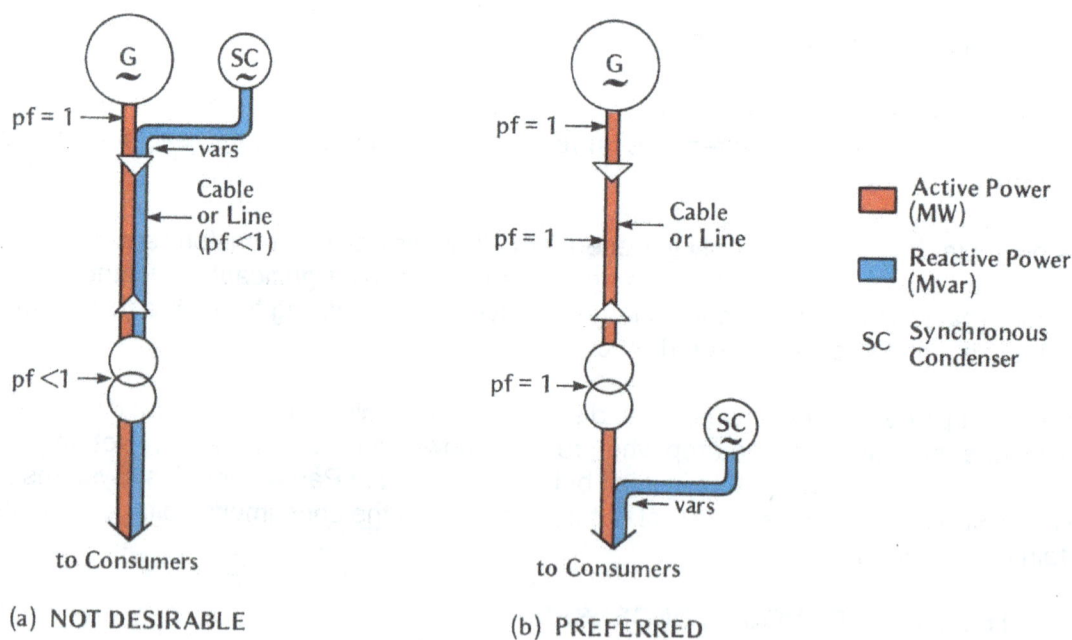

(a) NOT DESIRABLE

(b) PREFERRED

FIGURE 1.5
SITING OF SYNCHRONOUS CONDENSER

To arrange power-factor correction at the power station itself, as shown in Figure 1.5(a), relieves the generators of providing the reactive power and so uses them more efficiently, but it does not help the voltage drop in the transformers, cables or lines. It is preferable to site the synchronous condensers well downstream and as near the main consumers of reactive power as possible, as shown in Figure 1.5(b).

1.11   APPLICATION OF SYSTEM CORRECTION

Because system power-factor correction applies to a complete power network as distinct from the correction of individual pieces of consumer equipment such as motors, it presupposes that there is a system large enough to warrant correction. Offshore systems are by comparison with onshore networks very small indeed, and their network is local to the offshore installation. It follows, therefore, that the methods discussed in this chapter apply only to onshore systems.

1.12   EXTRA HIGH VOLTAGE ONSHORE SYSTEMS

In Part 4 'Electrical Systems' it is explained that, with the extra-high-voltage networks at 275kV, and more especially at 400kV, the capacitance of the overhead transmission lines is so large that it tends to swamp the effects of inductance, at least at the lower loads, and to cause a voltage **rise** instead of the normal drop due to inductive reactance. Therefore, far from having to provide extra capacitance in the form of synchronous condensers, exactly the opposite is done. The leading power factor of these lines at the lower loads has to be compensated by adding **inductance** in the form of reactors or by operating a synchronous condenser underexcited so that it provides the **leading** vars needed by the transmission line.

CHAPTER 2 CAPABILITY DIAGRAM

2.1 GENERAL

The principal limitations on the output of an engine-driven generator set are as follows:

    (a)   Current heating of the stator (armature).

    (b)   Power output of the prime mover.

    (c)   Current heating of the rotor (field).

    (d)   Stability of the rotor angle.

There are other limitations such as the heating due to iron losses, harmonic currents, negative-and zero-sequence currents, etc., but the four listed above have the most decisive limiting effect.

2.2 STATOR CURRENT

Consider first the stator current. The I^2R, or 'copper', losses due to the load current are, together with the iron losses, the main sources of stator heating. With a given cooling

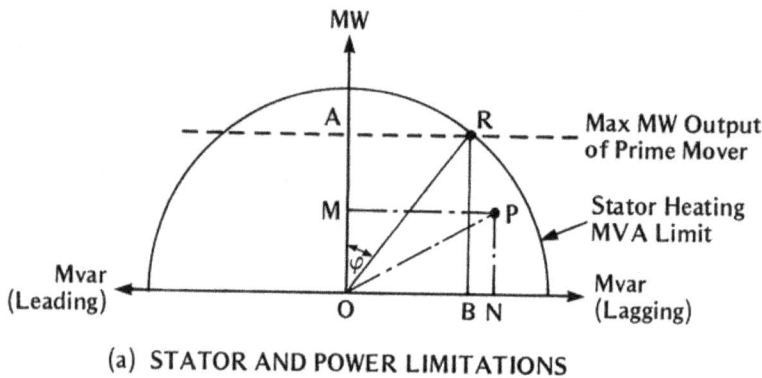

(a) STATOR AND POWER LIMITATIONS

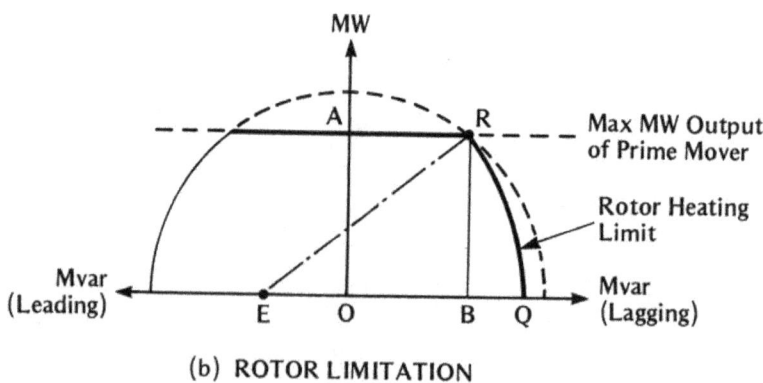

(b) ROTOR LIMITATION

FIGURE 2.1
GENERATOR CAPABILITY DIAGRAM (1)

system there is clearly an upper limit to such continuous stator currents no matter what their power factor may be. Stated another way, there is a limit of MVA beyond which the generator must not be allowed to go continuously, and this limit applies at all power factors.

In Figure 2.1(a), if the reactive (Mvar lagging) loading is taken as the x-axis and the active (MW) loading as the y-axis, then for any given loading P (PN being the active component and PM the reactive), the line OP represents the MVA of that load $(= \sqrt{PN^2 + PM^2})$, and the angle PÔM is the phase angle of the load. If a semi-circle is drawn about the origin O and with radius equal to the maximum permitted MVA, then only those loads (such as P) within that semi-circle are within the capacity of the generator. This is the first limitation - para. 2.1(a).

2.3   POWER OUTPUT

The electrical MW rating of an engine-driven generating set is limited by the mechanical output of the prime mover. Therefore, if a horizontal line is drawn across the MW axis at a level equal to the maximum output of the prime mover (OA), the top part of the semi-circle is cut off, since it represents MW power which is not attainable from the engine. Therefore the loading on the generator must be confined to points within the remainder of the semi-circle. This is the second limitation - para. 2.1(b): it is shown in Figure 2.1(a) as a dotted line.

If this horizontal line cuts the maximum MVA semi-circle at the point R, then the rating of the generator is RB megawatts and RA megavars. Angle RÔA (φ) is the rated phase angle, and the rated power factor is then cos φ.

2.4   ROTOR CURRENT

To achieve the rated MVA loading OR, a certain level of excitation is required. This calls for a certain rotor current, with its consequent I^2R losses which cause rotor heating. The cooling system takes this, together with the stator heating, into account. Any increase in excitation beyond this level - and hence in rotor current - will cause rotor overheating; so at first sight the load point P should remain to the left of the line RB.

However, there is a point E* on the other side of the origin such that the line ER represents the emf of the generator when operating at its rated load and power factor. ER then represents not only the emf but also the excitation - and so the rotor current - needed to produce it. Constant excitation at various maximum loads and power factors is therefore represented by an arc of a circle through R, centre E, shown by the arc RQ in Figure 2.1(b). This arc thus represents the maximum allowable rotor current at different maximum loads and power factors. To avoid overheating the rotor, therefore, the load point must lie to the left of the arc RQ (not to the left of line RB as first suggested above). This is the third limitation - para. 2.1(c).

---

*Though not strictly necessary to know for the purposes of this explanation, the position of point E is determined from the generator's synchronous reactance. If this is n per unit (= per cent ÷ 100), then the length OE is $\frac{1}{n}$ times the radius of the semicircle. Thus if n = 200% (fairly typical), E is halfway between O and the circumference.

2.5   STABILITY OF THE ROTOR

When the machine is generating, the rotor is driven ahead of the stator's rotating magnetic field at an angle depending on the actual active load - it is called the 'Power Angle', symbol λ. The opposing torque developed by the generator on the engine is due to the magnetic back-pull on the rotor's poles by the stator's magnetic field. The greater the power angle, the greater the back torque. The driving torque delivered by the engine is just balanced by this back torque from the rotor, and the rotor is stable.

When the loading is lagging reactive, armature reaction in the stator causes the field poles to become partly demagnetised (see Part 2 'Electrical Power Generation'). The consequent loss of air-gap flux reduces the net emf and so the terminal voltage; this is detected by the AVR, which causes increased excitation to restore the air-gap flux and so the emf.

With **leading** reactive loading the opposite effect occurs. The leading stator current causes the field poles to become **more** magnetised at first. The gain of air-gap flux **increases** the net emf and so the terminal voltage; this is detected by the AVR, which then decreases excitation to restore the air-gap flux.

If this process were allowed to continue and the leading load to increase further, the point would be reached where the excitation of the rotor poles would be reduced to nothing, and all the air-gap flux would be provided by the stator alone. There would then be no rotor poles upon which the stator could pull back. The prime mover would drive the rotor ahead out of synchronism, and the generator would go unstable, pulling out of step.

This situation would occur if the reactive loading reached a value equal to OE (leading), since at E the excitation is reduced to zero. Therefore the load point P must never be allowed to go to the left of the vertical line through E - that is, the leading Mvar must never be allowed to exceed the value OE. This is the fourth limitation - para. 2.1(d): it is shown in Figure 2.2(a).

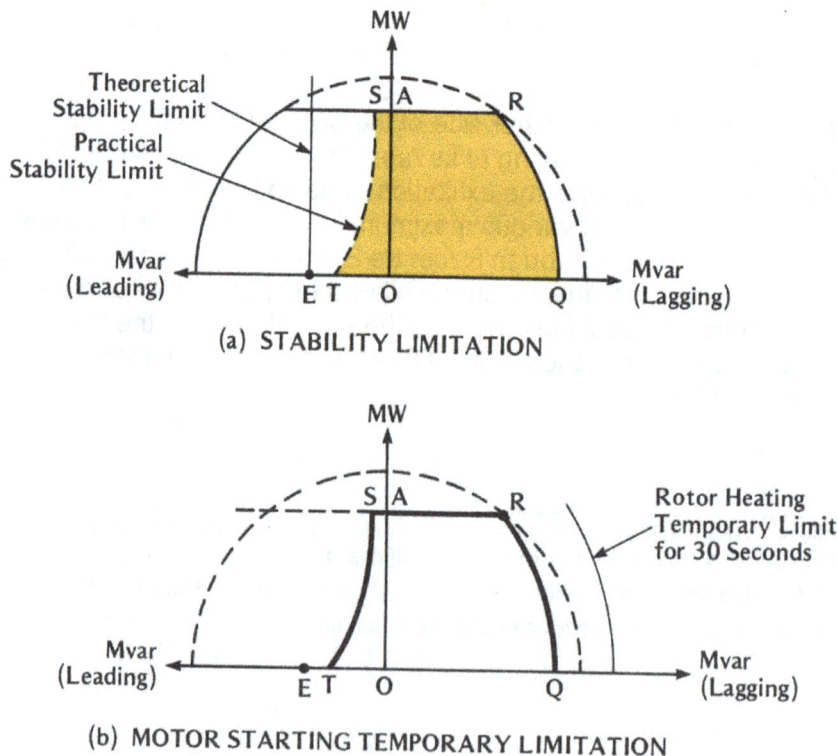

(a) STABILITY LIMITATION

(b) MOTOR STARTING TEMPORARY LIMITATION

FIGURE 2.2
GENERATOR CAPABILITY DIAGRAM (2)

Clearly such a situation must never be allowed to occur **or even to be approached,** as the generator would become difficult to control. There is therefore a limit to the amount of leading current (or MVA) which a generator may be allowed to produce. The theoretical limit set in the previous paragraph is therefore in practice too high. The practical limit will be appreciably less and is represented by the dotted curve ST in Figure 2.2(a); the calculation of this curve is complicated, and it is merely indicated here. Its shape will depend on whether the rotor is cylindrical or has salient poles. In practice, however, leading loads on platforms should never arise.

So in Figure 2.2(a) the theoretical semi-circle inside which the loading of the generator must lie is limited at the top and now at **both** sides, and the only 'usable' part is the coloured area if the generator is to run within its rating and remain stable.

The coloured part of Figure 2.2(a) is called the 'Capability Diagram' of that generator set. It provides a constant guide not only on the loaded state of the generator but also on its maximum allowable further loading. It shows whether any intended additional loading will still remain within, or go outside of, the rating of the generator set, and it therefore indicates whether or not a further machine should be started up. All that is required to determine the existing load point on the capability diagram are the generator MW and Mvar instrument readings.

2.6   TEMPORARY LIMITATION

A special case arises when large motors are to be started. They impose a large, but temporary, reactive loading while starting, which falls to a much smaller value when run up to speed. The extra excitation needed for this can if necessary be found from the AVR's field-forcing circuits, which provide extra field current above the normal maximum. This, although above the steady-current limit of the rotor windings (arc RQ), may be tolerated for a short time without damaging the rotor.

For this reason capability diagrams are sometimes furnished with one or more additional rotor current limitation arcs with a specified time limit of, for example, 30 seconds, as shown in Figure 2.2(b). This means that, within this time, the reactive loading may be increased to the indicated higher limit to allow the motor to start, provided that it falls back to within normal limits within the specified time when the motor has run up to its steady speed. On some systems, if the rotor current goes beyond the higher limit, or if it fails to return to within normal limits within the specified time, the generator is tripped.

If the total reactive load on starting falls outside even this higher temporary limit and automatic tripping is not fitted, it goes beyond the field-forcing limit of the AVR, and a prolonged dip of the system voltage will result, besides risking damage to the rotor.

2.7   USE OF THE CAPABILITY DIAGRAM

To use the capability diagram, the operator looks at the wattmeter and varmeter of the generator which is to be further loaded and he plots the point P on the diagram corresponding to the Mvar and MW standing load readings (see Figure 2.3). He notes, or calculates, the additional load in Mvar and MW which he intends to put onto the generator; these he adds to the Mvar and MW values of the point P. If the resulting point P_1 lies within the coloured area, the generator can accept the additional load. If not, he must be prepared to start an additional generator to share the total load. If the excess is marginal, he must use his discretion.

Particular care is needed if a large motor is to be started. Although the additional Mvar and MW at full load may be acceptable to the generator, the Mvar due to the large starting current may not. For example, a certain gas export compressor motor has an input of 5MW at 0.85 pf, giving a full-load demand of 2.6Mvar and 5MW, which might be acceptable on top of the standing load. But the **starting** current is approximately 1 500A at 0.25 pf, giving a starting demand of about 25Mvar and 7MW.

While even the 7MW might be acceptable to the generator, the capability diagram would show that the additional 25Mvar on starting almost certainly would not be, even taking into account any temporary margin allowed. The operator would need to put on line extra generators to accommodate this start, even if he took them off again once the motor had run up. The following example illustrates this point with a different motor.

When a second generator is put on line, it is assumed that both share the load equally; therefore the MW and Mvar loadings on each are half what they were with one generator only. This means that the 'working point' P has half the previous values and is therefore much nearer the centre. This leaves more room for additional loading, remembering also that the additional load itself on each generator is also halved.

Example

A 6.6kV generator is rated 15MW at 0.85 pf. At a certain moment it is carrying a standing load of 8MW and 6Mvar (represented by point P in Figure 23) as given by its switchboard instruments. It is desired to start and run a 3 600kW$_m$, 0.8pf water injection motor (efficiency 90%) on this generator. The motor's starting current is four times full-load current at 0.25 pf. The capability diagram, including the temporary limitation curve, is given in Figure 2.3.

Can this generator carry the extra load, and can the motor be started on it? If not, what action should be taken?

(In the following calculations all results have generally been rounded off to the nearest 10 units. Operation at a generated voltage of 6.8kV has been assumed.)

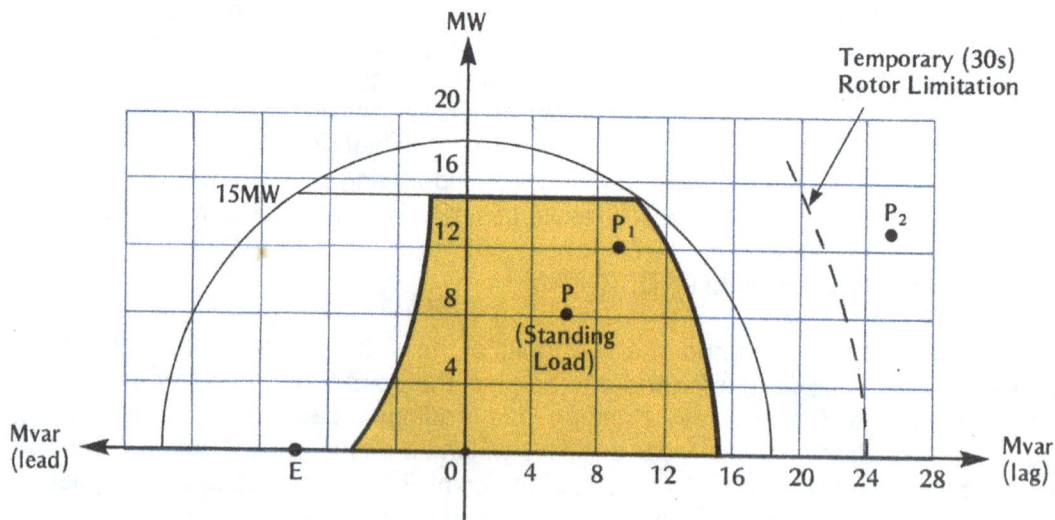

FIGURE 2.3
CAPABILITY DIAGRAM - EXAMPLE

Generator

Generator rating 15MW at 0.85 pf and 6.6kV (6.8kV operation)   (15MW)

$\therefore$ Full-load current $(I) = \frac{15\,000}{\sqrt{3} \times 6.8 \times 0.85} = 1\,500\text{A}$

$\cos \varphi = 0.85, \quad \therefore \quad \sin \varphi = 0.53$

$\therefore$ Full-load reactive power $= \sqrt{3} \times kV \times I \times \sin \varphi$

$= \sqrt{3} \times 6.8 \times 1\,500 \times 0.53$

$= 9\,360$ kvar (say 9.4 Mvar)

Motor (Running)

Motor's output is 3 600kW$_m$ at 0.9 efficiency

$\therefore$ Motor input is $\frac{3\,600}{0.9} = 4\,000 kW_e$ at 0.8 pf (4.0 MW$_e$)

$\cos \varphi = 0.8, \quad \therefore \quad \sin \varphi = 0.6$

$\therefore$ Full-load current $(I_{FL}) = \frac{4\,000}{\sqrt{3} \times 6.8 \times 0.8} = 425\text{A}$

$\therefore$ Full-load reactive power $= \sqrt{3} \times kV \times I_{FL} \times \sin \varphi$

$= \sqrt{3} \times 6.8 \times 425 \times 0.6$

$= 3\,000$ kvar (3.0 Mvar)

Motor (Starting)

Starting current $(I_{ST}) = 4 \times$ full-load running current (I_{FL})

$= 4 \times 425 = 1\,700\text{A}$ at 0.25 pf

$\cos \varphi = 0.25, \quad \therefore \quad \sin \varphi = 0.968$

$\therefore$ Starting active power $= \sqrt{3} \times kV \times I_{ST} \times \cos \varphi$

$= \sqrt{3} \times 6.8 \times 1\,700 \times 0.25$

$= 5\,010 kW_e$ (say 5.0 MW_e)

and starting reactive power $= \sqrt{3} \times kV \times I_{ST} \times \sin \varphi$

$= \sqrt{3} \times 6.8 \times 1\,700 \times 0.968$

$= 19\,380$ kvar (19.4 Mvar)

Combination

When running:	standing load is	8.0MW	and	6.0 Mvar
	motor load is	4.0MW	and	3.0 Mvar
	TOTAL:	12.0MW		9.0 Mvar

Plotted on the capability diagram, this gives point P_1, which is well within the coloured diagram limits and is therefore acceptable.

When running:			
standing load is	8.0MW	and	6.0 Mvar
motor load is	5.0MW	and	19.4 Mvar
TOTAL:	13.0MW		25.4 Mvar

Plotted on the capability diagram, this gives point P_2, which lies outside the diagram and even beyond the temporary limit. The starting of this motor is therefore not acceptable, even though the current could be carried continuously once running. Before starting, therefore, either the standing load must be sufficiently reduced or another generator set must be started and put on-line.

2.8 CAPABILITY DIAGRAM FOR SYNCHRONOUS MOTOR

Although the capability diagram so far described takes the form of a semi-circle, it can be continued below the horizontal axis to become a complete circle. In that case the y-axis, representing active power (or MW), is negative and so indicates negative active power supplied by the machine. This is equivalent to active power being **received by** the machine; that is to say, the machine is absorbing true power and is therefore motoring. The x-axis, representing lagging or leading machine power supplied, is not affected.

The two lower quadrants shown in Figure 2.4 thus represent the machine operating as a motor - that is, as a synchronous motor. The left-hand lower quadrant indicates such a motor running under-excited and therefore supplying some leading vars to the system, equivalent to **drawing** lagging vars from it.

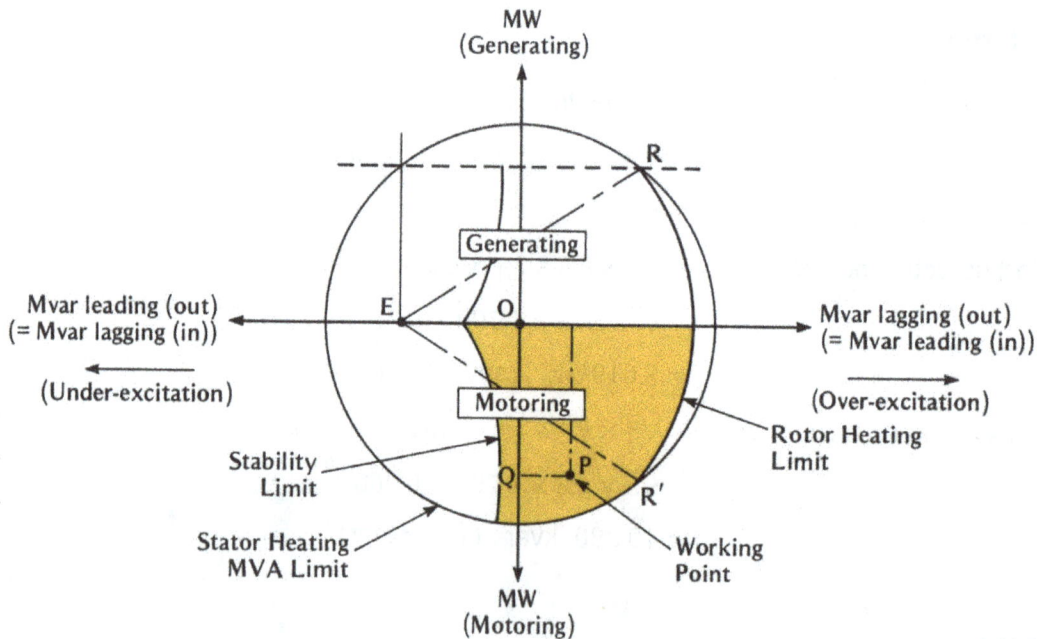

FIGURE 2.4
CAPABILITY DIAGRAM - SYNCHRONOUS MOTOR

The right-hand lower quadrant indicates a motor well excited and supplying lagging vars to the system - equivalent to **drawing** leading vars from it. (This mode of operation is further discussed in para. 2.9.)

It should be noted that the excitation can be adjusted so that the machine is drawing neither leading nor lagging vars (point Q) - that is, it is taking no reactive power at all, only active power. Such a motor is then running at unity power factor - a useful feature of the synchronous motor.

It is possible to go even further. The machine can be deliberately run as a motor overexcited - i.e. in the fourth quadrant - where it will draw active power from the system as it motors, but it will at the same time **supply** lagging vars, and it will therefore run at a leading power factor. If a mixed load of induction motors and a large synchronous motor is installed, the synchronous motor run in this manner can help compensate for the poorer power factors of the induction motors.

Below the horizontal the prime-mover output limitation clearly no longer applies, but the rotor and stability limitations apply as before. The 'working point' P must therefore fall within the coloured area of Figure 2.4 if the motor is to work within its design limits.

When determining the working point P, all losses (including friction and windage losses) must be added to the known mechanical power of the motor drive, since all go into the total power absorbed. The losses can be calculated from the efficiency of the motor at that particular loading.

There are no synchronous motor drives in most offshore or onshore installations, but they are used elsewhere onshore in larger plants where exact constant-speed operation is required.

2.9   CAPABILITY DIAGRAM FOR SYNCHRONOUS CONDENSER

It was stated in Chapter 1 that any synchronous machine when over-excited can be used as if it were a bank of static capacitors.

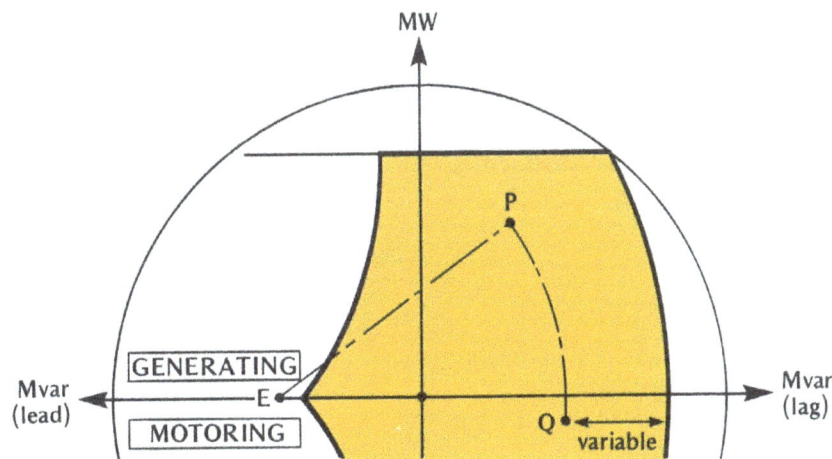

FIGURE 2.5
CAPABILITY DIAGRAM - SYNCHRONOUS CONDENSER

Figure 2.5 represents the capability diagram of such a synchronous machine. The full semicircle above the line depicts its generation mode, as discussed previously. The y-axis represents generated active power (MW). If extended downwards it represents negative active power - that is, motoring instead of generated power. The x-axis to the right represents lagging reactive power (Mvar) given out by the machine whether generating or motoring and is associated with the degree of excitation.

Imagine such a machine used as a generator and driven up to speed by its prime mover and synchronised onto the system, where it takes up its share of the active and reactive loads. The working point of the capability diagram of Figure 2.5 would be, say, P. Suppose then that the prime mover is unclutched, or that its fuel is cut off, but that the machine is well excited and its excitation remains unchanged. Mechanical drive to the machine then ceases, but it continues to rotate in synchronism with the system. It draws from the system only enough active power to keep itself going without driving any external load - it behaves as an unloaded motor, drawing just enough active power to make good its losses - a 'reverse-power' situation. Plotted on the capability diagram of Figure 2.5, the working point would now be Q, with slightly negative MW but delivering rather more lagging Mvar as before due to its unaltered excitation.

Such a machine would then be supplying reactive lagging power (megavars) but no active power (megawatts). **Supplying lagging** vars is the same thing as **receiving leading** vars, since one is the negative of the other. Therefore a machine operating as described above can be regarded as drawing leading reactive power - that is to say, it behaves as a static capacitor. The machine is then called a 'synchronous capacitor', although the old name 'synchronous condenser' remains in common use.

Moreover, the amount of leading reactive power drawn will be determined by the degree of over-excitation. Therefore the 'capacitance' is infinitely variable as required by the changing system load conditions, unlike that of a bank of static capacitors which can only be switched.

If a synchronous condenser is not available to correct the power factor of a system, a corresponding effect can be obtained by running any synchronous motors in the system in an over-excited state, as stated in para. 2.8. In this condition they will draw leading reactive current in addition to their active current. This will compensate for the large lagging currents drawn by many induction motors by providing a useful contribution to the lagging vars needed by those motors instead of calling on the generators to do so.

CHAPTER 3 SYNCHRONISING OF GENERATORS

3.1 GENERAL

The idea of synchronising is not new. Every time you change gear in a car you synchronise the engine to the road speed so that, when the clutch is let in, both shafts are running at the same speed and there is no jerk. Conversely, if you synchronise badly there is a jerk and possibly a lot of noise. The same applies with electrical machines when they are put in parallel.

Only offshore installations have main and subsidiary generators. Onshore there are only emergency generators, usually only one per installation. Consequently this chapter applies only to offshore installations.

3.2 D.C. GENERATORS

The simplest case occurs with d.c. generators.

Figure 3A represents two d.c. generators, both on open circuit but about to be paralleled by a switch. Each is separately excited such that machine 'A' has an open-circuit voltage V_A and machine 'B' V_B. Machine 'A' is assumed to be the 'running' generator, and machine 'B' is the 'incoming' generator which is to be paralleled to 'A'.

Before closing the switch which puts the two generators in parallel it is necessary only to ensure that their voltages are the same - that is, that $V_B = V_A$; then the switch may be closed, and no sudden current will flow - there will be no electrical 'jerks'.

If the voltages were different, suppose that V_A is greater than V_B. On closing the switch there will be a closed loop with the emfs V_A and V_B opposing one another. Since V_A is greater than V_B, there is a net clockwise emf in the loop, which will cause a clockwise current I_C to flow round it (shown in red), limited only by the resistances of the two armatures. This current appears suddenly as the switch is closed, putting a sudden load onto generator 'A', so causing it to slow with a jerk, and causing generator 'B' to motor, making it accelerate with a jerk. This circulating current, which occurs on closing the switch whenever V_A and V_B are not equal, is also called the 'synchronising current'. To avoid it and its consequent jerking effect on the system, the incoming machine voltage must first be matched to the voltage of the running machine - normally done by trimming the field of the incoming generator.

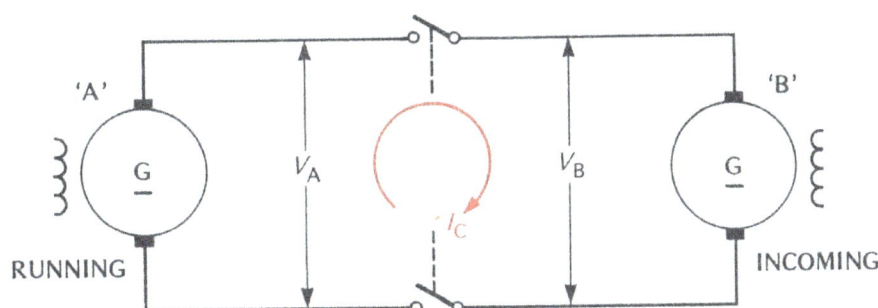

FIGURE 3.1
D.C. GENERATORS

3.3 A.C. GENERATORS

With a.c. generators the problem is more complicated. It can be seen in the d.c. case how a circulating current is caused by differing opposing voltages. In d.c. this is straightforward, but in a.c. a voltage difference can be caused **either** by differing voltage amplitudes **or,** for the same voltage amplitudes, by differing phase.

In Figure 3.2(a) the two voltages V_A and V_B are in phase with one another, but their amplitudes are different. At any instant such as time T, the instantaneous voltage of machine 'A' is TA and that of machine 'B' is TB. Therefore there is, at that instant, a voltage difference AB which will cause a circulating current to flow between the generators when the paralleling switch is closed. This is true at any instant other than a common voltage zero.

In Figure 3.2(b) the two voltages have equal amplitudes but are displaced in phase, V_B lagging on V_A. At any instant such as time T the instantaneous voltage of machine 'A' is TA and that of machine 'B' is TB. Although the two voltages are equal in amplitude, there is still an instantaneous difference of voltage AB which will cause a circulating current to flow between the generators when the paralleling switch is closed. Therefore, even though the voltage levels (as read by voltmeters) are the same, a difference of phase will still cause a circulating, or 'synchronising', current to flow between the machines, causing one to accelerate and the other to decelerate and to jerk them into phase with each other as the switch is closed.

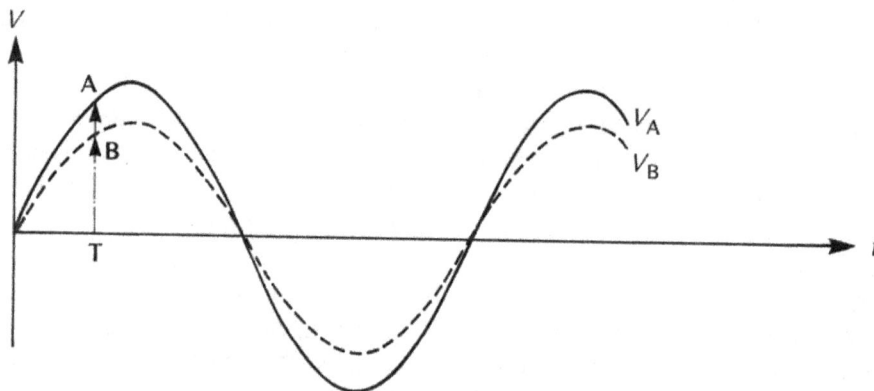

(a) VOLTAGE DIFFERENCE (Same Phase)

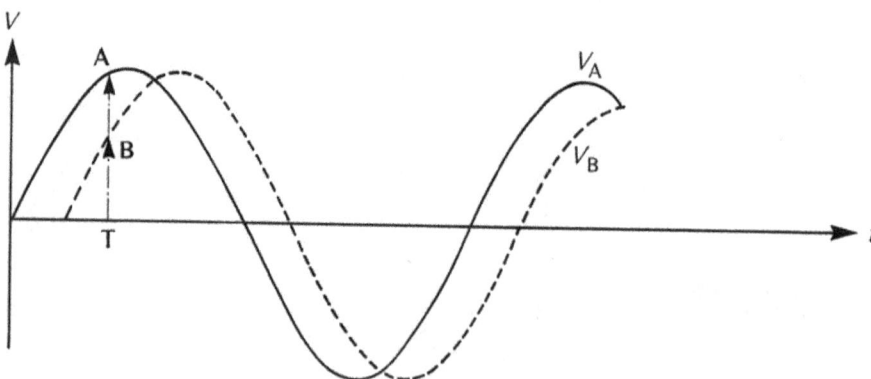

(b) PHASE DIFFERENCE (Same Voltage)

FIGURE 3.2
VOLTAGE AND PHASE DIFFERENCE

Therefore, to prevent sudden circulating currents occurring and to achieve smooth paralleling, the voltages of both machines must first be equalised **and** the machines then brought into phase. This is described in para. 3.4.

There is one further requirement. As when changing gear in a car, the two generator speeds must also be equalised before paralleling. If this is not done, the faster machine will be jerked back and the slower jerked forward, which could cause serious mechanical problems in large machines, as well as to the couplings, gear trains and prime movers.

If the two machines are running at different speeds before paralleling, this will show as different frequencies on the frequency meters. Therefore a preliminary to synchronising is to equalise as nearly as possible not only the machine voltages but also their frequencies, using the switchboard voltmeters and frequency meters.

3.4   SYNCHRONISING A.C. GENERATORS

It is assumed that one machine 'A' (the 'running' generator) is already in service on the busbars and is on load, and that a second machine 'B' (the 'incoming' generator) has been started and run up and is ready to be put in parallel with 'A' in order to share its load. Before this can be done the incoming generator 'B' must be synchronised with the running machine 'A'.

As already described, the first step is to match the incoming to the running voltage by reference to the voltmeters on the two generator control boards, and by using the incoming voltage regulator to trim it. Similarly the incoming frequency is matched to the running frequency by reference to the two frequency meters and by trimming the incoming speed regulator. Note that the **running** machine controls should not be touched - the incoming machine is always matched to the running, not vice versa.

It now remains to bring the generators into phase. Even after matching the frequencies by meter, the speeds will still not be exactly equal, and one machine will be slowly overtaking the other. As this occurs, their phase relationship will be steadily, but slowly, changing. The idea is to make this take place as slowly as possible and, as they momentarily pass through the 'in-phase' state, to catch them at that point, to close the paralleling switch and to lock them there.

There are two ways in which the correct phase may be detected - the first is by lamps, and the other is by an instrument called a synchroscope.

3.5   LAMP SYNCHRONISING

3.5.1 The 2-Lamp Method

Synchronising by lamps makes use of the circuit shown in Figure 3.3; two lamps in series are connected across the same phase of each generator. Only when the two systems are in phase is the voltage across the lamps continuously zero, and both lamps are out. At all other times there is a voltage difference, and the lamps glow. This is known as the 'lamps dark' method of synchronising.

The voltage phase vectors of both generators are shown. Machine No 1 is the 'running' and its vectors are in full line. Machine No 2 is the 'incoming' and its vectors are dotted. It is approaching synchronism with No 1.

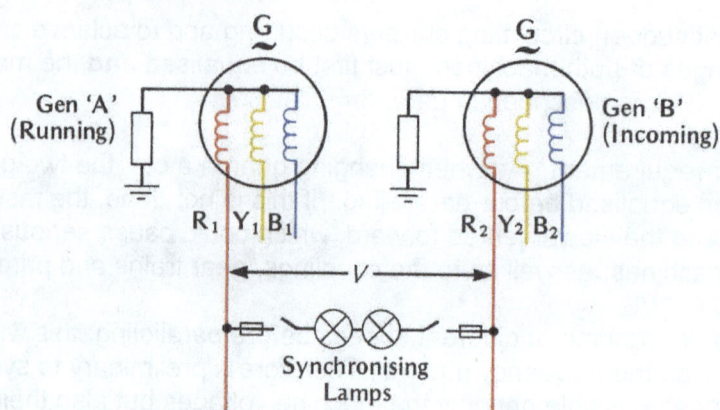

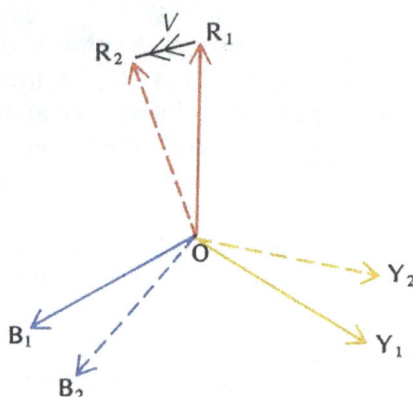

FIGURE 3.3
LAMP SYNCHRONISING (2-LAMP METHOD)

When the machine frequencies are nearly equal, the lamps are switched on and alternately glow and go out, giving a slow flashing appearance. The nearer the frequencies are to being equal, the slower the lamp flashing period. Therefore to achieve phase matching, the in-coming machine's speed is slowly trimmed until the lamps are flashing very slowly; then, as they are changing from bright to dark, the operator places his hand over the breaker control button or handle and, at the moment when the lamps go completely out, operates it to close the breaker. The lamps then stay out, but they should be switched off after completing the synchronising.

NOTE   The lamps could be connected to burn at their brightest, instead of being dark, when the systems are in phase, but this 'lamps bright' method is seldom used today. It is easier to detect the exact point of 'no light' in a lamp than to estimate when it is at its brightest. The 'lamps dark' method is almost universally found.

It is necessary to use two lamps in series because, when the systems are fully out of phase (lamps at brightest), the voltage difference is then double the system phase voltage.

3.5.2 The 3-Lamp Method

An alternative method, known as '3-lamp synchronising' is found on many platforms. It is shown in Figure 3.4.

The three lamps are connected as shown: No. 1 (yellow-to-yellow), No. 2 (blue-to-red) and No. 3 (red-to-blue). In the centre diagram the full lines refer to generator 'A' (R_1, Y_1 and B_1), and the dotted lines to generator 'B' (R_2, Y_2 and B_2). Machine 'B' is shown approaching synchronism with machine 'A'.

With the lamps so connected, the voltage across No. 1 lamp ($Y_1 - Y_2$) is small, and the lamp glows dimly. The voltages across No. 2 and No. 3 lamps ($B_1 - R_2$ and $R_1 - B_2$) are large, and both lamps are bright. As synchronism is reached (left-hand of the three lowest diagrams), No. 1 lamp goes out and the other two have equal brightness.

When the two generators are 120° out of synchronism (centre of the three diagrams) it can be seen that it is No. 2 lamp ($B_1 - R_2$) which has no voltage and goes out. 120° later (right hand diagram) No.3 lamp ($R_1 - B_2$) goes out.

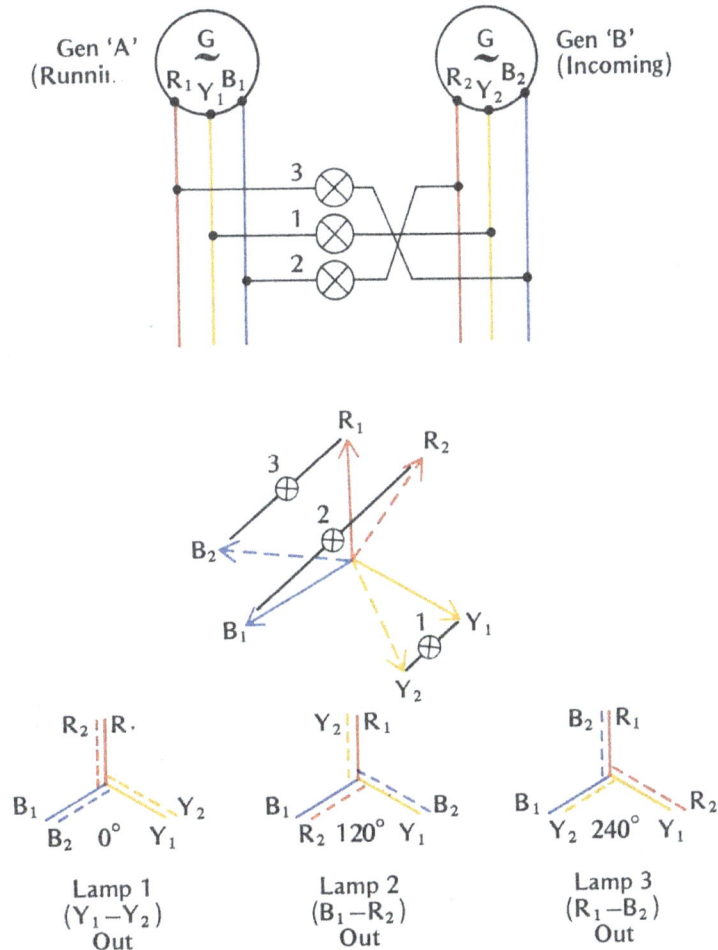

FIGURE 3.4
LAMP SYNCHRONISING (3-LAMP METHOD)

Thus, as generator 'B' catches up with generator 'A', each lamp goes out in turn, and at a decreasing rate, as synchronism is approached. Finally, at synchronism, No. 1 lamp remains extinguished long enough for the generator breaker to be closed.

The lamps are arranged either in a triangle with No. 1 at the top, or in a line with No. 1 in the centre. They may be lettered 'A', 'B' and 'C' instead of being numbered. Depending on whether the order of going out is clockwise or anti-clockwise with the triangular arrangement, or left-to-right or right-to-left with the in-line arrangement, the operator can determine whether the incoming generator is fast or slow - which cannot be done with the 2-lamp method.

3.6   SYNCHROSCOPE

The synchroscope method is normally used on those offshore installations where they are provided, but the synchronising lamps are often retained as a fall-back in case the synchroscope should fail. Therefore synchronising by lamp should be regularly exercised where this facility is available.

FIGURE 3.5
SYNCHROSCOPE

A typical synchroscope is shown in Figure 3.5. It is an instrument with a movement similar to that of a power-factor meter, but with the two windings fed from the running and incoming voltages. Whereas in a power-factor meter the current/voltage phase relationship is fixed and the pointer is stationary, in a synchroscope the phase relationship between the two voltages is constantly changing and the pointer rotates continuously, the direction of movement depending on whether the incoming machine is rotating faster or slower than the running. The face is marked with arrows denoting FAST or SLOW; these terms always refer to the incoming generator. When the pointer passes through the 12 o'clock position, the machines are momentarily in phase. (Some synchroscopes are marked '+' and '-'. The plus sign corresponds to FAST and the minus to SLOW.)

3.7   SYNCHRONISING AT A SWITCHBOARD

Most switchboards control two or more generators, and some have section breakers or interconnectors to other switchboards, any one of which may have to be synchronised with running machines. It would not be economic to have a separate synchroscope for each one, as it is used only infrequently.

The practice is therefore to have one synchroscope (usually with back-up lamps) in a central or conspicuous position on or near the switchboard together with selector switches whereby any chosen machine may be made the incomer. Selection may be by manual switch, key or plug. The running side is usually taken from the busbar. Where the switchboard handles high voltage the incoming and running voltage signals are taken through voltage transformers. The synchroscope is provided with fuses and an isolating switch, as it is not good practice to leave it in circuit when it is not in use.

To use the synchroscope, having selected which is to be the incoming generator, the voltages and frequencies are first matched as already described in para. 3.4. The synchroscope is then switched on; its pointer will be rotating. The incoming speed regulator is trimmed until the pointer is moving very slowly in the FAST direction. As it next approaches the 12 o'clock position, the hand is placed over the breaker control button or handle and, just before the pointer reaches 12 o'clock, it is operated to close the breaker. The synchroscope will then stop and remain locked in the 12 o'clock position as the generators remain in synchronism. Finally, the synchroscope must be switched off.

The reason why the incoming generator should be running the faster is that, when the breaker is closed, it will immediately take up a small part of the load. If it were running slower, that load would be negative - that is, the machine would 'motor' - and a reverse power situation would exist. The generator's reverse power protection might then cause the breaker to trip.

3.8   AUTOMATIC SYNCHRONISING

Most offshore switchboards are provided with an automatic synchronising feature. This consists of a number of relays (usually in a single case) which compare the incoming and running voltages and frequencies as well as their phase relation. Should any of these be outside limits, the incoming voltage regulator or speed regulator is automatically trimmed. Only when all three are within predetermined limits is a signal given automatically to the circuit-breaker to close.

Here again there is usually only one auto-synchronising unit to each switchboard; it is connected automatically to whichever machine is being started so long as the synchronising selector switch is set to AUTO.

Auto-synchronising is usually reserved for generators only. All other synchronising - for example across section breakers or interconnectors or on LV switchboards - is normally by hand.

3.9   CHECK SYNCHRONISING

In many instances, particularly with smaller generators and in the cases just mentioned, automatic synchronising is not used, and the exercise must be carried out manually by lamp or synchroscope. In such cases there is a danger that, if the manual synchronising is carried out unskilfully or by an operator under instruction, the switch could be closed at the wrong instant and severe damage could result to expensive machinery.

This can be prevented by 'check synchronising'. The equipment is similar to that used for auto-synchronising, but it does not automatically trim the incoming voltage, frequency and phase - it only monitors them. Nor does it carry out the final act of closing the circuit-breaker automatically; these all have to be done manually by the operator. However it does inhibit the breaker's manual closing circuit so that, unless all three synchronising conditions are satisifed together, the operator cannot close the breaker even though he presses the CLOSE button. If the breaker then fails to close, the whole synchronising process must be repeated.

Some check synchronising units sense only phase angle difference and do not monitor voltage or frequency differences. They rely on manual adjustment of voltage and frequency and only inhibit the closing of the breaker when the phase angle difference is excessive. It should be noted that voltage difference will cause circulating **reactive** current only. Although this is not desirable, it does not cause any mechanical shock and consequent damage to the transmission or the turbine since no active power is involved.

When, and only when, the check synchroniser is satisfied that the voltages, frequencies and phase difference are within acceptable limits (or, in the case of the 'phase only' type, that the phase difference is within limits), it closes a contact which 'arms' the circuit-breaker closing circuit, so permitting closure when the operator presses the CLOSE switch. The same contact on the check synchroniser also momentarily lights an IN SYNCHRONISM or READY TO SYNCHRONISE lamp, indicating to the operator that the breaker is ready for closing. Once this lamp has gone out again, he cannot close the breaker until it illuminates a second time.

Where check synchronising is fitted, it is brought automatically into circuit whenever a second or subsequent generator has been started and selected for switching on-line; it so serves as a protection against incorrect operation.

Check synchronisers may also be fitted across section breakers, interconnectors and LV incomers from transformers - in fact at any point in the network where it might be possible to close across two unsynchronised systems accidentally. They are also fitted across main generator incomer breakers even when auto-synchronising is provided. They come into action automatically if manual synchronising is selected.

Sometimes operators form the bad habit of holding the breaker control switch closed before synchronism is reached, and relying on the arming contact of the check synchroniser to complete the closing circuit. This is bad practice and must be avoided.

3.10   CLOSING ONTO DEAD BUSBAR

If it is required to connect an incoming generator, or LV transformer incomer, onto a dead busbar, the check synchroniser will not allow it to happen because, one side being dead, the two sides can never be in synchronism. In that case the check synchroniser must be temporarily 'cheated' while the connection is made. On most switchboards a special switch is provided for this purpose. It is spring-loaded to return to the OFF position so that the check synchroniser cannot be left permanently out of operation. This cheating switch may be tagged CLOSE ONTO DEAD BUSBAR or CHECK SYNC. OVERRIDE or other similar wording.

CHAPTER 4 ELECTRICAL CONTROL PANELS

4.1 GENERAL

The power systems of all offshore installations are remotely controlled from a central control point. This is usually an Electrical Control Room (ECR) containing the main control panel or console. It may be separate from the Platform Control Room (PCR) or combined with it.

The control panel usually takes the form of a console or desk, but in some cases it may be a vertical panel.

Onshore installations vary in their control arrangements according to their size. The control panel or console may be in a central control room, in a separate ECR or in a manned main substation. Whatever arrangement applies, the control panel provides similar facilities to those of an offshore installation with the exception of generator control and instrumentation.

As an offshore control panel is the more comprehensive, the following descriptions are aimed at typical offshore facilities. Excluding those parts relating to generators, however, they may be taken to apply to onshore arrangements.

FRONT VIEW

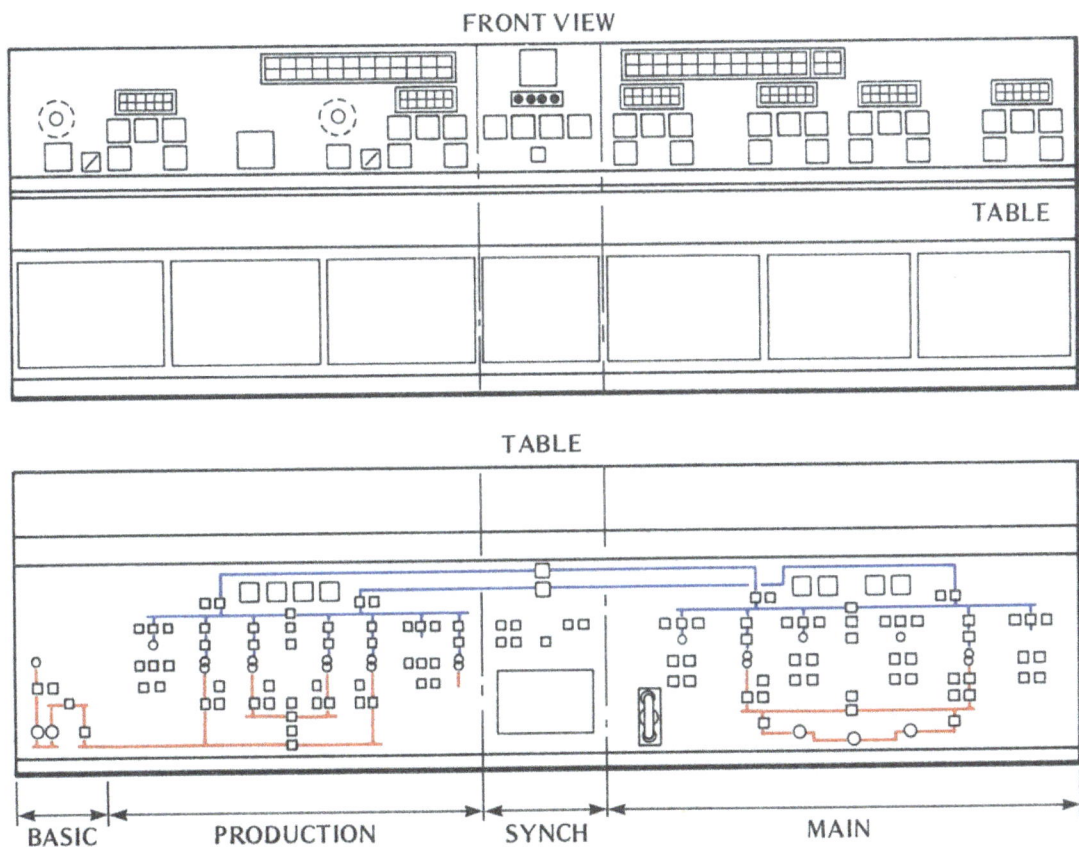

FIGURE 4.1
TYPICAL OFFSHORE ELECTRICAL CONTROL CONSOLE

4.2   CONTROL FACILITIES

Whatever form the panel takes, it provides the operator with certain facilities for monitoring and controlling the power network. The principal feature is a mimic diagram, which presents the whole of the bulk supply power network in single-line form. Symbols represent generators and transformers, and switches are placed in their correct positions in the diagram to indicate and control the various circuit-breakers in the system. There are also miniature indicating instruments placed in their proper positions on the diagram to show current or voltage in a particular connector. Lamps may show which generators are running, and other lamps may indicate fault conditions.

Each generator symbol may illuminate when the generator is running, and associated with it, either on the mimic itself or close beside it, are a number of indicating instruments and controls for the generator itself. Instruments usually include a voltmeter, ammeter, wattmeter, varmeter, power-factor meter and frequency meter. Controls may include start and stop pushbuttons, manual voltage control, manual speed control, controls for adjusting the set-points of the automatic voltage regulator and governor and Manual/Auto selector switches.

On or behind the mimic board there are usually alarm annunciator panels to show grouped alarms associated with each turbo-generator set (see Part 10 'Electrical Protection'), and sometimes also a further annunciator panel to show the progress of a gas-turbine's run-up sequence. There may also be kilowatt-hour meters and MW chart recorders included on or within the control desk equipment.

Figure 4.1 shows a typical control console. On such panels the HV system is usually coloured blue and the LV system red. The console shown here is in the form of a desk, the mimic diagram being placed on the sloping desk top (lower part of Figure 4.1), and the instrument panel forms a vertical back to the desk (top of figure).

On some offshore installations the control panel takes the form of a simple vertical board instead of a console.

4.3   THE MIMIC PANEL

Because of the large number of distribution circuits, especially at low voltage, a mimic panel with its controls makes no attempt to cover anything beyond the main items of the network and its bulk power supplies. The mimic is confined to the HV generator sets and the HV switchboards, including their incomer, section and interconnector circuit-breakers and also those circuit-breakers which feed LV switchboards through main transformers. It does **not** cover the feeders to individual HV motors.

On the LV side the coverage of the mimic extends down to the principal 440V switchboards, but not beyond. It includes their incomer, section and any interconnector circuit-breakers, but no distribution. It does however include any auxiliary LV generators and their associated circuit-breakers which it monitors, but does not control.

It is only these features that are monitored and controlled by the Electrical Authorised Person at the control panel. For the HV feeders to high-voltage motors control is exercised only from the PCR or their local motor control panels. For all LV feeders control is carried out from the local motor control panel or, in some cases, from the PCR. These limitations on the coverage can be seen on the mimic of Figure 4.1, where the section controlling the four Main generators is on the right, and that controlling the two Production generators on the left. The small section which monitors the Basic Services generator is on the extreme left.

4.4   SYNCHRONISING SECTION

In addition to the network mimics already described, control panels have a common section devoted to synchronising. Since various switching operations must be preceded by synchronising, and only one would be carried out at a time, a single common synchronising section is provided to which any of the two elements, running and incoming, can be connected. On most panels only the synchronising of generators is automatic and the synchronising section is not then used, but facilities for manual synchronising are usually provided in case the automatic equipment for generators becomes faulty and also for synchronising sections of the network. In that event the synchronising section is used, being brought into action by inserting a key in a switch beside the appropriate incoming generator or transformer control on the mimic.

The synchronising section consists of a synchroscope, with synchronising lamps provided as an alternative, a SYNCHRONISING ON/OFF switch and sometimes a READY-TO-SYNCHRONISE lamp; also two voltmeters and two frequency meters (incoming and running). The purpose of these is described in Chapter 3.

Control panels may have many other controls and switches, many of them key-operated, depending on the manufacturer and on the system in which they are used. For any particular platform it will be necessary to study the Procedures Manual for that platform.

4.5   DISCREPANCY SWITCHES

The remote control of circuit-breakers from the mimic panel is by means of devices called 'discrepancy switches'. They are set into the lines of the mimic and operate in the manner of 'gas-taps'; their semaphore handles point along the mimic line when the switch is on, and across it when off.

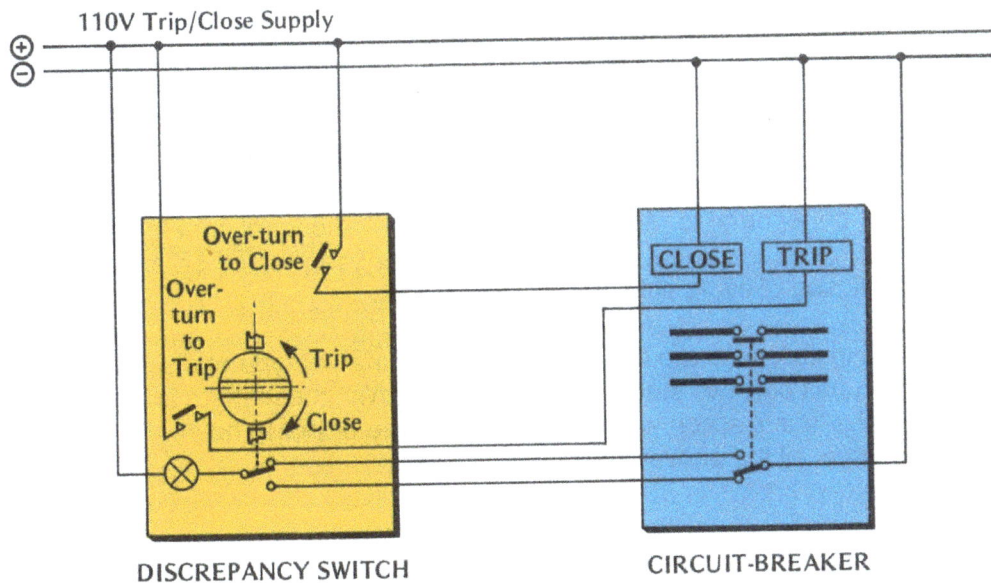

FIGURE 4.2
DISCREPANCY SWITCH

To operate the switch, the handle is turned 90 degrees counter-clockwise to open, or clockwise to close the breaker, as required. It is then pushed in and turned a further 30 degrees in the same direction as before but now against a spring. On release it will spring back to the new position. The final turn after pushing initiates the signal to the circuit-breaker to open or close. During the short period between the initial turn of the handle and the actual operating of the breaker (that is, while a discrepancy exists between the intended switch order and the actual breaker position), the switch becomes illuminated from within. As soon as the breaker operates and aligns with the new switch position, the discrepancy disappears and the light is extinguished. Thus the persistence of the light would indicate that the breaker has failed to operate, either due to malfunction or because of some interlock. Figure 4.2 shows the basic connections of this type of discrepancy switch.

It will be noted that it requires three positive actions to make a change: turn, push and over-turn against the spring. This reduces the risk of incorrect operation. Similarly an accidental push before the switch has been over-turned will not cause a change signal. This three-step arrangement also has the advantage that the switches can be used to try out a network rearrangement by turning them and studying the make-up without actually causing the switches to operate.

A further, and important, benefit is that, if a system disturbance causes one or a number of breakers to trip, those affected can be immediately identified by the illumination of their discrepancy lamps, because the tripped breakers are now misaligned with their 'on' switches. This can be 'acknowledged' by turning the switch to align with the tripped position, so extinguishing the discrepancy lamp.

Alternatively the original pre-trip set-up can be seen from the position of those discrepancy switches whose lamps are still illuminated, and the network can be restored, if desired, to its original state by pushing and turning against the spring those switches one by one until all lamps are extinguished.

Under normal conditions all lamps are out; this is known as the 'dark board' philosophy. Any lamp illuminating immediately draws attention to an abnormality.

As the lamp illuminates momentarily every time its switch is used, this provides a continuing test of it. If ever a lamp should fail to illuminate the instant its switch handle is turned, it should be reported to Maintenance. Normally all control panels have a 'Lamp Test' button which independently tests all lamps on the board together, and this should be used at the beginning of every shift.

Certain offshore installations are fitted with a slightly different type of discrepancy switch: this is one which has only two steps - turn and push. With these there is no 'over-turn' stage, and it is the 'push' action which creates the close or trip signal to the breaker. Apart from this difference, all the above description applies.

CHAPTER 5 BLACK START AND LOAD CONNECTION

5.1   INTRODUCTION

The expression 'black start' can be applied only to a completely self-dependent installation where the electrical power is generated within the installation and there is no access to external sources, such as in most offshore installations. It means, in fact, starting up a dead platform, from the point where no generators are running and only some battery supplies may be available.

This chapter, therefore, applies only to offshore installations. A complete supply failure in most onshore installations can occur only by reason of a widespread failure of the Supply Authority's system, and restoration of supply is not under the control of the authorised person. (This however is not the case with some onshore installations abroad which do not receive power from a local supply authority. The black start procedure in these cases will depend on the local arrangements.)

It is possible that from time to time an emergency or a plant failure offshore may lead to a partial or complete shutdown of the electrical system. At worst it might be necessary to abandon the installation for a time if the emergency were critical.

After a shutdown of electrical generation, essential services and safety systems are maintained operational for a time by batteries, but most of the battery systems become discharged after three or four hours. The exceptions are the navigational aids batteries which supply the secondary white and subsidiary red navigating lights and the standby foghorns; these last for a minimum of four days. Only in exceptional circumstances, however, are the diesel or gas-turbine starter batteries likely to be discharged.

It is possible, therefore, that after a shutdown lasting several hours the authorised person would be faced with a dead installation and with the batteries of several of the important d.c. systems discharged.

The electrical network of a typical offshore installation is shown in Figure 5.1; it must be emphasised that this system is typical only and does not represent any particular installation. Each system has individual differences. The 2.5MW gas-turbine generating sets are the 'secondary' or 'sub-main' generation, 'main' generation being provided by the 15MW sets.

The broad priorities for the restoration of supplies are:

1.   Start up and use the basic services generating set to provide power for living quarters, emergency lighting, basic life support, recharging the starting batteries of the secondary generating sets (if necessary) and for providing d.c. supplies for the safety and the communications systems. (For the action to be taken if the diesel starting battery is discharged, see para. 5.4.)

2.   Start the secondary generating sets when their batteries have been charged and energise the 'A' switchboards from them to provide power for full lighting, life support and start of production; also for firm supplies for the d.c. systems and for ventilation. Connect to basic services switchboard. Basic services generator may now be shut down.

3.   Close the interconnectors to 6.6kV switchboard 'B'. (No 'B' loads to be applied yet.)

4.   Run up main generator auxiliaries from basic services switchboard.

5.   Start up and synchronise the main generating sets to take over the 'B' switchboards and supply the full platform load including, where relevant, gas compression and re-injection, water injection and pipeline pumping.

6.   According to circumstances unload, disconnect and shut down the secondary generators.

Only the secondary and basic generating sets are equipped with independent battery starting; the main sets require an a.c. auxiliary supply, which comes from the basic services switchboard (see Stage 4 above).

Many main generating sets are equipped for burning only gas fuel and therefore are not available for service until gas is being produced from the process stream. The secondary sets are always suitable for dual fuel, reverting to diesel fuel automatically on the failure of the gas supply. They can therefore always be started on liquid fuel before production commences.

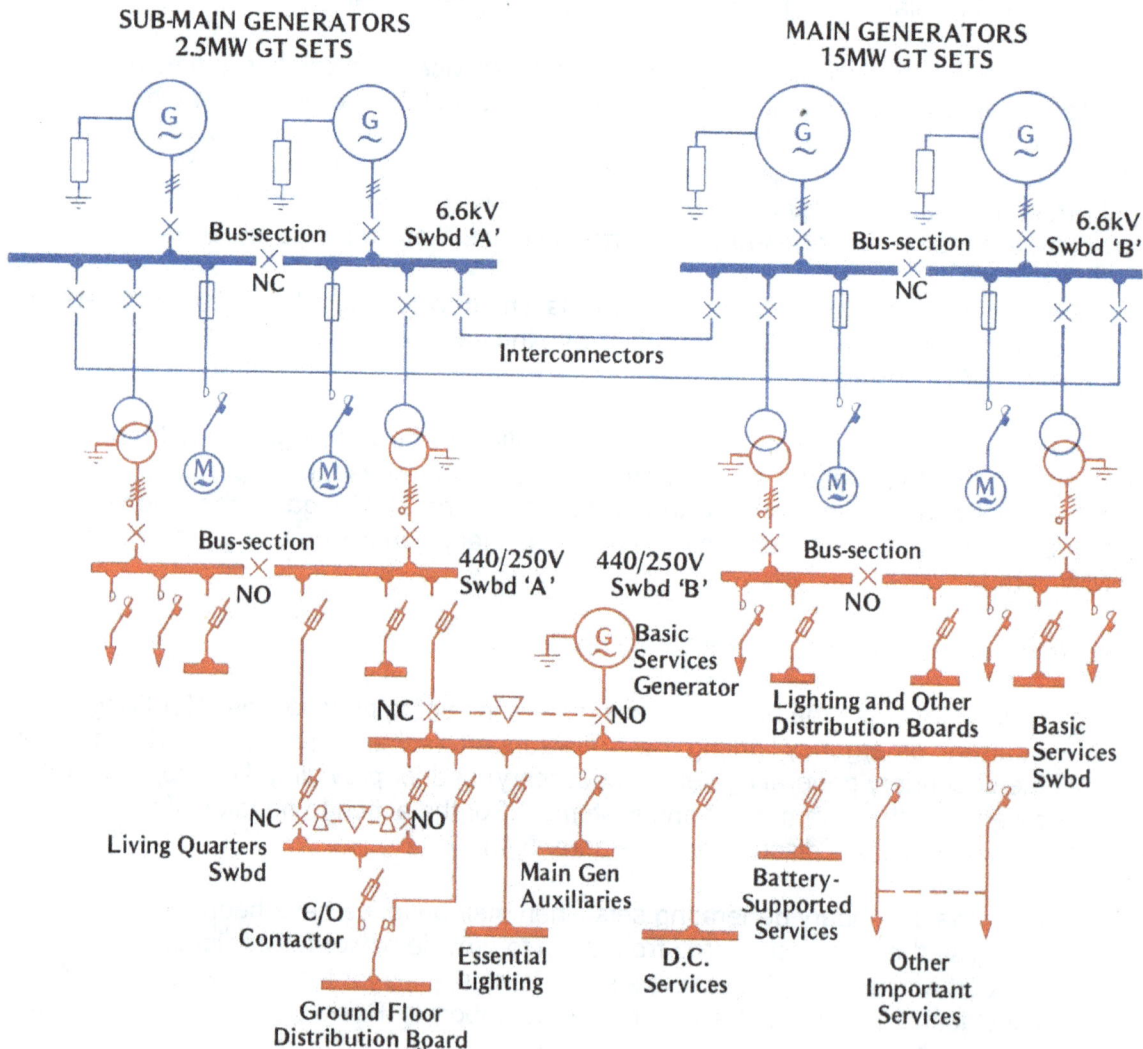

FIGURE 5.1
TYPICAL OFFSHORE POWER SYSTEM

A partial failure of generation or a major system fault should not lead to a prolonged complete loss of supply. The circuit protection isolates faulty equipment, and automatic load-shedding or undervoltage protection reduces the load on the system. The secondary generation, if not already running, starts automatically to restore the supply to HV and LV switchboards 'A'. Only a failure of the auto-start feature or an offshore emergency associated with the shutdown of all electrical plant should cause a complete loss of main supply, and even then the basic services generator is normally available. However, a complete disconnection of all supplies, including the basic services generator, can follow certain cases of offshore emergency, in which event a full black start will be required after the emergency is over.

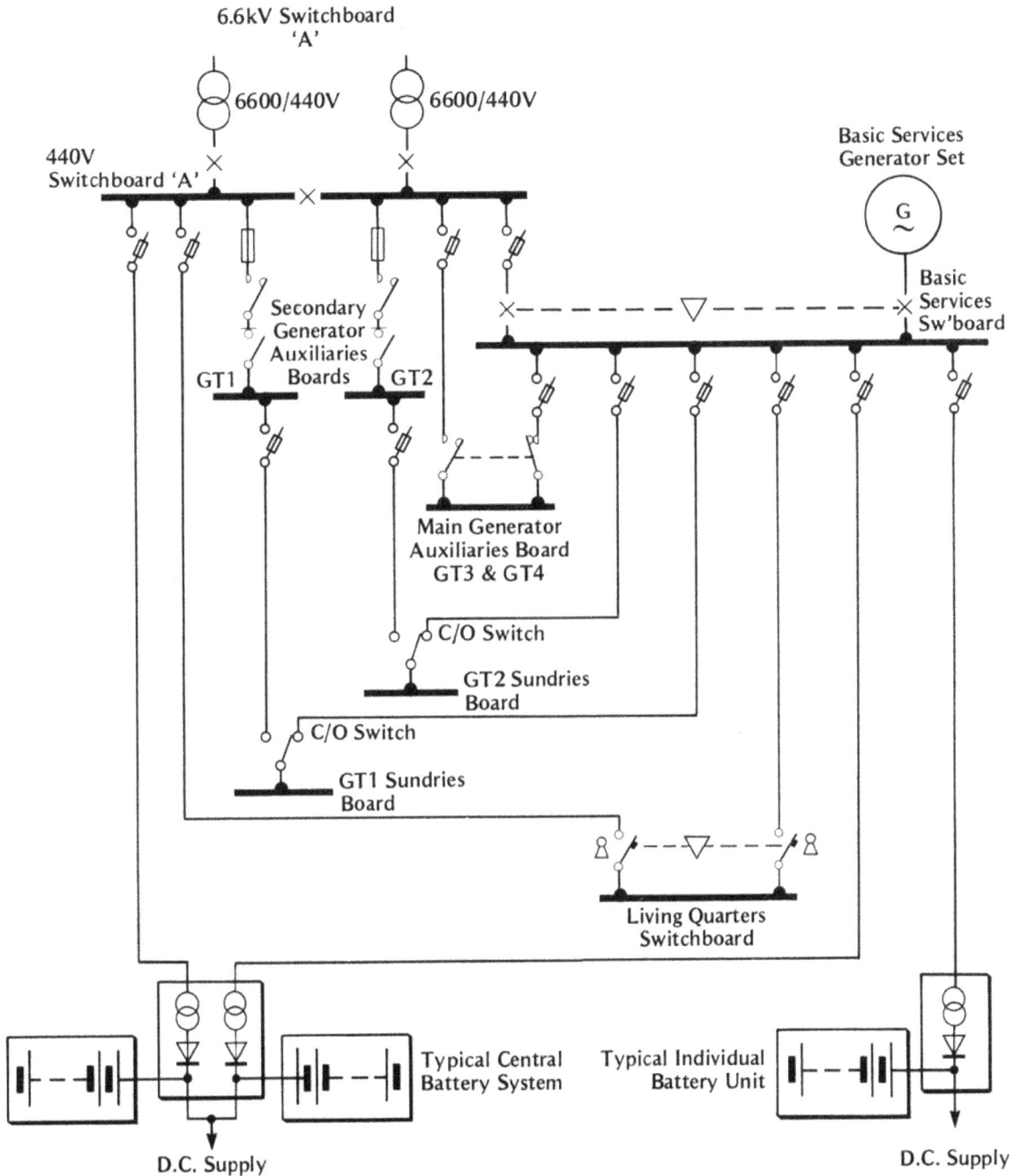

FIGURE 5.2
ALTERNATIVE FEEDS FOR OFFSHORE START-UP SUPPLIES

5.2 NETWORK CONDITIONS AFTER A COMPLETE SHUTDOWN

After a complete shutdown all generating sets are stopped and the associated generator circuit-breakers are open. All motor starter contactors will have opened automatically either on undervoltage protection in HV circuits or by inherent undervoltage drop-out in LV starters. The interconnector circuit-breakers forming part of 6.6kV switchboards 'A' and 'B' will have been tripped by undervoltage protection.

Certain services are fed through changeover switches which, when operated, provide an alternative supply from the basic services switchboard. Examples are the GT1 and GT2 Sundries boards, seen in Figure 5.2, which feed the single starting battery chargers for the secondary generating sets GT1 and GT2. Thus, when the basic services generating set starts, it supplies only lighting and small power to the offices and Radio Room in the living quarters, but at the same time the supplies to charge the starting batteries of the secondary generating sets are assured, should they be partially discharged.

As explained previously, the batteries associated with some of the d.c. systems may be completely discharged if the shutdown has been prolonged.

5.3 ACTION AFTER A COMPLETE SHUTDOWN

To illustrate a possible course of action, it is assumed that supplies are to be restored progressively after a complete shutdown during which some of the batteries associated with the main d.c. systems have been discharged.

If it is suspected that any part of the electrical network has been damaged, it must be fully isolated and a thorough examination carried out.

To prevent premature start-up, the auto-start facility for the secondary generating sets should first be disabled. For the precise method of carrying out this and the other control operations described later, reference should be made to the Procedures Manual for the particular installation.

If the shutdown had been due to high-level gas in a safe area or to an emergency resulting in the manual initiation of Total Platform Shutdown, a complete electrical black-out would have followed, including the stopping of the basic services generator. Before restarting, confirmation must first be obtained that it is safe to start up generation. If approval is given, and if the safety-system shutdown has not been reset, operate the appropriate key-switches on the Fire and Gas Protection Panel to override the generation shutdown or the 'start-up inhibit', allowing the electrical system to be progressively re-energised, as illustrated in Figure 5.3 and described in detail below.

5.4 RESTORE BASIC SERVICES SUPPLY

The basic services diesel generating set is first started from its local control panel. When the output voltage and frequency are correct, the manual operation of a control on the 440V basic services switchboard closes the generator circuit-breaker and at the same time opens the normal incoming breaker from 440V switchboard 'A' (if not already open). This assumes that the starting battery for the basic services diesel engine is not discharged. If it is, then either a charged battery or a portable engine-driven charging set must be brought in from outside. Some diesel sets can be started from an alternative hydraulic system using a hand-pumped accumulator.

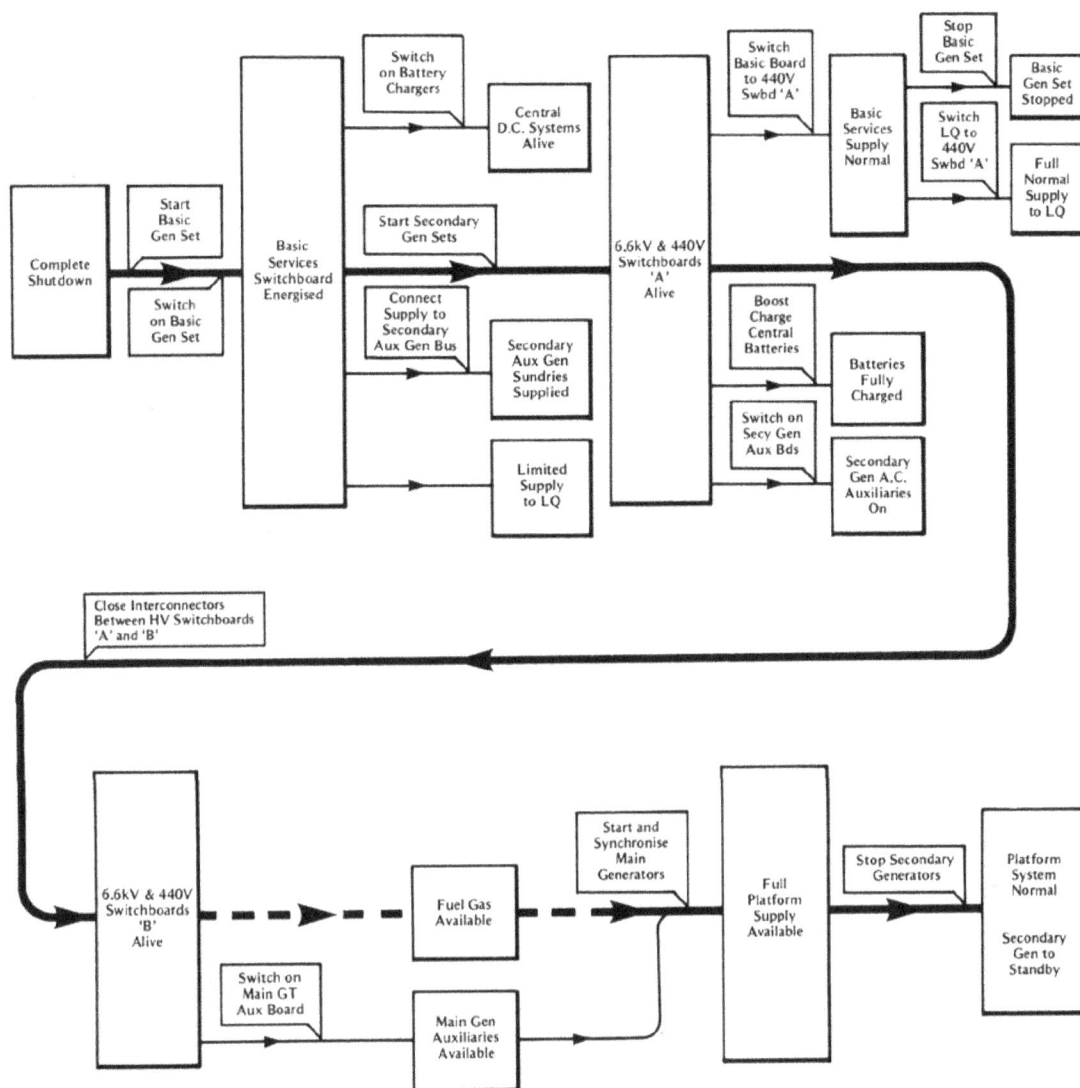

FIGURE 5.3
TYPICAL START-UP SEQUENCE

Supplies are now available to feed all loads connected to the basic services switchboard. These include one of the two charging rectifiers associated with each main d.c. system, the starting and purge battery chargers for the basic services and secondary generating sets and the diesel fire pump, and also supplies for the main generators' auxiliaries.

Limited power is now available for the living quarters and for basic life-support services such as sewage plant, one hot water heater, one water-maker, one air compressor and the heating circulating pump. Platform emergency lighting, a diesel transfer pump, lifeboat davits and some firefighting systems are also fed from the basic services switchboard, and motors may be started as required. All battery chargers fed from the basic services switchboard must now be switched on.

NOTE If the starting batteries of the secondary generators are still charged, it may be possible to ignore the basic services generator altogether and go straight to the starting of the secondary generators (para. 5.5).

5.5   START UP SECONDARY GENERATORS

Provided that their starting batteries are charged and diesel fuel is available, the secondary generating sets can be started from their local panels or from the Electrical Control Panel. When running correctly, the first to reach full speed is connected to the dead busbars of HV switchboard 'A', and the other auto-synchronises onto the bars. If the starting batteries are discharged, it will be necessary to wait for them to be recharged using the supply from the basic services generator. The 6.6kV switchboard 'A' can now be loaded up and the inter-connectors completed to the 440V switchboard 'A', which can now also be loaded.

The basic services switchboard is changed over to its normal interconnector supply by operating the manual control to trip the basic services generator circuit-breaker. This operation automatically closes the incoming circuit-breaker from 440V switchboard 'A'. A short interruption in the basic services supply will take place which may cause some starter contactors of motors fed from this board to drop out. Those motors must be individually restarted. The basic services generating set is then shut down.

Supplies are now available from the secondary generation at the following positions:

- 6.6kV Switchboard 'A'
- 440V Switchboard 'A'
- Basic Services Switchboard

A full supply to the living quarters from 440V switchboard 'A' can now be provided by opening the incoming circuit-breaker at the living quarters switchboard to disconnect the feeder from the basic services switchboard. This releases the interlock key which is then used to enable the incoming circuit-breaker from the 440V switchboard 'A' to be closed. Supplies are now available for full life support, ventilation and for production to commence; Production Staff should be informed accordingly.

Although the secondary generating sets start and initially run using only battery-powered d.c. auxiliaries, their a.c. auxiliaries must be brought into service as soon as possible by energising the secondary generation auxiliaries boards (Figure 5.2). This is done by closing the associated contactors at the 440V switchboard 'A'. When these two boards are alive, the connections to the two sundries boards are transferred from the basic services switchboard to the 440V switchboard 'A' by their changeover switches.

Batteries that are completely or partially discharged must now be fully charged as quickly as possible. When the 440V switchboard 'A' is energised, power is available to feed the second of the two chargers forming part of the power supply unit of each of the central d.c. battery systems, and these should be switched on at the chargers. Where an automatic changeover from 'Float' charging rate to 'Boost' is not provided, 'Boost' charge should be selected manually on one of the two chargers at each power supply unit, and its associated battery bank brought to a state of full charge. This procedure should then be repeated using the second charger to boost-charge the other bank of batteries.

Having ascertained that GT1 and GT2 are running in a satisfactory manner, that is to say that they are sharing active and reactive load equally and that the system voltage and frequency are correct, the interconnector circuit-breakers at the 6.6kV switchboards, first at the switchboard 'A' end and then at switchboard 'B', can be closed to restore supplies to 6.6kV switchboard 'B'.

5.6   START UP MAIN GENERATORS

When gas fuel is available, one or both main generating sets may be started. Before this is done, the contactor on the main generation auxiliaries switchboard that controls the circuit from 440V switchboard 'A' (Figure 5.2) must be closed; this transfers the main generation auxiliaries board from the basic services switchboard to the 440V switchboard 'A'. The various auxiliaries can now be run up.

The main generating sets are now started and synchronised onto the busbars of 6.6kV switchboard 'B' either by hand or automatically. The 6.6kV switchboard 'B' and its associated 440V switchboard 'B' can now be loaded up.

Depending on the loading, the secondary generating sets may now be shut down, the main sets now feeding 6.6kV switchboard 'A' through the interconnectors as well as feeding all 'B' boards. With their controls set for auto-start the secondary sets then act as standby for the main units. Both interconnectors between the 6.6kV 'A' and 'B' switchboards are left closed.

The offshore installation is now supplied from the main generating sets with the secondary generation on automatic standby, and the electrical system has been restored to its normal operating condition.

5.7   GENERAL

It must be remembered that the above description and sequence of events apply in detail only to the typical network shown in Figure 5.1. All installations have their own variations, and the detailed procedures, although similar in principle, differ from one to another.

The black starting of an offshore installation depends very much on the readiness of well charged batteries in good condition, especially those needed to start generating machinery. Whereas certain batteries, such as those for navigational aids, will have discharged after a prolonged blackout, others, such as those used for diesel starting, are used only intermittently and should remain charged even through a blackout.

It is therefore important that all batteries and their systems shall be in a healthy and well maintained state at all times, and that their charge shall not be subject to the smallest leak.

CHAPTER 6 LOAD SHARING

6.1 GENERAL

When two or more a.c. generators have been synchronised as described in Chapter 3, and they are feeding a common load, ideally they share that load between them in proportion to their sizes. However, to allow for flexibility in operating the system, each generating set is provided with field excitation and prime-mover governor controls whose settings affect not so much the voltage and speed of the sets but rather the share of the load taken by each when operating in parallel.

6.2 THE D.C. CASE

To see what happens when a power source is connected to an external circuit, consider first the simple d.c. system of Figure 6.1(a), where a battery is shown as the source of d.c. power.

The battery develops an emf of E volts and has an internal resistance of r ohms. Only the terminal voltage V is available to be measured, using a voltmeter. This voltage is used to drive the current I_L through the load resistance of R ohms. When the battery is not supplying current, there is no internal voltage drop and $V = E$. When current I_L is flowing, the direction of flow inside the battery is from the negative terminal to the positive and, in passing through the internal resistance r, it causes a voltage drop $I_L.r$ of opposite polarity to the battery emf. Thus the terminal voltage V is equal to $E - I_L.r$.

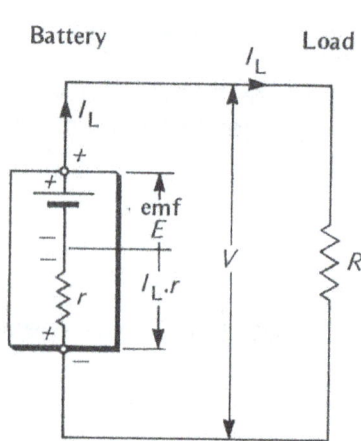

$$V = E - I_L.r$$

$$I_L = \frac{V}{R}$$

(a) BATTERY SUPPLYING LOAD

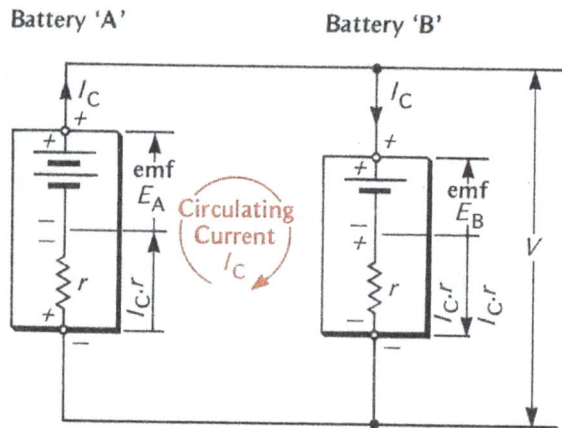

$$I_C = \frac{E_A - E_B}{2r}$$

$$(E_A - I_C.r) = (E_B + I_C.r) = V$$

(b) TWO BATTERIES IN PARALLEL (Unloaded)

FIGURE 6.1
D.C. SOURCES OF POWER

Figure 6.1 (b) shows two different batteries connected in parallel without an external load. The emf E_A of battery 'A' is assumed to be higher than the emf E_B of battery 'B'. The difference between these emfs causes a circulating current I_C to flow out of battery 'A' and into battery 'B', thereby discharging 'A' and charging 'B'; it is limited in value by the internal resistances of the two batteries in series. (In the example shown these resistances r are assumed to be the same.) Because the batteries are connected in parallel, they have a common terminal voltage V. In this particular example, V has a value mid-way between E_A and E_B; this is because the common circulating current causes equal voltage drops in each battery. The drop in battery 'A' is **subtracted** from the emf E_A and the drop in battery 'B' is effectively **added** to the emf E_B, since it is being charged by battery 'A'.

Thus for battery A: $$V = E_A - I_C.r$$

and for battery B: $$V = E_B + I_C.r$$

6.3   A.C. GENERATORS

With a.c. sources of power, the principal ones involved in load-sharing are gas-turbine-driven generators. The electrical indicating instruments of one such generator are shown in Figure 6.2(a), and it is the interpretation of the readings of these instruments which enables the operator to adjust the output of each set to achieve correct parallel operation.

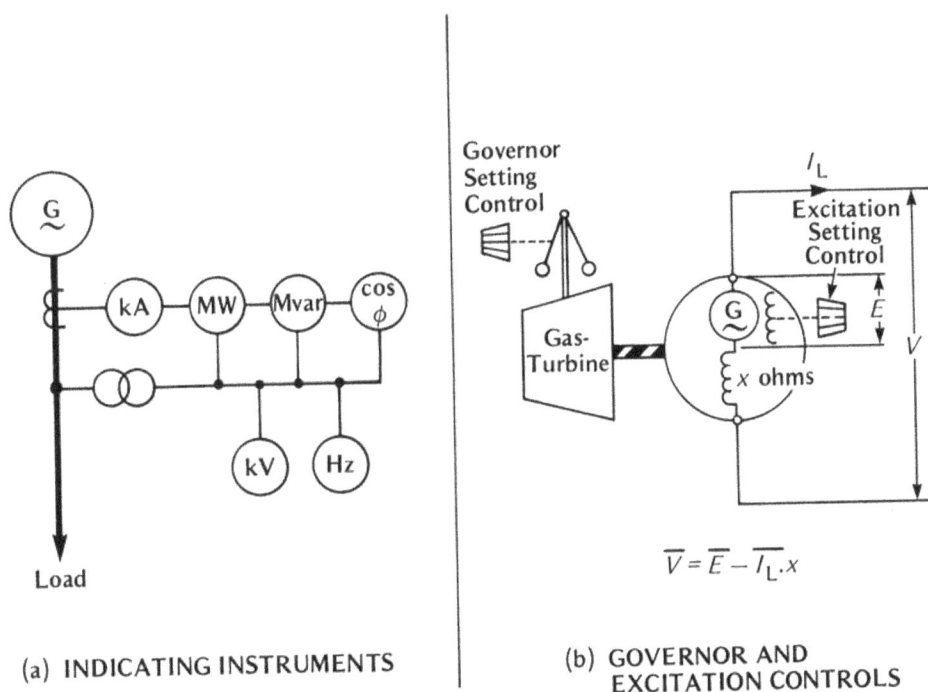

(a) INDICATING INSTRUMENTS

(b) GOVERNOR AND EXCITATION CONTROLS

$$\overline{V} = \overline{E} - \overline{I_L}.x$$

FIGURE 6.2
A.C. SOURCES OF POWER

The governor and excitation controls of a generating set are represented in Figure 6.2(b). The emf E of the generator is controlled by setting the excitation current in the field coils; an excitation setting or 'volts adjust' control is provided to vary the emf produced by the generator. The governor or 'speed' setting varies the mechanical power supplied to the generator by the gas-turbine. Like the battery, the generator has an internal impedance, z ohms. This consists of both resistance (r) and inductive reactance (x), but the resistive component is so small that it may be disregarded and only the reactance x is considered.

6.4   CONTROL OF GENERATOR LOADING

Consider now the effect of operating the 'volts adjust' and the 'speed adjust' controls of a generator which is running in parallel with another. Each operation will be considered separately and independently.

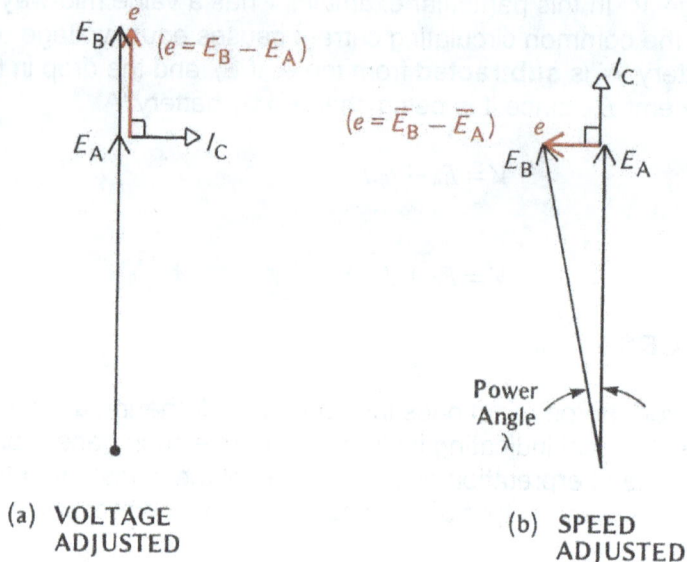

(a) VOLTAGE
     ADJUSTED

(b) SPEED
     ADJUSTED

FIGURE 6.3
LOAD SHARING CONTROL

6.4.1 Voltage Adjustment

In Figure 6.3 E_A and E_B represent the generated emfs of two generators 'A' and 'B' running in parallel, and for simplicity assume that E_A and E_B have the same length and direction initially. If the excitation of generator 'B' is now increased by operating the voltage adjust control in the RAISE VOLTS direction, the voltage vector E_B becomes longer than E_A but is still in the same direction. The difference between E_B and E_A is then a net emf e which causes a circulating current I_C to flow from machine 'B' to machine 'A'. Since these machines have reactance but negligible resistance, the circulating current is limited by the combined reactances of the two machines and lags $90°$ on e.

This is shown in Figure 6.3(a). Since I_C is at right angles to E_A and E_B, generator 'B' is producing, and generator 'A' is receiving, lagging wattless or reactive power, as would be shown by their respective varmeters. (The varmeter of generator 'A' would read in the negative sector unless prevented by a stop.)

Thus adjustment of the voltage control causes lagging reactive power, or vars, to circulate between the generators from the one with the higher to the one with the lower excitation.

If both machines are already on load and producing vars for the system, raising the voltage setting on one generator increases its var-loading and decreases the var-loading on the other, so varying the sharing of reactive power between the sets, as indicated by different readings on their varmeters. Lowering the voltage setting has the opposite effect.

Note that, although the control may be marked VOLTS ADJUST, with parallel generators it has little effect on the system voltage but becomes principally a reactive (or var) load sharing control.

6.4.2 Speed Adjustment

Consider next the effect of operating the 'speed adjust' control of one of the generators. Suppose the speed control of generator 'B' is moved in the RAISE SPEED direction.

This causes the fuel valve of generator B's prime mover to open more, so admitting more fuel and increasing the engine torque. It will drive the generator rotor more strongly in the forward direction of rotation. Since the rotor carries the field system, the emf of that generator advances relative to that of the other. In Figure 6.3(b), if E_A is the emf of generator 'A' and E_B that of generator 'B', then E_B is advanced in phase because of the rotor position, but its length is not altered since there has been no change in the excitation. The angle between the new rotor position and the old is called the 'power angle', symbol λ.

As before, the difference between the two emfs E_B and E_A is e; only this time it is a vectorial difference. The difference emf e, as before, causes a circulating current I_C to flow from machine 'B' to machine 'A'. As these generators have reactance but negligible resistance, the circulating current is limited by the combined reactances of the two machines and lags $90°$ on e.

Since I_C is now almost parallel with E_A and E_B, generator 'B' is producing, and generator 'A' is receiving, in-phase wattful or active power, as would be shown by their respective watt-meters. (The wattmeter of generator 'A' would read in the negative sector unless prevented by a stop.)

Thus adjustment of the speed control causes active power, or watts, to circulate between the generators from the one with the higher speed setting to the one with the lower. If generator 'A' were producing no other active load, receiving this power from generator 'B' would cause it to 'motor', and a reverse-power situation would be set up.

If both machines however are already on load and producing watts for the system, raising the speed setting on one set increases its watt-loading and decreases the watt-loading on the other, so varying the sharing of active power between the sets as indicated by different readings on their wattmeters. Lowering the speed setting has the opposite effect.

Note that, although the control may be marked SPEED ADJUST, with parallel generators it has little effect on the speed (and so on the system frequency) but becomes principally an active (or watt) load-sharing control. When used to its limit, it can off-load a generator completely onto any others remaining on-line - a normal operation before taking a generator off-line.

6.4.3 Summary

Notwithstanding that the controls may be marked VOLTS ADJUST and SPEED ADJUST, when generators are running in parallel these controls have little effect on voltage or speed. They are used respectively to control the sharing of reactive and active loading between generators (as indicated by their varmeters and wattmeters), and each can be operated quite independently of the other.

Thus:

> to adjust sharing of reactive load - operate the VOLTS ADJUST controls
>
> to adjust sharing of active load - operate the SPEED ADJUST controls.

Although the sharing of active or reactive load can be varied by operating the controls of one machine only, it is better practice to operate the controls of one generator in the RAISE direction and **simultaneously** those of the other in the LOWER. The two effects combine, and this method has the advantage of causing least disturbance to the system voltage or frequency.

Thus, to increase the share of active load on generator 'A', the SPEED ADJUST control of set 'A' is turned in the 'raise' direction and at the same time the corresponding control of set 'B' is turned in the 'lower' direction. If this is done with skill, the re-sharing of active load will have been done without disturbing the frequency at all. If however there has been a small increase (or decrease) of frequency, this can be trimmed out by operating both controls in the **same** direction. A similar sequence is used for adjustment of reactive load-sharing.

Figure 6.4 gives an example of the instrument readings of two parallel-running generators at a given instant.

Examination of the instruments shows that the common voltage (6.8kV in practice) is correct and the common frequency (60Hz) is also correct. But the varmeters and wattmeters show considerable imbalance of both reactive and active load-sharing.

Generator 'A' is producing 9.5 Mvar and generator 'B' 6.7 Mvar (total 16.2 Mvar). If they were in balance they would each be giving 8.1 Mvar. The action to take, therefore, is to lower the voltage setting on 'A' and simultaneously raise the voltage setting on 'B' until both varmeters read 8.1 Mvar. This simultaneous action will avoid disturbing the system voltage, but if there is a small error after balancing the reactive loading, it can be trimmed out by raising (or lowering) both voltage controls **together** so as not to disturb the now correct sharing of reactive load.

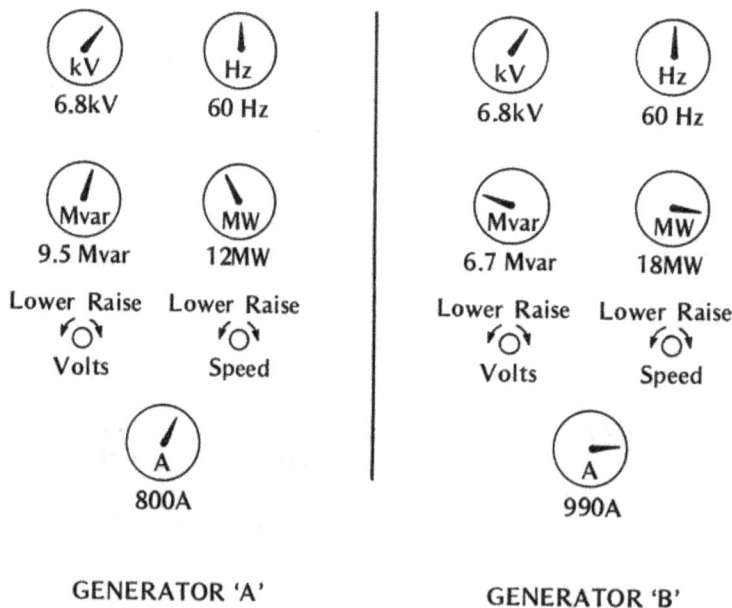

FIGURE 6.4
GENERATOR INDICATIONS (LOAD BALANCE)

Similarly the wattmeters show that generator 'A' is producing 12MW and generator 'B' 18MW (total 30MW), a considerable imbalance. If they were in balance they would each be giving 15MW. The action therefore is to raise the speed setting of 'A' and simultaneously to lower that of 'B' until both wattmeters read 15MW. This simultaneous action will avoid disturbing the system frequency, but if there is a small error after balancing the active loading, it can be trimmed out by raising (or lowering) both speed controls **together** so as not to disturb the now correct sharing of active load.

After both the active and reactive loads have been brought into balance, the MVAs are also automatically in balance and the two ammeters should give the same readings. Also the power-factor meters, if fitted, should give equal readings. Where power-factor meters are provided, they should not be used for balancing so long as both wattmeters and varmeters are fitted.

Some installations however do not have varmeters. In that case balancing is achieved by first equating the wattmeter readings, using the speed controls as described above. Then the voltage controls are used until the two ammeter readings are the same. It is essential that these two steps be taken in that order.

6.5 PRINCIPLES OF PARALLEL OPERATION

Although the governor control is usually designated 'speed adjust' - and indeed on single machines it does control speed - on parallel machines it is primarily an active load-sharing control, and speed control is only a secondary function. When used for active load-sharing, it does nevertheless slightly affect speed and hence system frequency. After the load share of one machine has been raised, the system frequency may have to be trimmed back by reducing the speed settings of both machines together until the system frequency is again within limits.

Similarly, although the Automatic Voltage Regulator (AVR) control is usually designated 'voltage adjust' - and indeed on single machines it does control voltage - on paralleled generators it is primarily a reactive load-sharing control, and voltage control is only a secondary feature. When used for reactive load-sharing, it does nevertheless slightly affect voltage and hence the system voltage level. After the reactive load share of one machine has been raised, the system voltage may have to be trimmed back by reducing the voltage settings of both generators together until the system voltage is again within limits.

The reasons for the above statements are given in paras. 6.5.1 to 6.5.4 below.

6.5.1 Droop

All control devices must have a 'droop' in order to give stable control, even though that droop may be very small. Droop is defined as the drop in the controlled output of a system for an increase in the controlling factor. For example, the speed of a governed generating set is in general not constant for all loads but falls slightly as the load increases, as explained in Chapter 5 of Part 2 'Electrical Power Generation'. The fall of speed from no load to full load is called the 'speed droop' and is expressed as a percentage of the full speed. A 4% droop is typical for a mechanical governor, but electronic governors can have much smaller droops, down to ½% or less. The theoretical and ideal limit of 0% (which can never in practice quite be reached) is called 'isochronous'. In practice the small droop of electronic governors may actually have to be increased artificially when generators are run in parallel, as explained in para. 6.5.8.

Similarly AVRs can never be absolutely perfect in operation. Although they aim to maintain the generator's output voltage constant at all loads, nevertheless there will be some fall of voltage as the reactive load increases (see Part 2 'Electrical Power Generation' Chapter 4).

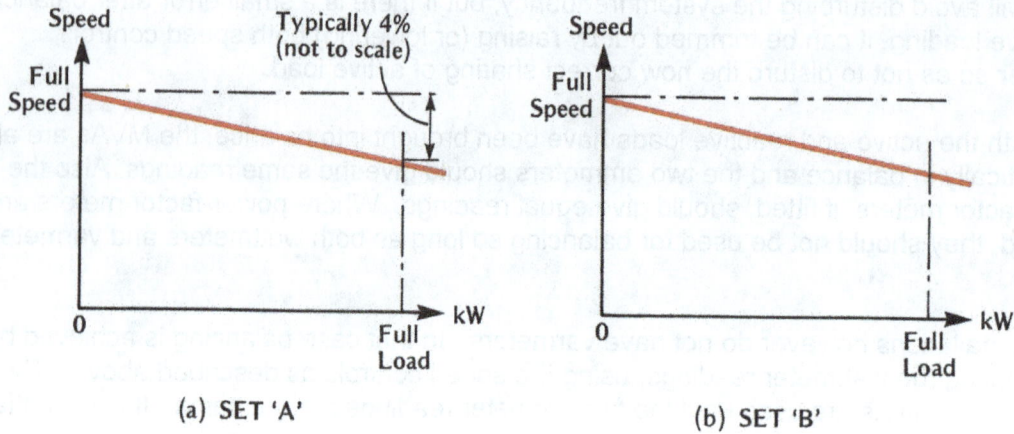

FIGURE 6.5
GOVERNOR DROOPS

Figure 6.5(a) illustrates a typical governor characteristic, somewhat exaggerated for clarity to show the droop in speed as the active (kW) load increases. Figure 6.5(b) shows the governor characteristic of a second, parallel set, and it is assumed that it is identical with the first.

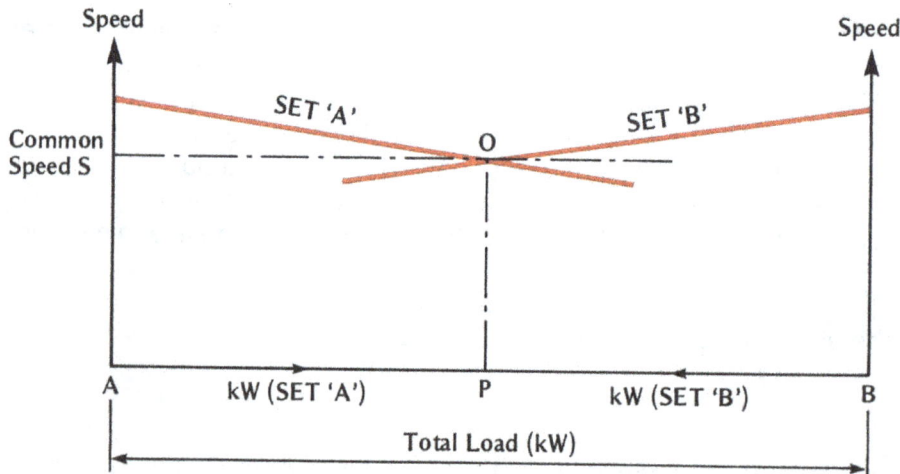

FIGURE 6.6
COMBINED GENERATING SETS

Both characteristics have been combined in Figure 6.6. The characteristic of set 'A' is repeated from the previous figure, whereas that of set 'B' is drawn reversed, with its speed axis on the right. The combined figure is thus quite symmetrical.

Since, when paralleled, the sets are locked in synchronism, their speeds are exactly the same; therefore the operating points of both are where the lines cross at O. AS is the common speed, PA is the load on set 'A', and PB is the load on set 'B'. AB is then the total load on both and, by symmetry, PA = PB ½AB. That is to say, the load on set 'A' is the same as that on set 'B', each being half the total load.

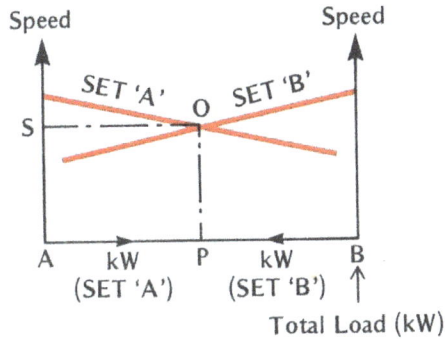

(a) TOTAL LOAD BEFORE INCREASE

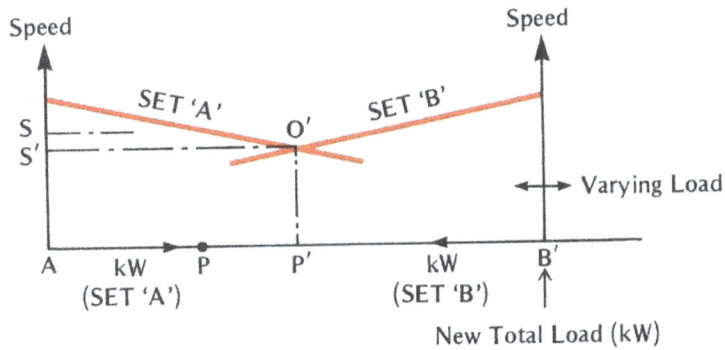

(b) TOTAL LOAD AFTER INCREASE

FIGURE 6.7
INCREASE OF TOTAL LOAD

If at any instant AB is the total load, Figure 6.7(a) (which is a repeat of Figure 6.6) shows that, by symmetry, each set takes half the load and that the common speed is AS.

If now the total load increases to AB' (Figure 6.7(b)), the slopes are unaltered but point B moves out to point B'. The figure remains symmetrical, but wider. Each set still takes half the new load (P'A and P'B'), but the crossing point O' is now lower, and therefore the common speed AS' is a little less than before.

Thus, if two identical generating sets have governors with similar characteristics, once set to share the load equally they continue to do so at all loads, although the common speed falls with increasing load.

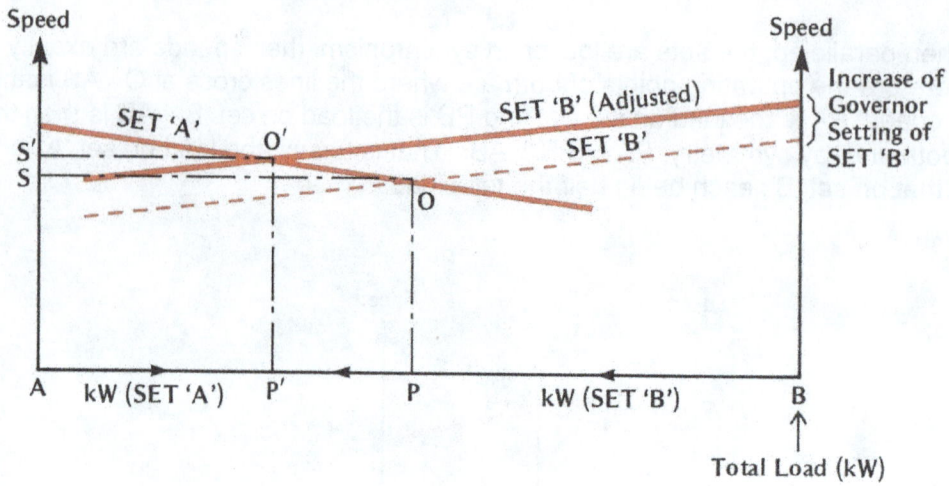

FIGURE 6.8
ADJUSTMENT OF SPEED SETTING

6.5.2 Trimming and Governor Setting

If the combined load is AB as before and the governors are equally set, the lines cross at O and the sets share the load equally (PA = PB = ½AB). If now the governor setting of set 'B' is raised, the whole line of set 'B' is raised to a position shown in full in Figure 6.8, though its droop angle is unaltered. The crossing point is now O'. The load is divided P'A to set 'A' and P'B to set 'B' - that is, they are no longer equal, the machine with the higher governor setting taking a greater share of the total load AB. Also, since O' is a little higher than O, the common speed AS is increased to AS'. This confirms the earlier statement that with paralleled sets, although the speed control is primarily used for sharing of active load, its use in this manner does nevertheless affect speed slightly, and it may be necessary to

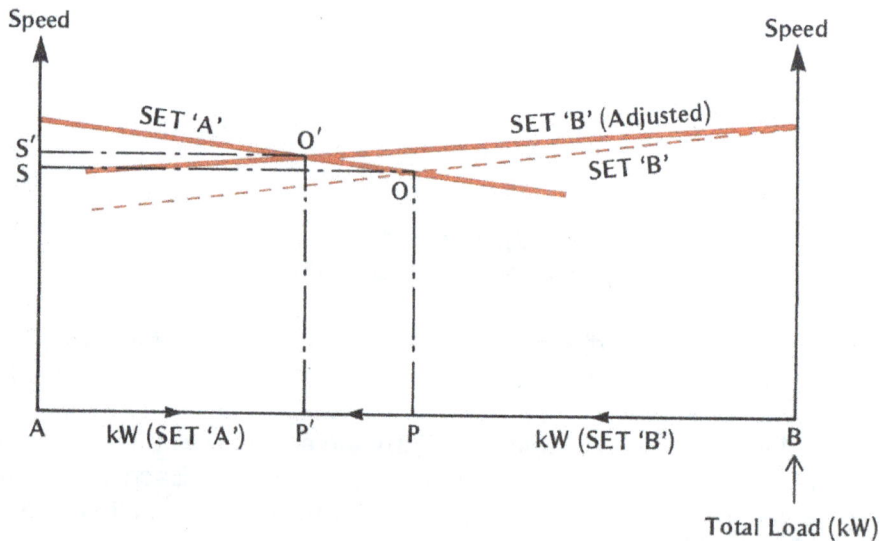

trim both sets down together to restore system frequency.

FIGURE 6.9
ADJUSTMENT OF DROOP

Thus, if the governor setting of one ('B') of two identical generating sets is raised, that machine takes a greater share of the load and the common speed is slightly raised, as shown in Figure 6.8. The same will occur if the governor setting of the other machine ('A') is lowered. Machine 'B' takes a greater share of the load, but the common speed will be slightly lowered. If both are done together, it is possible to adjust the sharing of load without affecting the speed at all.

6.5.3 Trimming and Governor Droop

Whereas the droop of a mechanical governor is built into the mechanism and cannot usually be adjusted, that of an electronic governor, such as is used on most offshore installations, is usually adjustable from maximum down to almost zero, the minimum position being labelled 'isochronous'.

If, instead of raising the governor speed setting of set 'B' as in Figure 6.8, the droop setting were reduced, Figure 6.9 shows that the crossover point shifts from O to O', the load-sharing is altered from PA : PB to P'A : P'B and the speed is raised from AS to AS'.

This achieves the same effect as altering the speed setting, but in an unsymmetrical manner, and it must never be used as a means of adjusting the sharing of load. Further, sets with different droops must not be run in parallel if it can be avoided. If it cannot - for example with two sets of different makes - then the operator must be prepared to find the load-sharing balance upset as load changes, and he must adjust the speed setting of one of the sets whenever it is necessary to restore balance.

6.5.4 Stability

No governor is perfect; all are subject to operating tolerance, although the electronic types are better than the mechanical in this respect. Ideally the characteristics are straight lines crossing at O, with equal load-sharing.

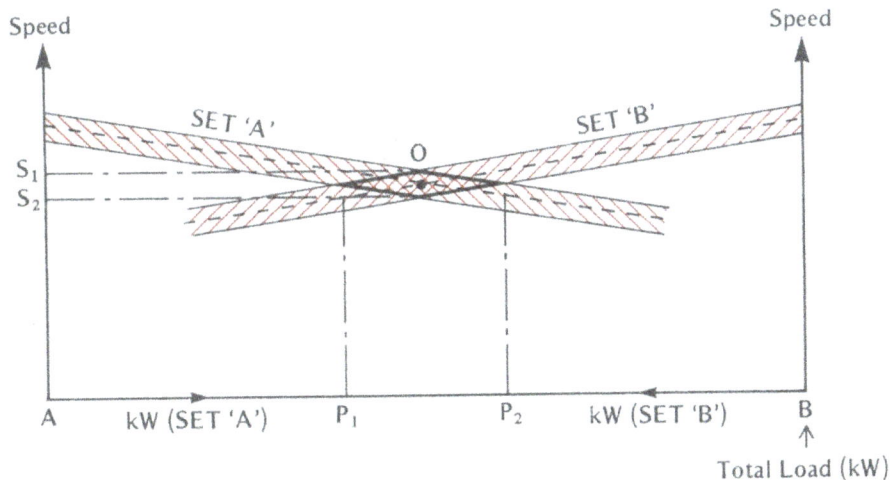

FIGURE 6.10
EFFECT OF GOVERNOR TOLERANCE

In practice the characteristic may lie anywhere within a band either side of the nominal lines - i.e. within the shaded areas of Figure 6.10. These bands may be of, say, ± 2% width for a mechanical governor when new (or more when worn) down to ± ½% for a good electronic governor.

The crossover point O could therefore lie anywhere within the central diamond formed by the two bands, and the common speed anywhere between S_1 and S_2. The load-sharing point may lie anywhere between P_1 and P_2, and the sharing of load anywhere between $P_1A : P_1B$ and $P_2A : P_2B$. So long as the diamond is small, this is probably acceptable. Wear on a mechanical governor, however, results in wider tolerance bands and consequently greater speed- and load-sharing errors.

If governors are used which have a very small droop (for example electronic governors set to low droop to achieve good speed regulation) even small tolerances can result in the diamond becoming very elongated, as shown in Figure 6.11. The sharing of load can then vary widely and can swing uncontrolled between P_1 and P_2, often without warning. The system is then unstable.

It is therefore good practice not to use low-droop or 'isochronous' modes to obtain good frequency regulation when generating sets are in parallel, although the isochronous mode may be used for a single machine. Some generator systems with electronic governors have automatic changeover from ISOCH to DROOP when a second set is paralleled; in other cases the operator has to make the changeover by hand switch. Unstable load-sharing is often a signal that the change is needed or perhaps has been forgotten.

Normally the level of droop used in the DROOP mode is set during commissioning and should never have to be altered by the operator.

It is good practice however not to change over between ISOCH and DROOP when going from single-machine to parallel-machine running, but to leave the control permanently in DROOP to ensure stable running at all times. This practice accepts that there would be loss of frequency with increasing load if the governor were the only control provided. However, most offshore installations have an overriding frequency control unit which holds the overall system frequency effectively constant irrespective of governor droop. This is described in sub-para. 6.5.6.

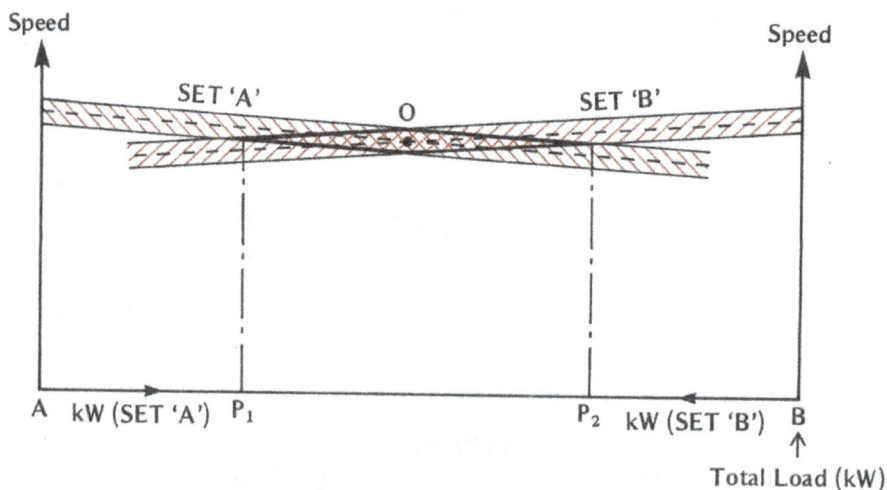

FIGURE 6.11
TOLERANCE WITH LOW DROOP LEADING TO INSTABILITY

6.5.5 Automatic Load Sharing

When identical sets, or sets with governors having identical droop characteristics, operate in parallel, they should, as in Figures 6.7(a) and (b), continue to share load equally, without further adjustment by the operator, once they have been set up to do so. If however the sets or their governors are not identical, some outside trimming is required to monitor any departure from equal sharing and to adjust accordingly. Equipment for this purpose is known as 'Automatic Load-sharing'. It is wholly static, is set up during commissioning and should need no adjustment by the operator. It operates on the auto speed electronic amplifiers of the various turbine governors, adjusting their set-points as necessary. This trims the sharing of active load as shown in Figure 6.8.

This equipment also senses system frequency. If it has deviated from the correct value (60Hz), the unit trims **all** governor set-points together, thereby restoring the frequency to its nominal value without upsetting the load balance. It thus overcomes any general loss of speed due to the droops of the various governors.

Another purpose of this equipment is for use when starting an incoming generating set by remote pushbutton and allowing it to synchronise automatically. After synchronising, the running machine will still be carrying its original load, but the incoming machine, though paralleled, will still be taking no load.

Without such equipment the operator, after synchronising, would have to go to the turbine control panel or Electrical Control Panel (ECP) and adjust the speed setting of the incoming set for it to take up its share of the load. He then trims back both sets to bring the frequency back within the limits - a manual operation, even though the starting and synchronising may have been remote and automatic.

The automatic load-sharing equipment, if switched on, makes the operation wholly automatic. It immediately senses a difference of loading between the loaded running machine and the unloaded incoming machine. It therefore automatically loads up the new set to its correct share and then trims the combined frequency.

It must be emphasised that automatic load-sharing equipment deals only with **active** load and operates on the sets' governors. It exercises no control over the reactive load-sharing, which depends on excitation and is controlled by the characteristics of the AVRs (see para. 6.5.9).

A typical automatic load-sharing equipment is shown in Figure 6.12, where the principal load-sharing elements are coloured yellow. The frequency control is coloured blue.

Busbar voltage is sensed by a voltage transformer, and the current being supplied by each generator is sensed by current transformers in its output line. Output current and voltage signals are taken, for each generator, to a Power Detector unit which converts the a.c. watts (active power) into a proportional d.c. signal.

Connecting both power detector units is a 'comparator loop' which in effect averages the d.c. signals from the two power detectors.

Suppose the active loading on the two machines is not in balance and that No. 1 machine is taking the greater share. Then the actual power from No. 1 generator exceeds the average, and a d.c. 'error signal', being the difference between actual and average, is passed to an Amplifier Unit which in turn passes it to the governor motor, directing it to reduce the fuel valve opening.

Similarly, the actual power from No. 2 generator is less than the average; its error signal is in the opposite sense and will cause its governor fuel valve to open more.
The combined effect of these two actions is to transfer active load from No. 1 to No. 2 machine and to bring the two loadings towards balance. When this point is reached both loadings are the same, and both are equal to the average. There is then no longer any error signal from either machine, and the two governor motors stop moving. Balance has been automatically restored, and both generator sets are now sharing the load equally.

The above description has dealt with two generators in parallel, but it applies equally to any number. The comparator loop is simply connected in parallel to all power detector units and averages out the active loading on all running generators. All machines are then automatically controlled until their outputs become equal to the average and so to each other. If a machine is not running, the 10V d.c. power supply is removed from its amplifier, and a relay disconnects its power detector from the comparator loop, so that the detector's zero signal does not influence the average carried by that loop. This can be seen in Figure 6.12 where the opening of the generator supply breaker disconnects the 10V d.c. supply.

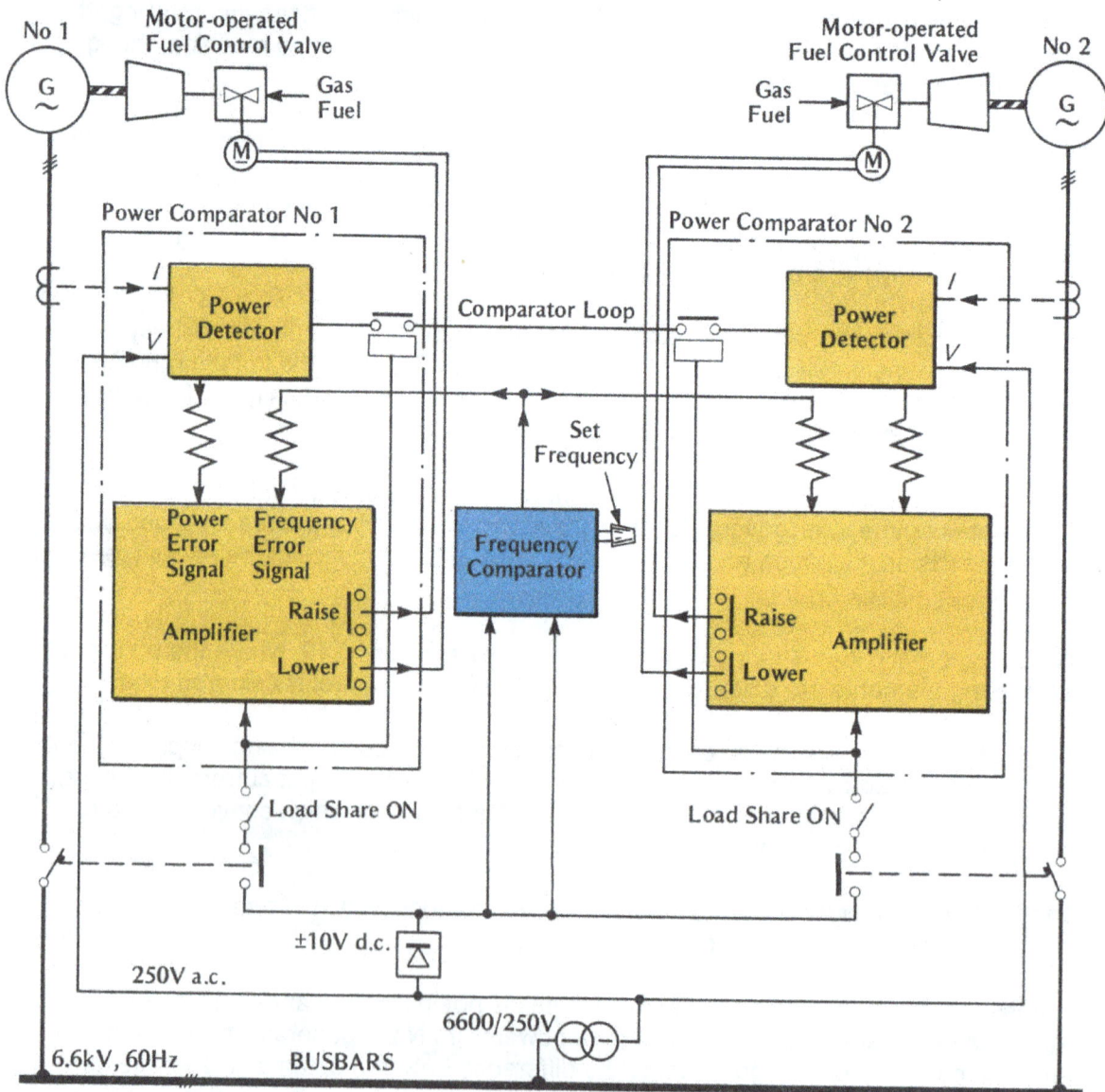

FIGURE 6.12
AUTOMATIC LOAD SHARING EQUIPMENT

Each amplifier can also be switched on or off manually. The switch is usually on the control panel and is marked LOAD SHARE OFF/ON. It must be switched off when the generator is in manual speed control, as explained in sub-para. 6.5.7. Otherwise it is normally left on.

6.5.6 Automatic Frequency Control

The automatic load-sharing equipment is also provided with a Frequency Comparator. This has nothing to do with the sharing of load as such, but it makes use of the load-sharing amplifiers and their connections to the governor motors. It appears in the centre of Figure 6.12.

The frequency comparator (coloured blue) is fed from the busbar VT and compares the busbar frequency with the set frequency; this datum (60Hz) may be set manually or by remote-controlled motor. The difference is passed on as a d.c. signal, proportional to the frequency error, to all amplifiers together. Those associated with the machines which are running will pass signals each to its governor motor, which then all run in the same direction to adjust the fuel valves of all machines together to reduce the frequency error. If they do not all move equally, the automatic load-sharing circuits already described will bring them into line.

It has already been shown that, for stable parallel running, a droop is necessary in the governor load/speed characteristic. Without the frequency comparator there would be a steady loss of system frequency as active load increased, but the comparator automatically compensates for this, holding the frequency within tight limits at all loads and providing so-called 'isochronous' governing.

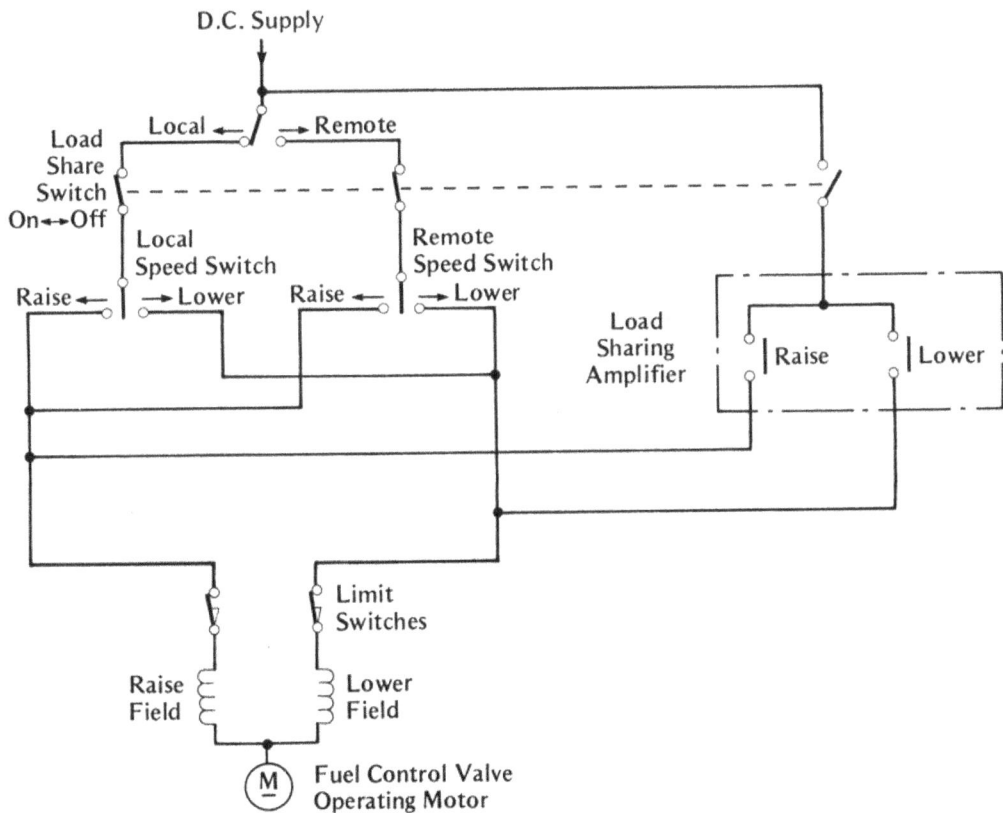

FIGURE 6.13
EXTERNAL CONTROLS FOR LOAD SHARING

6.5.7 External Controls

Figure 6.13 shows the external controls necessary when automatic load-sharing is provided. Control may be automatic or manual, and manual itself may be local or remote from the ECR. A LOCAL/REMOTE switch selects this.

The LOAD SHARE switch already referred to selects either manual or auto. The auto control must be disconnected when in manual because otherwise, if a manual change of speed setting were made on one generator, the active loading would become unbalanced and the automatic unit would immediately operate to counter it.

6.5.8 AVR Droop

As stated earlier, a voltage control system is as subject to errors as any other closed-loop system. The principal error, like that of the speed control system, is 'droop' - that is, the amount by which the voltage is **not** restored to its set level as the load increases. It was shown in Part 2 'Electrical Power Generation' that an a.c. generator suffers internal voltage drop mainly owing to the **reactive** component of the load current flowing through its internal reactance - that is, owing to kvar loading only. The AVR seeks to compensate for this by increasing the excitation, but it is only partially successful in doing so. There is always a small shortfall due to the inherent droop of the AVR, but with modern electronic types this is normally less than ½% between no load and full kvar load.

For purposes of stable reactive (kvar) load-sharing between parallel-running generators it is necessary to have some droop. This is for exactly the same reasons as have been explained for the sharing of active (kW) load between paralleled sets.

The natural, inherent droop of an AVRs is around ½%, and this is not enough for stable running of paralleled generators, although it is excellent for controlling the regulation of single-running sets. Therefore, when generators are paralleled, artificial droop must be injected into the AVR of each set to give it the necessary sloping characteristic - usually about 4% from no load to full kvar load. If generator sets are not in the DROOP mode when being paralleled, it is possible that they may go unstable and throw reactive load backwards and forwards between each other. This condition is indicated by violent fluctuations of the kvar and power-factor meters; the latter may even swing into leading.

Artificial droop is injected into an AVR in the following manner.

It must be remembered that it is necessary to measure only the **reactive** component of the load current, and to use this to bias the measured output voltage (which the AVR is trying to sustain) so that the AVR senses not only the output voltage but also an additional element proportional to the reactive part of the load current. As a result the AVR senses a voltage which is actually a little too high, and it reduces the excitation accordingly. The generator output voltage will therefore fall slightly with increasing kvar loading, so producing the desired artificial droop. The quantities are arranged to make this droop about 4% at full kvar load.

A switch is usually arranged in the biasing circuit to short out the bias when it is not required - i.e. when the set is running singly. The switch may be manual or may operate automatically on the closing of a second or further main breaker. In some practices, however, it is left permanently open - i.e. in permanent DROOP mode.

To single out the reactive component of the line current is very simple in a balanced 3-phase system, as different phases are used for current and for voltage measurement to give the necessary quadrature effect. This is shown in Figure 6.14.

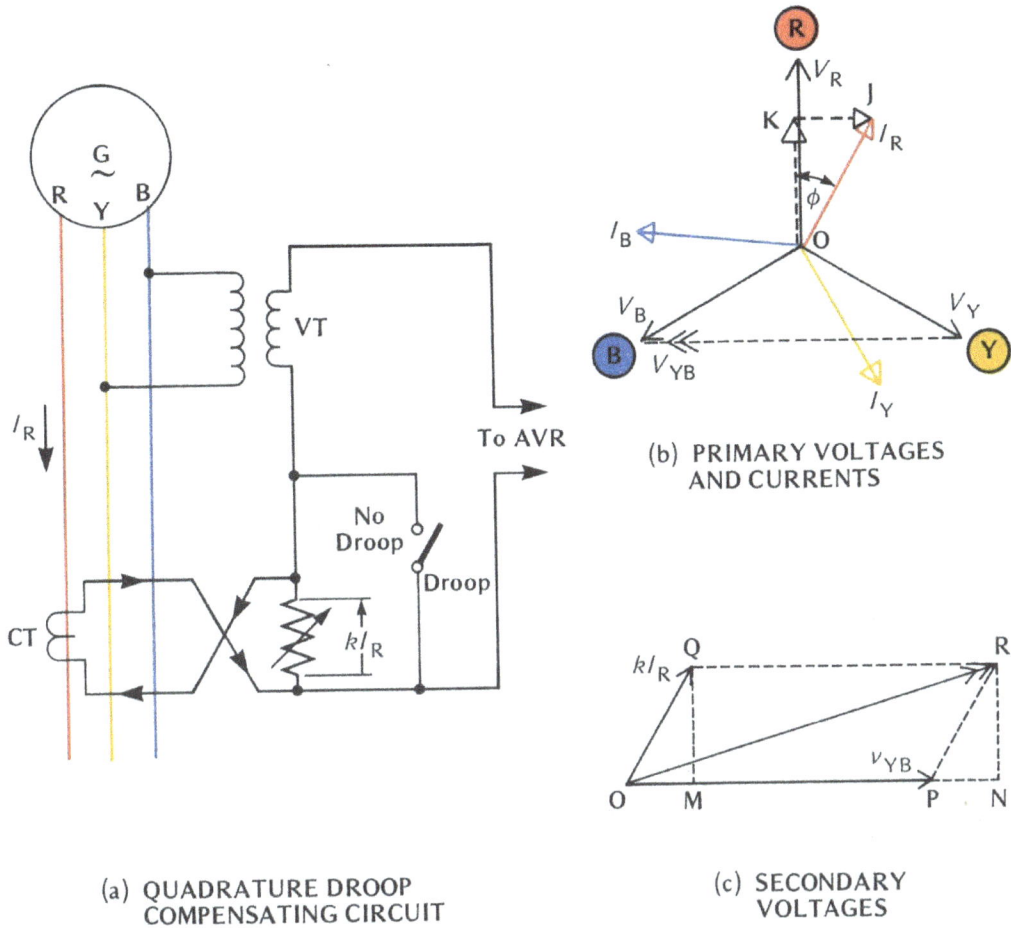

(b) PRIMARY VOLTAGES
AND CURRENTS

(a) QUADRATURE DROOP
COMPENSATING CIRCUIT

(c) SECONDARY
VOLTAGES

FIGURE 6.14
QUADRATURE DROOP COMPENSATION

Figure 6.14(a) shows a 3-phase generator, the three output lines being red, yellow and blue. Sensing voltage for the AVR is taken through a single-phase voltage transformer from Y and B phases (or between the Y and B phases of an existing 3-phase transformer), and a current transformer is inserted in R phase. If the load current lags on the voltage (normal condition) the three voltage and current vectors appear as in Figure 6.14(b), the current vectors lagging by an angle φ on the voltages.

In Figure 6.14(b) vectors OY and OB represent the main yellow and blue phase voltages V_Y and V_B, and vector YB is therefore the yellow-to-blue line voltage V_{YB}. Vector OJ represents the red line current I_R. This can be resolved into components OK and KJ respectively in phase and in quadrature with red phase voltage V_R.

Figure 6.14(c) represents the situation in the VT and CT secondary circuits.

Yellow-to-blue line voltage in the VT secondary, v_{YB} is the reverse of that in the primary and is represented by the vector OP. The CT secondary feeds a burden of a simple resistance; therefore the voltage across this resistance is directly proportional to and in phase with the line current I_R - let it be called 'kI_R', represented by the vector OQ (Figure 6.14(c)), parallel with the current vector OJ (Figure 6.14(b)). (The vector OQ is not opposite in direction to OJ of Figure 6.14(b) due to the reversed leads from the CT secondary.)

If the CT resistor is placed directly in series with the VT secondary so that their voltages add vectorially, the situation is then as shown in Figure 6.14(c), where OP is the VT secondary voltage and OQ the voltage across the resistor due to the CT secondary current. OR is the vector sum of the two and is of magnitude $\overline{v_{YB}} + \overline{kI_R}$. This is what is sensed by the AVR.

If a perpendicular, QM, is dropped from Q to the horizontal and another, RN, from R, then the length OM represents the **quadrature component** of the voltage in the resistor, corresponding to KJ in Figure 6.14(b), which was the quadrature component of the actual red line current I_R. Because OQ is small compared with OP, the lower triangle is flat and elongated, so that the length OR is little different from ON. But ON = OP + PN = OP + OM, which is the sum of the VT secondary line voltage and an element proportional to only the quadrature component of the line current.

Therefore the AVR, in sensing a voltage OR, is in fact effectively sensing the line voltage **increased** by a small element proportional to the reactive, or quadrature, component of the line current. It causes the generator excitation to be reduced, so dropping the line voltage by this amount and therefore introducing artificial droop which increases as the line current - or, more accurately, the reactive or kvar component of the line current - increases. The droop is a straight line inclined downwards.

By making the value of resistance variable, the amount of percentage droop, and so the downward slope, can be varied. By shorting out the CT bias altogether all artificial droop is removed and the AVR operates in a 'No Droop' mode, or 'isochronously' (to extend the meaning of this word from speed to voltage).

This process is known as 'Quadrature Droop Compensation'. It has the effect of ensuring a balance of reactive load-sharing between paralleled generators. If generator 'A' is taking a greater share of reactive load than is generator 'B', then the AVR of generator 'A' receives a higher input voltage than 'B', because of the compensation, so it reduces the excitation of 'A'. Similarly the AVR of generator 'B' receives a lower input voltage and increases the excitation of 'B'. These combine to reduce the reactive loading on 'A' and to increase it on 'B', as explained in para. 6.4.1, until balance is restored.

6.5.9 Reactive Load Sharing

The effect of governor setting on the common speed of paralleled generating sets and on their active load-sharing - that is, on their kW loadings - was discussed earlier and shown in Figures 6.6 to 6.11.

Exactly the same conditions apply to the effect of AVR setting on the common voltage of paralleled generators and on their reactive load-sharing - that is, on their kvar loading. This is shown in Figure 6.15, which corresponds to Figure 6.6 for active load-sharing but where 'kvar' replaces 'kW' and 'voltage' replaces 'speed' on the earlier figures.

The droop characteristics may be presented in exactly the same way as for governors, the cross-over point determining the common voltage AV and the sharing of reactive load PA : PB. In the symmetrical case they share it equally (PA = PB = ½AB) - half the total reactive load.

Altering the voltage setting of one generator's AVR (as Figure 6.8 for a governor), or changing the droop (as Figure 6.9), has effects similar to those described for a governor. Too low a droop (as Figure 6.11) tends to produce instability, causing the reactive load to swing between one generator and the other, showing as swings on the kvar meters and the power-factor meters, which may even go into leading.

Droop in the AVR characteristic is provided automatically by 'quadrature droop compensation' as described above. When generators are operated singly, this droop may be removed automatically in some systems, or by manual switch in others, in order to give the best voltage regulation. As already stated, it is good practice to leave the droop permanently in, accepting a small drop of system voltage as loading increases. The droop, however, is adjustable (by varying R); it is optimised on commissioning and should not thereafter need to be touched.

Automatic load-sharing equipment operates on active load-sharing only. It does not operate on reactive load-sharing, for this follows automatically from the quadrature droop circuits.

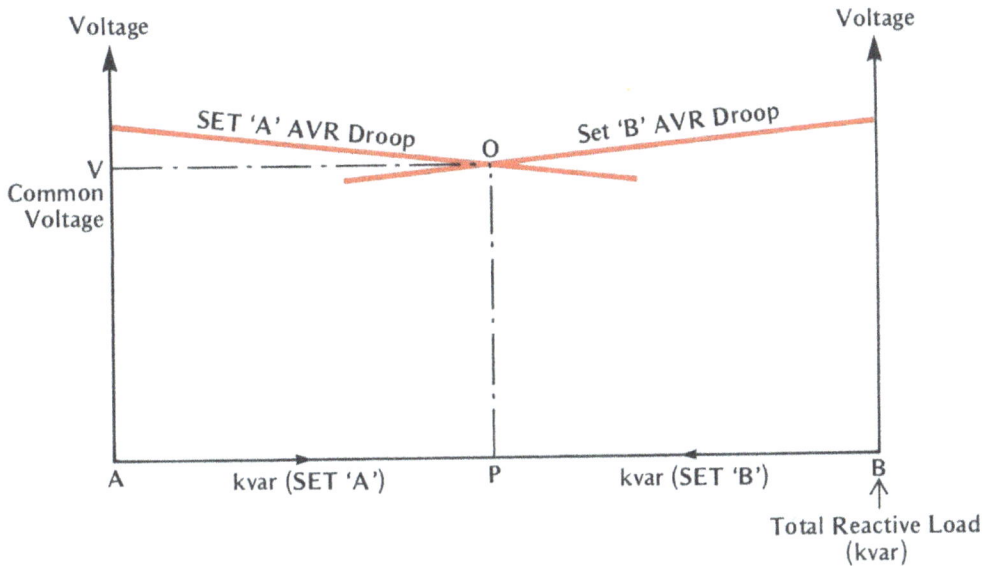

FIGURE 6.15
COMBINED GENERATORS - REACTIVE LOAD SHARING

CHAPTER 7 LOAD SHEDDING

7.1 GENERAL

In an onshore installation the duty Electrical Operator has no control over his supply source and is only concerned with distribution. He can only proceed on the assumption that his source is infinite and that he can supply his full load at all times. This is not the case with offshore plant.

Consequently this chapter refers only to offshore installations, where the operator can control his supply source and has knowledge of its limitations.

In Chapter 1 it is stated that ideally a power system should be run with firm capacity - that is, with enough generators connected - so that the sudden loss of the largest of them would not leave the remainder overloaded. If this is done, and a generator fails, it is not necessary to reduce the load.

This however is an ideal state of affairs. In certain conditions it may not be possible to have spare generating capacity connected. When an offshore installation is in full production and the load is high there may not be enough generators installed to provide this margin, or perhaps one of the large sets is unserviceable due to breakdown or maintenance. In either case the loss of a running generator would produce a crisis condition, with the remainder overloaded. It would then be necessary as a matter of urgency to remove the overload before the overcurrent protection of another generator operated and caused it to trip, so aggravating the situation. Indeed this could lead to a catastrophic loss of all main generators.

7.2 ACTION

Such a crisis can only be prevented by quick operator action as soon as he realises that the whole system is in a state of overload and may become unstable. Indeed, if such a possibility existed, he should already be aware of it, and the loss of a running generator would be the signal for immediate action.

This action consists of removing load from the system in order to bring the total remaining load within the capacity of those running generators that are left; it is called 'load-shedding'. To be effective the amount of load shed at each step should be appreciable - that is, of several megawatts - and the operator should make use of the megawatt-meters on his control desk or panel. In this context 'load' means active load only, i.e. megawatts. No account need be taken of the reactive load (megavars), as this may be assumed to be a more or less constant proportion of the active load. In other words, the power factor may be assumed to remain reasonably constant both before and after load has been shed. The generator ammeter should only be used as an indication of when the generator current is approaching its trip value.

The need for load-shedding may arise from any of three causes:

(a) **Acute**: Sudden loss of a running generator while there is an insufficient margin of connected generators on the system.
(b) **Threatening**: Need to shut down a running generator while there is insufficient margin of connected generators on the system and no more are available.
(c) **Threatening**: Growth of load on the system when there is no more generating capacity available to be connected.

In the 'acute' case (a) action must be immediate; there will be no time for consultation, and therefore a prearranged plan must have been prepared. In the 'threatening' cases (b) and (c) there will, in general, be time to consult with the Control Room Supervisor and load can be shed under more controlled conditions.

7.3   SHEDDING CONTROL

Only certain operations are open to the authorised person at the electrical control desk. A typical desk is shown in Chapter 4, Figure 4.1, from which it can be seen that the operator can start and stop generators and control the high-voltage incomer, bus-section, transformer feeder and interconnector switchgear. He can also control the 440V incomer and section switch-gear but cannot control from the electrical control desk the switchgear that feeds individual high-voltage motors or any of the 440V distribution. Therefore from that position the operator can only remove very large parts of the load by isolating whole sections of the system, which in general would be far too drastic as a load-shedding exercise, except as a last resort to save a total shutdown.

Since it is necessary to shed fairly large units of load, this in practice means the high-voltage motors, which range in size between about 300kW and over 9 000kW. Shedding loads smaller than this would hardly be effective.

It should be remembered, however, that, just because a motor is rated at, say, 3 600kW, it may not be taking full load. Therefore the shedding of such a motor may not always achieve the full results expected.

All high-voltage motors on an offshore installation are part of the process, injection or gas compression systems and are probably controlled from the process control desk in the Platform Control Room or from the gas compression control panel in the Gas Compression package. Shedding such motors, when not done automatically, is therefore carried out by the Control Room Supervisors on the advice of the Electrical Authorised Person. Where the platform and electrical control rooms adjoin each other this presents no problem, but on some installations they are remote from each other, and communication must be by telephone. It is obvious that instructions must be absolutely clear, and the exercise should be regularly practised (but not implemented).

7.4   PLANNING

As load-shedding involves stopping certain aspects of processing, the first units to be shed must be those which upset the process least; this usually meant the water injection units which have motors up to 3 600kW, whose disconnection would contribute in great measure to restoring stability to the system. (This statement was true at one time, but priorities for shedding are constantly changing.)

In general a systematic plan of shedding must be worked out with the Production Supervisor, and each successive stage must be clearly specified, stating from what position the tripping must be carried out, and by whom. As this plan may have to be put into action at a moment's notice and by whoever is manning the control rooms at the time, it is essential that it be regularly exercised (short of actual tripping) by all personnel who may have to carry it out.

The authorised person on duty must at all times be aware if his system is in such a condition as to be vulnerable to the sudden loss of a generator, and he must be ready to put the load-shedding plan into immediate operation. He must inform the Control Room Supervisor whenever this is the case, and he must make sure that his relief on the next shift knows it.

7.5   AUTOMATIC LOAD SHEDDING

7.5.1 Direct Shedding

On some offshore installations the pre-planning will have been carried out during the design stage, and the whole sequence of load-shedding is then predetermined and can be carried out automatically.

In a simple case such as an installation which has only two main generator sets, the act of disconnecting or the self-tripping of one of them while they are in parallel automatically sheds certain large loads. Such a system is shown in Figure 7.1. If either generator comes off the board the two very large gas compression motors are immediately tripped (if running), as well as three of the six 2 500kW water injection pump motors, as indicated in red. Moreover, when only one generator is running and connected, none of these motors can be started.

Compared with the sequence load-shedding system described in para. 7.5.3, this direct method of shedding is a 'blunt instrument': it takes no account of the actual loading on the generators or of the load being taken by the motors to be shed, or indeed of those which are running. All are tripped simultaneously. If further shedding is needed, it must be carried out manually.

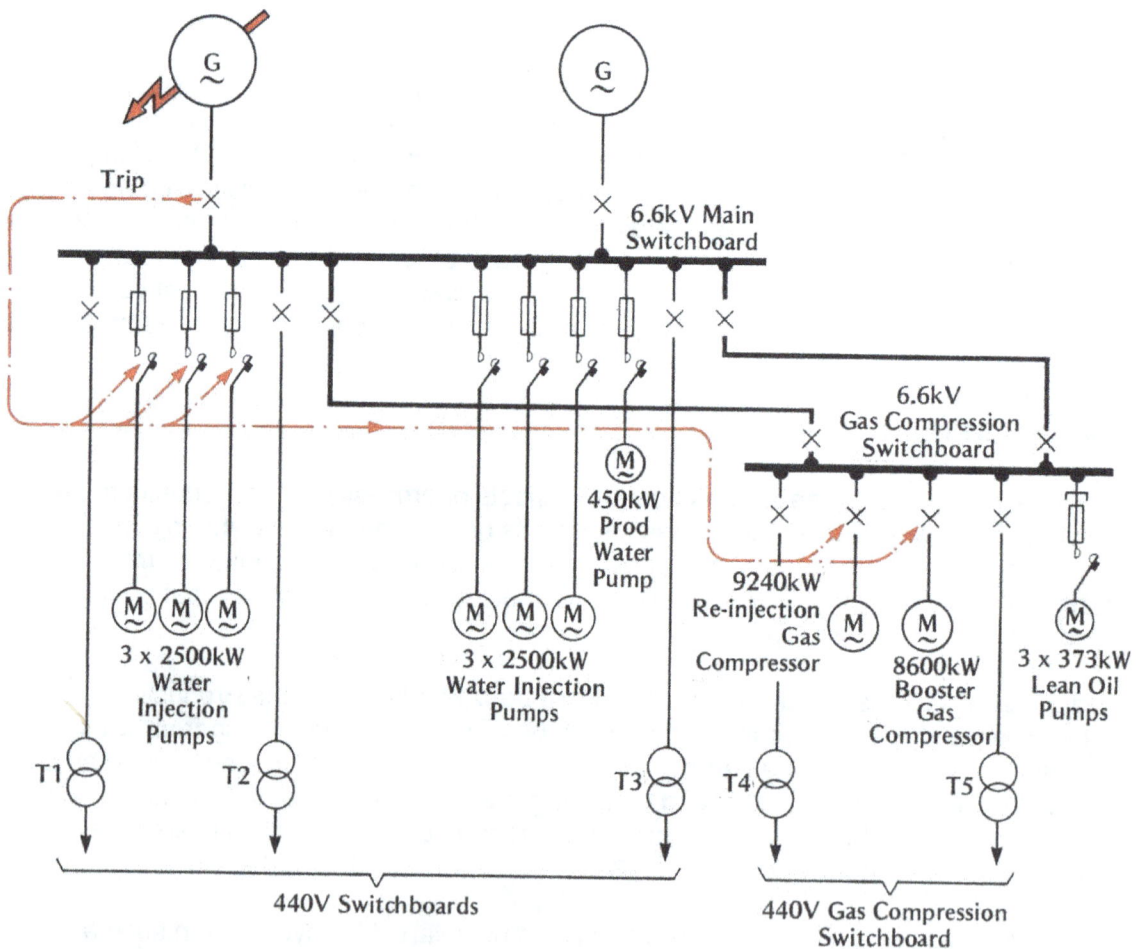

FIGURE 7.1
AUTOMATIC DIRECT LOAD SHEDDING OF HV MOTORS (TYPICAL)

7.5.2 Sequence Shedding

On certain offshore installations with more complicated networks, such as the typical one shown in Figure 7.2 which has four 15MW main and two 2.5MW secondary generator sets, a more sophisticated system is used which sheds large loads in sequence as the situation demands and which also gives warning of impending sheddings.

The shedding sequence is programmed and is put into effect by a number of logic circuits inside a special Load-shedding Panel. It takes account of the number of generators running and connected at any instant and the actual total load on them. The difference is the power margin currently available. It also calculates what would be the margin if the largest running generator failed. Such a margin might well be negative, indicating that the loss of the generator would overload the remainder. The logic also measures the actual loading on all the running motors selected for sequential tripping, and it thus determines how many should be tripped if a generator should fail.

In Figure 7.2 the various motors which may be shed are numbered in sequence (1 to 6) and are indicated in red. Group 1 consists of seven 3 600kW water injection motors on the Main board which are shed in a predetermined sequence. Motors 2 to 4 are the three very large gas compression motors (9 240kW, 7 140kW and 7 140kw), also on the Main board.

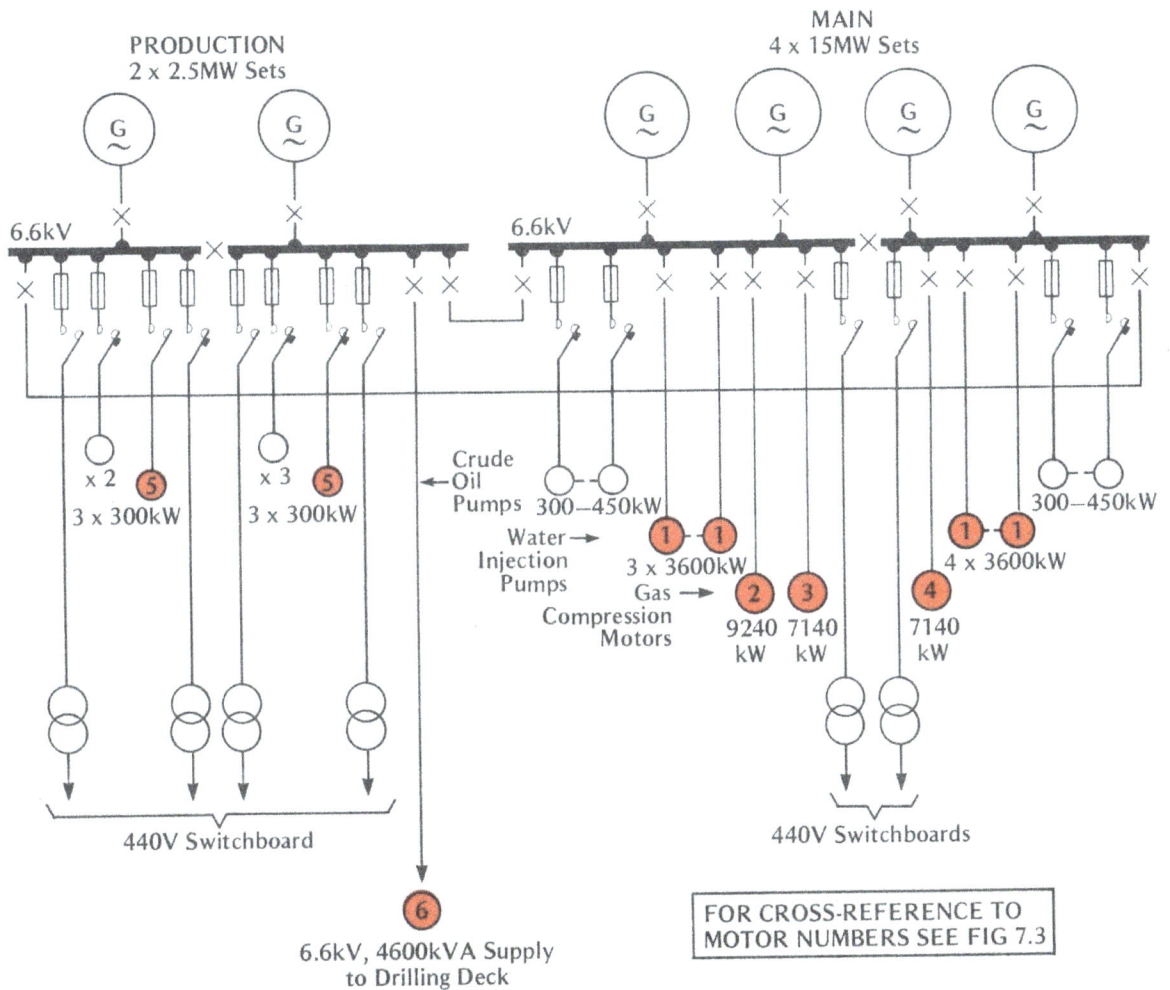

FIGURE 7.2
AUTOMATIC SEQUENCE LOAD SHEDDING OF HV MOTORS (TYPICAL)

Group 5 consists of six 300kW oil transfer and booster pump motors on the Production board which are shed together, giving a unit of 1 800kW. Finally, No.6 is not a motor but is the feeder from the Production board to the drilling package transformer, rated 4 600 kVA.

A typical load-shedding panel is shown in Figure 7.3. On the left-hand side is a mimic diagram of the high-voltage power supply system with lamps to indicate which generators are running and connected and which section breakers and interconnector breakers are closed. The generator indicator lamps also have a pushbutton operation, whose purpose is explained below. There is also an OFF/NORMAL/TEST switch; in the NORMAL position load-shedding is in operation, and in the OFF position it is not in use. The use of the TEST position is explained below.

On the right-hand side of the panel, displayed vertically, is a list, starting with the least important, of all those HV and other motors which may have to be shed. Against each are two lamps: a yellow warning lamp tallied WILL BE STOPPED and a red tripped lamp HAS BEEN STOPPED. When the logic has selected the motor or motors next to be shed, their yellow lights illuminate; if and when the motors are actually tripped, the yellow light goes out and the red light comes on. An alarm is also given in the platform control room.

The Control Room Supervisor should always be informed when any yellow lamps illuminate, or if there is any change in them, so that he is forewarned that the electrical system is 'vulnerable' and that therefore he could lose certain items of process plant if a generator should fail.

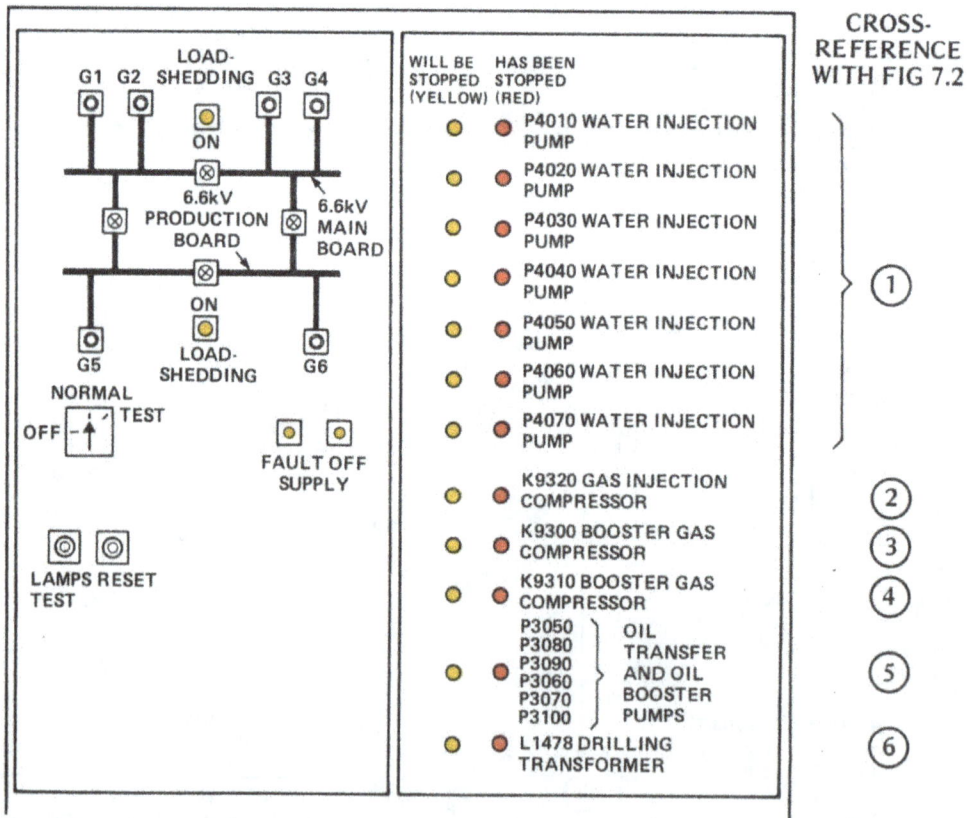

FIGURE 7.3
TYPICAL AUTOMATIC SEQUENCE LOAD SHEDDING PANEL

To enable the logic calculations to be made, the network is assumed to be in one of two states:

(a)   both HV section breakers closed and one or both interconnectors between HV switchboards closed at both ends;

(b)   both interconnectors open and the Production board section breaker closed.

State (a) means that the whole network is electrically a single unit and that up to all four 15MW sets and both 2.5MW sets (if running) are paralleled. State (b) means that each HV switchboard is on its own; in that case automatic load-shedding will operate on the Production board only.

Provided that the panel is switched on and that condition (a) above is met, LOAD-SHEDDING ON lamps burn against both Main and Production busbars on the mimic. If condition (b) only is met, only the Production LOAD-SHEDDING ON lamp burns. The two load-shedding lamps are of course extinguished when the main switch is set to OFF, but if one or both of these lamps is not lit when load-shedding should be in operation, the operator should determine the cause.

Other features of the load-shedding panel are a FAULT SUPPLY lamp which illuminates if either of the two alternative 24V d.c. auxiliary supplies fails, and an OFF lamp which indicates when the supply to the tripping relays has failed. All these lamps, as well as the yellow and red shedding lamps on the right-hand side, are restored by the RESET button when the fault has been corrected or the shed load reconnected.

Inside the panel is a MW chart recorder which makes a continuous trace of the total megawatt loading on the system.

7.5.3 Testing

When it is desired to test the internal circuits and operation of the automatic sequence panel, the selector switch is turned to the TEST position. The mimic should be showing lights corresponding to those generators that are running and connected to the busbars, and also lights indicating which bus-section breakers and pairs of interconnector breakers are closed. It should also be showing one or both LOAD-SHEDDING ON lamps. On the right-hand side yellow lamps should be burning against the names of those motors next selected for tripping (if any).

For testing the shedding which would follow a generator failure the mimic combined lamp/pushbutton of the generator selected as faulty is pressed and held. This simulates loss of that generator. If the largest running set has been chosen, all the selected yellow lamps on the right-hand side change to red, indicating that the motors concerned would have tripped (but do not actually trip under these test conditions). Further yellow lights will also appear indicating the next stage of shedding. When the pushbutton is released the selector switch springs automatically back to the NORMAL position, and the red shedding lights change back to yellow.

The above description is that of a typical load-shedding panel of one particular make and of how it operates on that particular network. It is given as an example only. The networks of other offshore installations may differ, and the load-shedding panels, where fitted, are tailored to suit them and may be of different construction.

7.6   LIVING QUARTERS

On most offshore installations the relatively less important kitchen and laundry loads can be shed separately from the other living quarters' loads by a pushbutton on or near the basic services switchboard. They can be restored by pressing an adjacent RESET button. It should be noted that the pushbutton, though it may be labelled LIVING ACCOMMODATION SHED, does in fact trip only the kitchen and laundry contactors and nothing else.

This limited shedding is provided for use when the basic services generator, with its small output, is running and feeding the basic services switchboard. It is used when the loading on that generator, as given by the local ammeter, is approaching its limit.

CHAPTER 8 SYSTEM INSTABILITY, MOTOR STARTING & RE-ACCELERATION

8.1 SYSTEM INSTABILITY

When a large electrical system suffers a serious disturbance, all motors which are running at the time begin to slow down. When normal voltage is restored, the motors may then be running at a large slip and drawing heavy reactive current. In an offshore installation, where the generators have finite capacity, the combined effect of all the running motors in this state - and especially the largest ones - could be to pull down the voltage of the whole system. This may occur, despite the removal of the disturbance and the operation of the AVRs, to the point where many loaded motors may not be developing sufficient torque to re-accelerate their loads, and all of them would run down to a stall. In such a situation the whole system has become unstable and will be brought to collapse unless steps are taken to deal with it.

In an onshore installation a similar situation could exist, depending upon the Supply Authority's local generation capacity and distribution arrangements. This is further dealt with in para. 8.3.

Normal overcurrent protection should operate in such cases to disconnect motors running in a 'large slip' condition, but it may not be quick enough to prevent the system going unstable, especially where very large motors are involved. As a back-up to prevent system collapse it is usual to fit undervoltage relays connected to the busbars. A sustained drop of voltage following a large system disturbance will cause the relay to operate and disconnect all the motors on that busbar. The removal of their combined large reactive currents will reduce the voltage drop and enable the system to recover. After this, the disconnected loads must be reconnected, in sequence, by hand so that they do not all draw their starting currents at the same time.

8.2 MOTOR RE-ACCELERATION

A more sophisticated solution to the problem may sometimes be met, which renders the process automatic. Selected motor control panels may be fitted with 're-acceleration units'. A typical unit senses loss of voltage and is able to distinguish between a very quick restoration in, say, 200 milliseconds, a longer delayed restoration in, say, 4.5 seconds and a still longer delay.

A typical arrangement would operate as follows:

(a) For a quick voltage restoration (200ms or less) no action is taken. Within this short time the loss of speed and consequent re-acceleration current drawn by the motors is considered small enough to be acceptable. The motor contactor trip is either delayed or, if the contactor has tripped, it is immediately reclosed.

(b) If voltage restoration takes more than 200ms but not more than 4.5s, selected motors are automatically reconnected in a pre-arranged sequence over a period of 60s or so. Thus their re-acceleration currents do not occur at the same time.

(c) If voltage has not been restored within the 4.5s all motors remain disconnected and run to a stop. None re-accelerates automatically. They must be restarted, in sequence, by hand.

Such re-acceleration systems are manufactured under various trade-names, of which 'Holec' is one. In that system certain motor control panels are fitted with special sockets into which a Holec unit may be plugged; each unit has set on it the delay time before re-acceleration starts as in (b) above - that is, before the motor contactor is reclosed.

8.3   ONSHORE INSTALLATIONS

Where an installation takes its total normal supply from an onshore Supply Authority, it is sometimes justifiable to think of the supply input as coming from a set of 'infinite busbars', that is, as being absolutely secure.

If that supply should fail, however, this concept can no longer apply. An onshore installation may have motors just as large as those found offshore, although they will be performing different functions. Such large motors would present a totally unacceptable reactive load if any attempt were made to accelerate them simultaneously. Arrangements vary with local circumstances, but Supply Authorities invariably insist upon arrangements to limit reconnection after a supply interruption or failure.

CHAPTER 9 QUESTIONS AND ANSWERS

9.1   QUESTIONS

1.   Specify the major differences between onshore and offshore electrical installations.

2.   What knowledge does a Electrical Authorised Person need to be able to control any system successfully and to meet its load requirements at all times without risk of overload?

3.   Each item of plant has a 'CMR'. What do these initials stand for?

4.   What is meant by the 'firm capacity' of a supply network?

5.   In an onshore installation, where the supply is taken from a local Supply Authority, does the term 'firm capacity' have the same significance?

6.   In a typical offshore installation, what loads are fed by the basic services generator?

7.   When is a generator called upon to supply the largest amounts of reactive power?

8.   What are the four principal limitations on the output of an engine-driven generator set?

9.   What factor limits the actual MW output of a generator?

10.   What factor controls the stability of the rotor angle (power angle)?

11.   When may above-normal lagging reactive loads on a generator sometimes be tolerated?

12.   Why is power-factor correction desirable in an onshore installation?

13.   With d.c. generators what condition must be fulfilled before two generators can be connected in parallel?

14.   What conditions must be fulfilled before two a.c. generators can be connected in parallel?

15.   When synchronising two a.c. generators, with one already on load (running) and the other not yet on load (incoming) which machine's controls are adjusted?

16.   How can the correct moment to parallel the two generators be detected?

17.   In the 2-lamp synchronising method why is the 'lamps dark' method used, rather than the 'lamps bright'?

18.   In the 3-lamp synchronising method two lamps are bright and one is out at the moment of synchronism. Which lamp is out?

19.   When a synchroscope is used, which way should the needle be rotating, and at what pointer position are the two generators momentarily in phase?

20.   What is automatic synchronising?

21. If automatic synchronising is not provided, what precaution is usually taken to prevent accidental damage?

22. What is a mimic diagram on a control panel?

23. A control panel usually has only one synchronising panel for use for all synchronising operations. What facilities does it provide?

24. If a system fault causes a circuit-breaker to trip, how is this event signalled at the control panel?

25. The expression 'black start' can only be applied to one sort of installation. Which?

26. What is the first action required for a black start?

27. On loss of their normal supply some services receive an alternative supply from the basic services switchboard. These include the starting battery chargers for the secondary generators. Why do these battery chargers have priority?

28. If the basic services generator starting battery is discharged, how can the first step in a black-start sequence be carried out?

29. When two or more generators are running in parallel and feeding a common load, what is the ideal load-sharing arrangement?

30. What two generator set controls are concerned with load-sharing between generators?

31. For two generators in parallel and generating the same emf, if the excitation of one is increased what happens to their reactive load-sharing?

32. For two similar generators running in parallel and sharing the load equally, what is the effect of altering the 'speed' control of one of them in the raise-speed direction?

33. If two similar generators are running in parallel and feeding a common load, with system voltage and frequency correct and with equal sharing of active load, but with the varmeter of generator 'A' reading 9.5 Mvar while that of generator 'B' reads 62 Mvar, what action is necessary?

34. What is 'droop' in a control system loop?

35. What is the technical name for the theoretically ideal but actually unattainable limit of 0% droop in a speed governor?

36. If two identical generators, with identical governors, are set to share their common load equally, what happens if the load increases?

37. Why should generator sets with different governor droop characteristics never be run in parallel if it can be avoided?

38. If the situation described in Q.37 is unavoidable, what operator action is necessary?

39. It is often necessary for dissimilar generator sets to be run in parallel. What equipment is usually provided to cater for this situation?

40. What is another important use of automatic load-sharing equipment?

41. When automatic load-sharing equipment is provided, what external controls are necessary?

42. When generators are running in parallel, what may happen if there is insufficient droop in their voltage control loops?

43. How is reactive load-sharing made stable?

44. How does quadrature droop compensation work?

45. Why does automatic load-sharing equipment operate on active load-sharing only?

46. To ensure that an offshore network is never overloaded, what state of affairs should, ideally, apply at all times?

47. If this ideal situation cannot be maintained, what must be done to prevent a cumulative situation in which all remaining generators trip on overload?

48. What three situations may necessitate load-shedding?

49. How do these situations differ in the action that must be taken?

50. In any planned load-shedding, which loads should be shed first?

51. There are two different kinds of automatic load-shedding. What are they?

52. To prevent the total collapse of an offshore electrical power installation after a serious, but fairly brief, disturbance, what steps can be taken?

9.2 ANSWERS

(Figures in brackets after each answer refer to the relevant chapter and paragraph in the text.)

1.   Offshore installations have their own generators and a number of life-support loads that onshore installations do not need. (1.1)

2.   Thorough knowledge of his system and the loading limits of each item of plant, plus close co-operation with other departments. (1.2)

3.   Continuous Maximum Rating. (1.4)

4.   If any one item of generation or distribution plant can be lost because of a fault without affecting the network's ability to supply its load, it is said to be loaded within its firm capacity. (1.5)

5.   Within the installation its significance is confined to distribution equipment only. (1.5)

6.   Battery-supported services, minimum life-support services, essential lighting and main generation auxiliaries. (1.6)

7.   When very large drive motors are started. (1.6 and 1.7)

8.   Current heating of the stator; power output of the prime mover; current heating of the rotor, and stability of the rotor angle. (2.1)

9.   The mechanical power output of its prime mover. (2.3)

10.   The generation of leading stator current. (2.5)

11.   For short periods while large motors are being started. (2.6)

12.   Because, while both watts and vars must be paid for, productive output results only from the watts (active power). Power-factor correction reduces the unproductive vars (reactive power). (1.8)

13.   Their output voltages must be the same. (3.2)

14.   Their output voltages and frequencies must be the same and they must be in phase. (3.3)

15.   Those of the 'incomer'. (3.4)

16.   By using lamps (either two or three according to the installation) or a synchroscope, or by the IN SYNCHRONISM lamp where check-synchronising is provided. (3.4 and 3.9)

17.   Because it is easier to tell when a lamp is fully 'out' than to estimate when it is brightest. (3.5.1)

18.   The top lamp if they are arranged in a triangle, or the centre lamp if they are arranged in a line. (3.5.2)

19.   Clockwise (FAST). At 12 o'clock. (3.7)

20. Automatic synchronising is a method of ensuring correct synchronising of a generator without the possibility of human error. Relays compare the voltages, frequencies and phase relation of the generators, trim them accordingly, and close the circuit-breaker at the correct moment. (3.8)

21. The use of a 'Check Synchroniser', which leaves all operations to be carried out manually, but inhibits the final operation of paralleling if conditions are not right. (3.9)

22. A single-line bulk supply network diagram, with symbols representing generators and transformers, and with switch controls placed in their correct network positions to indicate and control the network circuit-breakers. Instruments of various types are also provided. (4.2 and Figure 4.1)

23. Usually a synchroscope, synchronising lamps, two voltmeters and two frequency meters (running and incoming) and the necessary switches. (4.4)

24. Its associated discrepancy switch illuminates. (4.5)

25. An offshore platform. (5.1)

26. After checking that it is safe to do so, start the basic services generator to provide essential lighting. (5.1)

27. It is necessary, in the black-start sequence, to start the secondary generators as soon as possible, and their starting batteries must be fully charged for this. (5.2)

28. Either a charged battery or a portable engine-driven battery charger must be provided. If manual/hydraulic start equipment is provided, it should be used. If the secondary generator's starting batteries are charged it may be possible to start them first and ignore the basic services generator. (5.4)

29. They should share both the active and the reactive loads in proportion to their sizes. (6.1)

30. The governors of the prime movers for the sharing of active load, and the settings of the generators' excitations for the sharing of reactive load. (6.2 and Fig 6.2)

31. Lagging reactive power (vars) circulates between the generators from the one with the higher excitation to the one with the lower. With both feeding a common external load, the sharing of reactive load is changed so that the one with the higher excitation supplies a higher proportion of the var-loading. (6.4.1)

32. In-phase active power (watts) circulates between the generators from the one with the higher power input to the one with the lower. With both feeding the common external load the active load-sharing is changed so that the one with the higher power input supplies a higher proportion of the watt-loading. (6.4.2)

33. An equal sharing of reactive load would imply a reading of 8.1 Mvar for each generator. Lower the voltage setting of generator 'A' and simultaneously raise that of generator 'B' until both varmeters read 8.1 Mvar. (6.4.3)

34. A drop in the controlled output of a system for an increase in the controlling factor. (6.5.1)

35. Isochronous. (6.5.1)

36. They continue to share it equally, but their common speed falls slightly. (6.5.1)

37. The load-sharing balance becomes upset as load changes. (6.5.3)

38. The speed adjustment of one of the sets must be altered whenever it is necessary to restore balance. (6.5.3)

39. Automatic load-sharing equipment, which senses active load-sharing and system frequency, and controls the various governors accordingly. (6.5.5)

40. When a new generator is started, auto-sychronised and brought on-line, the automatic load-sharing equipment, if switched on, will automatically cause the incoming generator to take up its correct share of the total load. (6.5.5)

41. It must be possible to select either automatic or manual control, and the manual control itself may be either local or remote. (6.5.7)

42. They may go unstable and throw reactive load backwards and forwards between each other. Their varmeters and power-factor meters may show violent fluctuations and the latter may even swing into leading. (6.5.8)

43. By 'Quadrature Droop Compensation'. (6.5.8)

44. By adding to the AVR's normal line-voltage sensing input a small element proportional to the reactive (quadrature) component of the line current, so injecting artificial droop into the AVR characteristic. (6.5.8)

45. Because reactive load-sharing follows automatically from the quadrature droop circuits. (6.5.9)

46. There should be enough generators running and on load, so that the sudden loss of the largest of them would not leave the remaining machines overloaded. (7.1)

47. The overload on the remaining generators must be removed as a matter of urgency, preferably in a planned way. (7.2)

48. (a) Sudden unexpected loss of a generator with insufficient generating capacity remaining.

    (b) Need to shut down a running generator while insufficient capacity remains and no more generators are available.

    (c) Growth of load on the system with no more available generating capacity. (7.2)

49. (a) Is 'acute': there is no time for consultation and immediate action is essential;

    (b) & (c) are 'threatening': there is usually time for consultation and an orderly shedding of load. (7.2)

50. Those large loads that upset the production process least, e.g. water injection pumps, which have motors of up to 3 600kW. (7.4)

51.   (a)   The crude ('blunt instrument') type, in which large blocks of load are shed with no regard paid to the actual loading on the generators or of the load actually being taken by the motors that are shed.   (7.5.1)

      (b)   The more sophisticated 'sequence' type, which sheds blocks of large loads in a planned sequence and also indicates which loads are next at risk. (7.5.2)

52.   Large potentially reactive motor loads are sometimes provided with re-acceleration units, which sense a loss of voltage and distinguish between a very rapid restoration and a more delayed one. According to a pre-planned arrangement a given motor may be either reconnected immediately, reconnected as part of a controlled sequence over a period, or disconnected altogether.   (8.2)

PART 7 CONTROL DEVICES

CHAPTER 1 INTRODUCTION

1.1 GENERAL

In any sizeable electrical installation the control systems incorporate, apart from simple switches, various devices - mostly electrically actuated switching devices - which are required to carry out automatic, remote or other functions without the benefit of manual control. Traditionally these have been electromechanical devices, but there is now an increasing tendency for 'solid-state' techniques to be used for such purposes.

The kinds of device commonly found in installations - onshore and offshore - are described below, together with some account of the functions they normally perform.

1.2 TYPES AND APPLICATIONS

1.2.1 Relays

Relays operate contacts in response to an electrical input of relatively low power (the term is also used, though inaccurately, for devices controlled by other inputs, such as temperature). Typical uses are remote control, electrical isolation between control circuits, protection of equipment against potentially damaging conditions, and the interlocking of switchgear to prevent inadvertent misuse.

1.2.2 Sensors

Sensors are comparable to relays, but operate contacts (or deliver an electrical output) in response to a non-electrical input. Typical are temperature sensors, used to control the temperature of, say, an oil reservoir (thermostat) or to indicate excessive heating in a transformer or motor; pressure sensors, used to indicate excessive or low pressure in an oil-immersed transformer; and differential-pressure sensors used to indicate excessive pressure-drop, for example in an air filter. There is a wide variety of uses in connection with control, measurement and protection.

1.2.3 Transducers

Transducers* convert inputs, or combinations of inputs, in one form to corresponding (usually proportional) outputs in another form - e.g. temperature to voltage or electrical power to direct current. This enables relatively simple, standard instruments or other devices to respond to complex or non-electrical quantities. As well as being increasingly used for measurement purposes, they have applications in many aspects of control and protection, including, for example, generator speed control, generator load-sharing, and reverse-power detection. (Transducers are, however, often incorporated into complete control systems rather than existing as separate entities.)

1.2.4 Rectifiers

Rectifiers (diodes) create unidirectional current paths, and are used to convert alternating current to direct current, to block direct currents in undesirable directions and to provide discharge paths for stored energy in inductive circuits. Special diodes are used for other purposes such as d.c. voltage stabilisation.

*Transducers should not be confused with transductors, which are amplifying devices using saturating magnetic cores - now largely obsolete.

1.2.5 Thyristors

Thyristors, or Silicon Controlled Rectifiers (SCRs), are in effect diodes with a control function added. They are applied, in both d.c. and a.c. circuits, to voltage control in battery chargers, speed control of d.c. motors (particularly for drilling), and to inverters and a.c. static switches.

1.2.6 Transistors

Transistors are essentially continuously variable control devices but are very often used as switches, both in low-power electronics and in power circuits. To date the principal power applications are in small d.c. power supplies, audio amplifiers, oscillators and inverters.

1.2.7 Integrated Circuits

Integrated circuits contain no elements that are not known (at least in principle) as discrete components, but they make many complex electronic systems practicable, or economic, by virtue of the number of elements that can be accommodated in one small component and also by virtue of their high reliability. They find application in all kinds of control and low-power electronics, with particular benefits in automatic controllers and in logic, memory and other 'computer-type' circuits.

CHAPTER 2 RELAYS

2.1   CLASSIFICATIONS

A relay may be classified in three ways:

(a)   **Type**. Nearly all relays are operated either electromagnetically or electronically. The most common electromagnetic types are attracted-armature, induction disc and reed.

(b)   **Characteristics**. Relays can be divided into those which have no precisely defined characteristics and simply operate 'instantaneously' when an input is applied, and those which are calibrated, in terms either of operating level (current, voltage, etc.), or of operating time, or of both. If both, the level and time calibrations may be independent, or may be related by an inverse-time characteristic. Different characteristics may be combined within one relay or relay unit.

(c)   **Application**. Any type of relay may be used, within limits, for many different purposes, although it is usual to employ the simplest type which meets the requirements. The basic function implied by the word 'relay', and the simplest in practice, is to repeat in one circuit the effect of a signal in another circuit - for the purpose of isolation or interlocking, for example. More complicated functions, calling for special calibrated relays, include the many different forms of protection. These relays are designated according to their specific purpose - e.g. overcurrent relay, undervoltage relay, negative-phase-sequence relay etc.

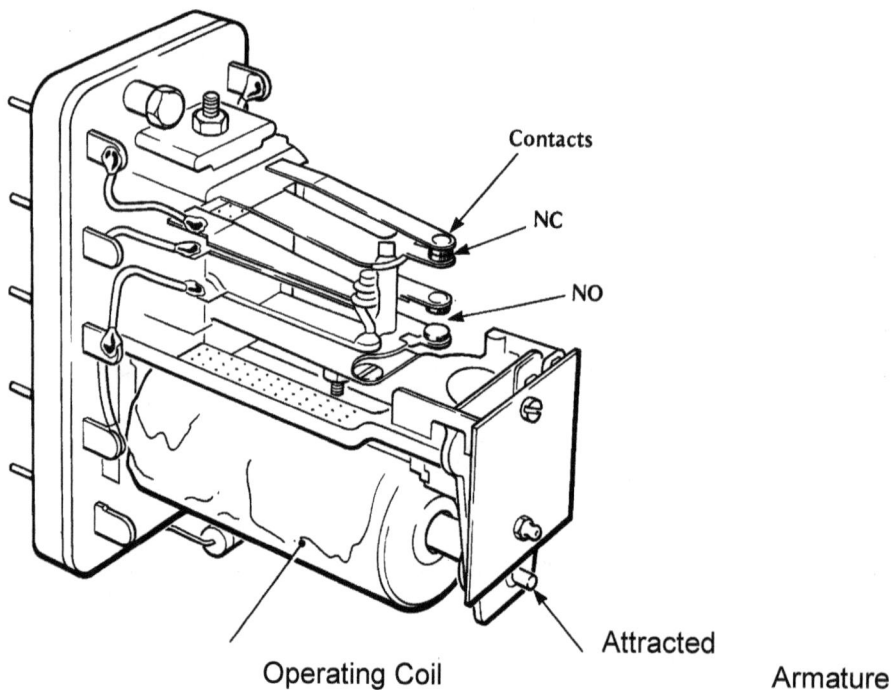

FIGURE 2.1
A CONTROL (INTERPOSING) RELAY ATTRACTED ARMATURE TYPE WITH TWO NC AND TWO NO CONTACTS

2.2   TYPES AND CHARACTERISTICS

For the purposes of description, relays may be grouped here as:

Control relays: those that perform no specific function other than operating their contacts in response to an input, with or without a specific time delay (often referred to as 'auxiliary relays').

Protection relays: those that are calibrated to operate in accordance with closely defined characteristics for specific purposes.

2.2.1 Attracted Armature Control Relays

Control relays are mostly of the simple attracted-armature type, having two states: energised or de-energised. The control supply is typically direct current at 24V or 110V. Figure 2.1 shows a typical construction.

A.C. relays are also used with a slightly modified magnetic structure, including a shading ring around part of the pole face to reduce vibration due to the alternating flux.

Individual contacts may be arranged to be opened ('normally closed') or closed ('normally open') by the energising of the operating coil (see Figure 2.2); the types of contact may be mixed on one relay. They may also be combined to act as changeover contacts.

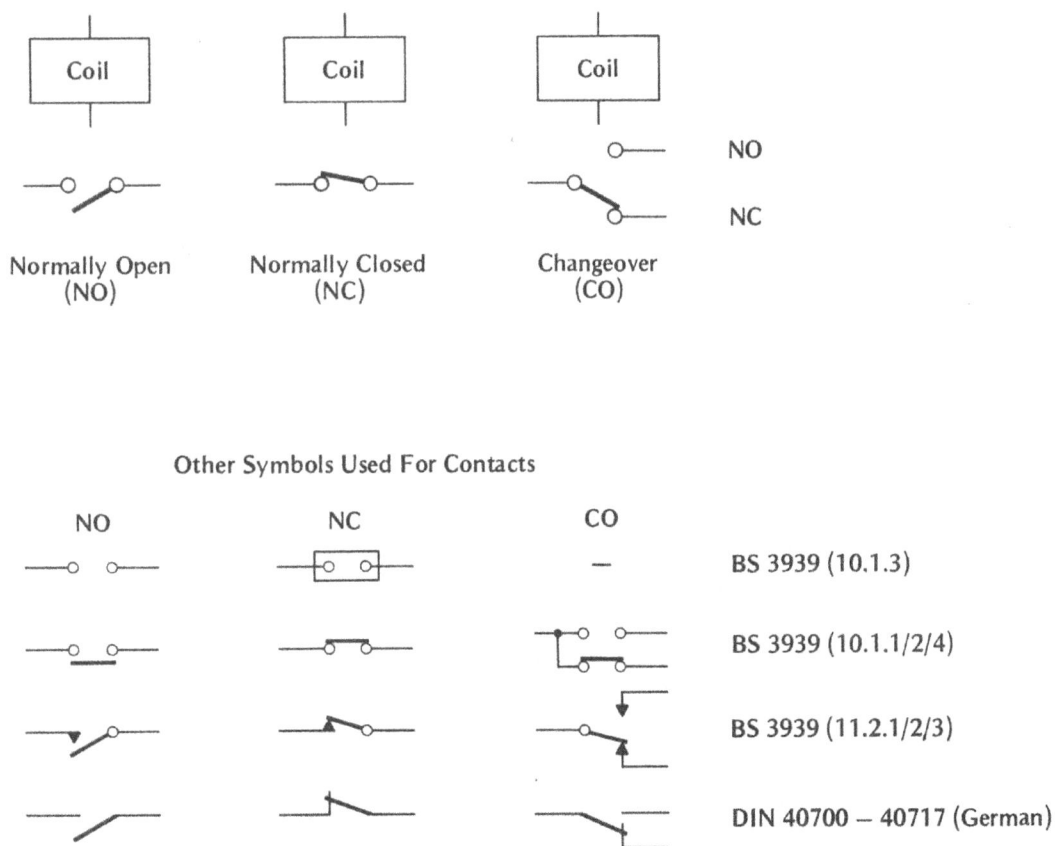

FIGURE 2.2
RELAY CONTACTS SHOWN IN THE DE-ENERGISED STATE

It is important to note that the designations 'normally open' (NO), 'normally closed' (NC) and 'changeover' (CO) define the types of contact fitted to the relay and refer to the states of the contacts when the relay is de-energised, or in its 'shelf' state. They do not necessarily relate to the usual state of the relay (which might be energised or de-energised) when it is in use in a circuit. A corresponding convention applies to circuit diagrams, in which contacts should always be shown for the de-energised condition of the relay regardless of its normal function. Normally-open contacts are sometimes referred to as 'make' contacts, and normally-closed as 'break' contacts, indicating what happens when the relay becomes energised.

Usually, when a relay is energised, the normally-closed contacts open before the normally-open contacts close ('break-before-make'). For special requirements they may be specified as make-before-break, so that for a very short period as the relay operates all the contacts are closed simultaneously.

'Instantaneous' operation in a relay means that it is not deliberately designed to introduce a delay, the operating time being normally a few tens of milliseconds. For particular purposes it is possible to increase the operating time by large amounts by simple expedients such as mounting a copper 'slug' on the magnetic core.

A control function may in some cases require a much longer time delay than the operating time of a simple relay; possibly it may need to be adjustable. Time-delay relays operate on a variety of principles, including thermal elements, clockwork escapements, induction discs, dashpots, pneumatic cylinders and synchronous motors. Modern relays increasingly make use of electronic timing circuits.

Relays are often provided with 'flags', which indicate clearly when the relay has operated and remain showing, even though the relay is subsequently de-energised, until reset by hand. It is very important that, when some mishap in a system has resulted in a trip, dropped flags should not be reset until a written record has been made of which flags have fallen, so avoiding a loss of valuable information. This applies especially to protection relays - see para. 2.2.2.

2.2.2 Protection Relays

A control relay, as described above, operates with an 'on-off', or 'digital', input of sufficient magnitude to actuate its contacts. By contrast a protection relay for use against overcurrent or other potentially damaging conditions in an electrical machine or system responds accurately to the level of its operating signal and is actuated when the signal exceeds, or falls short of, a preset value. Such a relay is sometimes referred to as a 'measuring relay'. Depending upon whether it responds to an excess or a shortfall, it is termed an 'overcurrent' ('overvoltage', 'overfrequency' etc.) relay or an 'undercurrent' ('undervoltage', 'underfrequency') relay.

Many relays of this type provide a time delay, which may be fixed (definite) or 'inverse'. With an inverse-time characteristic the delay decreases as the input signal increases, so that protection becomes more rapid as the severity of a fault increases.

The same nomenclature and conventions in regard to contacts apply to protection as to control relays (see para. 2.2.1); that is, 'normally open' and 'normally closed' define the states of the contacts when the relay is de-energised, and the contacts are so shown in circuit diagrams. Most protective relays are fitted with flags which indicate when they have operated and remain showing until they are reset by hand, even though the relays themselves revert to their normal states as soon as the fault is removed (see para. 2.2.1).

Protection relays are mostly of three basic types:

(a)   **Attracted-armature**. This type is used when 'instantaneous' operation is required, and it can be energised by either direct or alternating current. It is fundamentally similar to the attracted-armature control relay referred to in para. 2.2.1, but unlike the control relay it is calibrated in terms of operating current or voltage. The calibration depends upon the restoring force applied to the armature by gravity or by a spring. The operating current level is set by an adjusting screw at the top of the relay which controls the armature back-stop or adjusts the control spring - see Figure 2.3.

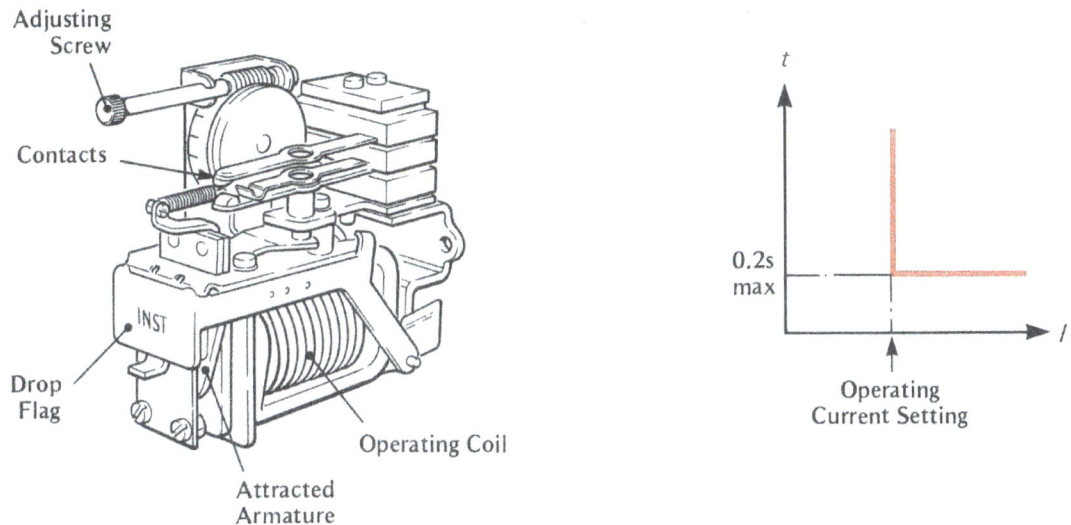

FIGURE 2.3
INSTANTANEOUS OVERCURRENT RELAY

For overcurrent protection - a common application - the relay is normally fed from a current transformer at a nominal current of 1A or 5A. In 3-phase systems three relays are assembled in one unit, or two in a 3-wire circuit (see Part 10 'Electrical Protection'), each coil being fed by a separate current transformer. Like other 'instantaneous' relays, the instantaneous overcurrent relay takes a finite time to operate, usually not more than about 0.2 seconds, and its overall characteristic is shown, somewhat idealised, in Figure 2.3. Other common uses are for undercurrent, undervoltage, overvoltage and earth-fault protection.

(b)   **Induction Disc**. The induction disc relay functions by the interaction of the magnetic flux which is generated by an energising coil and passed through the disc, and of the eddy currents which are produced in the disc by the same flux or by a second coil. The mechanism is described in Part 1 in relation to instruments, such as the integrating kilowatt-hour meter (e.g. the domestic 'meter'). For protection purposes this type of relay has the advantages that its operating time can be controlled over a wide range by means of eddy-current braking magnets and that a wide variety of functions can be obtained by using different arrangements of operating magnets and coils. The actual operating current level can be varied by adjusting a light restraining hairspring.

A variation of this relay is the 'induction cup', which operates on basically the same principle.

This type of relay is used in a number of forms; the principal ones are as follows:

(i) **Inverse-time Overcurrent (OCIT)**. This has a single shaded-pole driving magnet energised by alternating current from the associated current transformer, producing a torque which varies with the square of the current. When the current exceeds a predetermined value the driving torque overcomes the resistance of the restraining spring and the disc starts to rotate until eventually a moving contact attached to the spindle (or actuated by it) strikes a fixed contact (Figure 2.4).

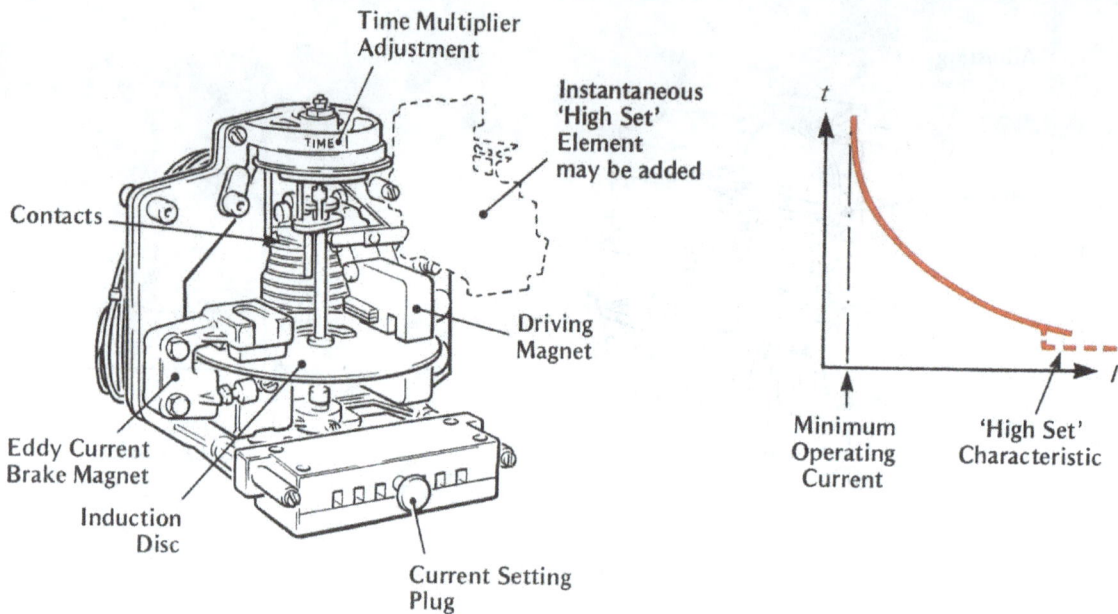

FIGURE 2.4
INVERSE TIME OVERCURRENT RELAY

The motion of the disc is opposed by the drag exerted by the permanent braking magnet, and this gives rise to an appreciable and consistent time delay. The greater the coil current relative to the minimum operating current, the faster the disc has to rotate before the braking torque balances the driving torque, and the shorter is the operating time. This results in the kind of inverse-time characteristic illustrated in Figure 2.4, with a long delay at currents barely greater than the minimum operating current but only a relatively short delay at high overcurrents.

Adjustment of both the operating current and the delay time-scale is provided for in order to enable a standard relay to accommodate variations in current transformer (CT) ratios and line currents and to facilitate discrimination in regard to other protection devices in the system (see Part 10 'Electrical Protection'). Current adjustments are made by selecting taps on the driving coil, usually by moving a plug between a number of holes at the front of the relay; typically the range covered is from 50% to 200% of the normal operating current (1A or 5A). The time delay is adjusted by moving the 'fixed' contact, or by altering the starting position of the disc, and so altering the travel of the disc necessary to close the contacts; this means that a particular adjustment alters all times on the inverse-time characteristic in the same ratio. Figure 2.5 shows a typical resulting family of characteristics scaled in terms of multiples of the current selector plug setting and the time multiplier set by the contact adjustment.

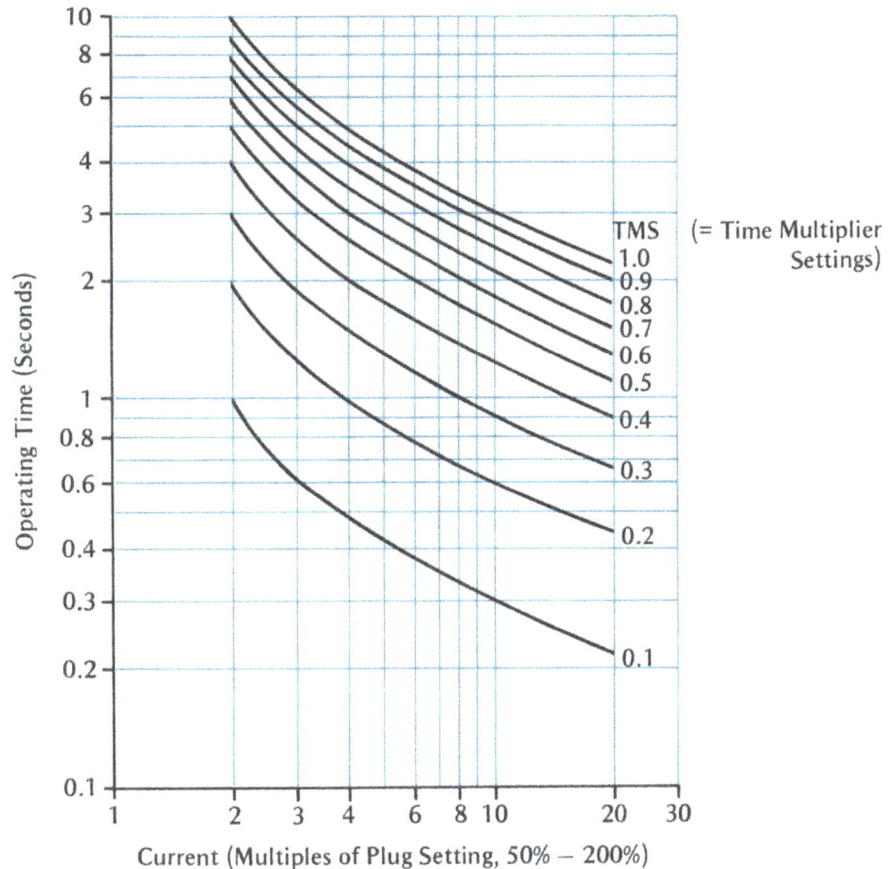

FIGURE 2.5
TYPICAL RELAY SETTING CURVES

The exact manner in which the current and time adjustments are set for any particular circuit is explained in detail in Chapter 3 of Part 10 'Electrical Protection'.

An inverse-time relay may be equipped with an additional instantaneous element in the same casing set to a high current value, referred to as a 'High Set' element. This gives it the feature of a combined 'inverse-time and high-set instantaneous' relay, the instantaneous feature overriding the time delay only on the most severe faults. An example of this additional feature is shown dotted in Figure 2.4. The modification to the time/current characteristic is indicated in that figure by the dotted section of the curve.

(ii) **Very Inverse and Extremely Inverse Overcurrent.** There are two variations of the inverse-time overcurrent relay: they are referred to as 'very inverse' and 'extremely inverse'. The differences lie mainly in the shape of the time/current characteristic, and examples of each are shown in Figure 2.6, where they are compared with the characteristic of a normal type. There are no recognised special abbreviations for these variations of OCIT relays.

Both these variations have characteristics which are steeper than that of the normal inverse-time type. Advantage is taken of this when there is a long chain of circuit-breakers with inverse-time relays and it is desired to achieve sufficient discrimination between their tripping times for a given fault current. This is indicated in principle in Figure 2.7, although the full explanation is more complicated.

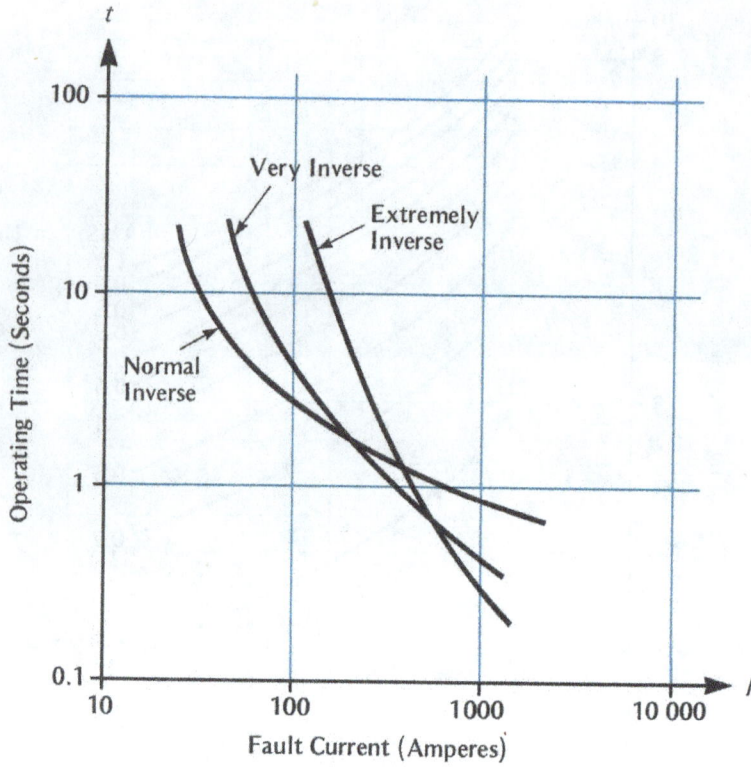

FIGURE 2.6
VERY INVERSE AND EXTREMELY INVERSE OVERCURRENT
RELAY CHARACTERISTICS

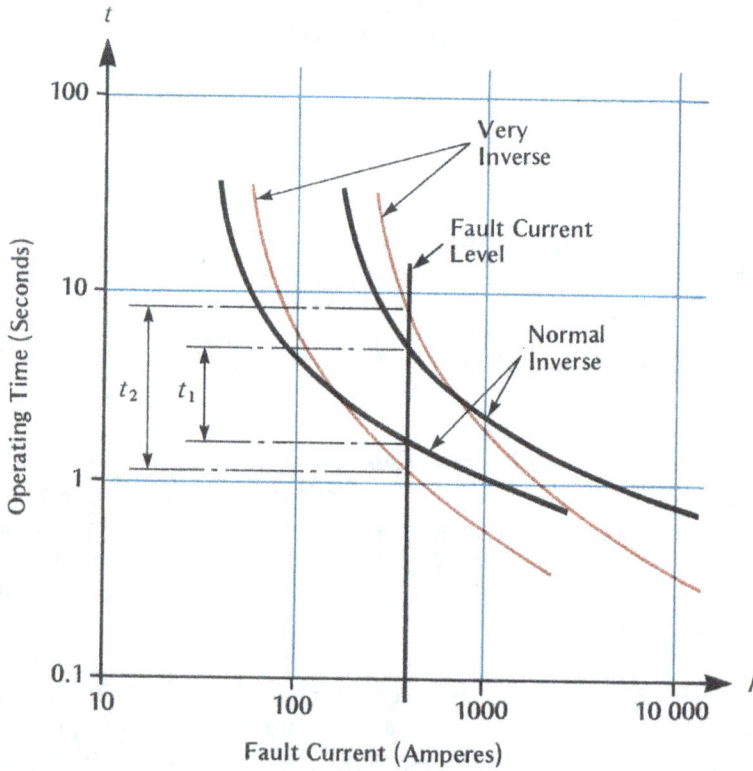

FIGURE 2.7
COMPARISON OF DISCRIMINATION USING NORMAL AND VERY INVERSE
OVERCURRENT RELAYS

The two black curves are the characteristics of two adjacent relays in the chain, with the same time settings and with their current plug settings appropriate to the fault levels at those points. The two red curves are the corresponding characteristics of two very inverse-time relays installed at those points in place of the normal relays.

It can be seen that, for a given fault current, the difference (t_2) in operating time between the two very inverse relays is greater than the difference (t_1) between the two normal inverse-time relays. Therefore, if the discrimination time between circuit-breakers in a distribution time is 'tight', the use of very inverse overcurrent relays could offer a solution. For example, this might occur where there is a relatively long chain and the tripping time delay at the supply end would otherwise be unacceptable.

The further variation of the 'extremely inverse' relay merely exaggerates this feature. It is often employed when it is necessary to discriminate with a fuse, which also has a steep characteristic at the lower current levels (see Part 10 'Electrical Protection', Figure 3.12). The longer time delay at the lower end also permits large 'switching-in' currents such as might occur when reclosing a circuit which has loads still connected. For example heaters or refrigerators may remain connected even after a prolonged interruption of supply. The in-rush currents of large transformers can be similarly passed.

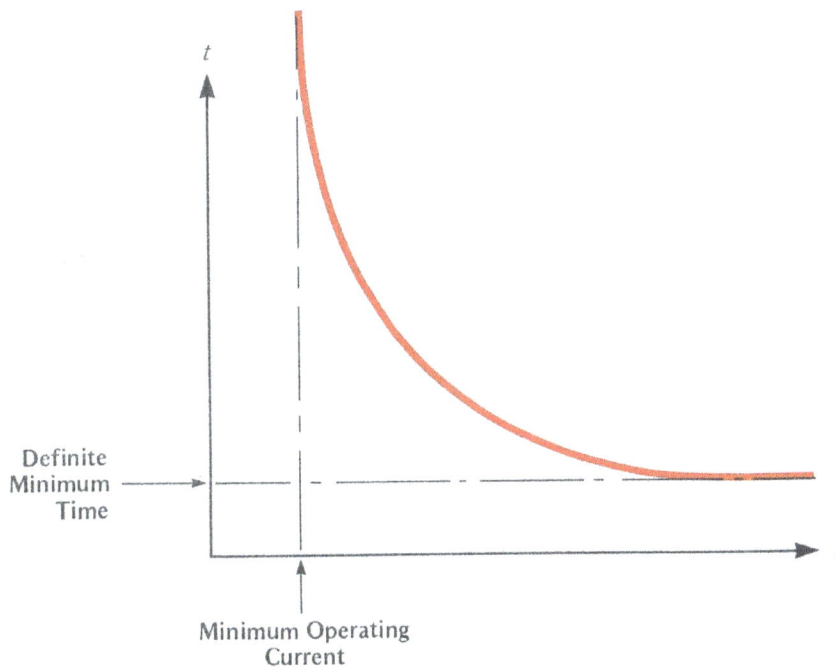

FIGURE 2.8
INVERSE DEFINITE MINIMUM TIME OVERCURRENT
RELAY CHARACTERISTIC

(iii) **Inverse Definite Minimum Time Overcurrent (OCIDMT).** The exact shape of the inverse-time characteristics is controlled to some extent in design by appropriate design of the driving electromagnet. A common variation is the Inverse Definite Minimum Time characteristic, shown in Figure 2.8, in which a lower limit is set to the delay time as the current increases.

(iv) **Other Single-quantity Relays.** The type of relay described above for inverse-time overcurrent protection is also applied to a variety of other functions for which an inverse-time characteristic is appropriate, such as earth-fault and overvoltage protection. Voltage relays have higher-resistance windings, in some cases with series resistors for adjustment, and are normally fed from voltage transformers.

(v) **Voltage-restrained Overcurrent.** The voltage-restrained OCIT relay is designed to overcome a difficulty encountered in protecting generators, namely that even a severe fault causes a relatively low overcurrent after the initial transient period, because of the high synchronous reactance which is normal in synchronous generators. This is explained in Part 2 'Electrical Power Generation', Chapter 2. Such a fault is therefore not cleared before an undesirably long delay has elapsed. In this modification an additional driving element is coupled to the disc (or a separate disc may be mounted on the same spindle) energised by the line voltage and arranged to produce a restraining torque in opposition to that produced by the current coil. When the fault causes a drop in system voltage, this additional restraining torque is weakened, so making the relay more sensitive and reducing the time delay. Allowance is made in the current setting for the increased restraining torque and for the consequent longer time delay occurring when operating at normal voltage. This relay is further explained in Part 10 'Electrical Protection', Chapter 3, Figure 3.7.

British Standard 3939 does not give a Device Function Number for this variation, but it has become the practice to use number 51(V).

(vi) **Voltage-controlled Overcurrent.** This relay achieves a similar effect to that of the voltage-restrained type, but by a different method. The normal solid-copper shading rings in the pole faces of the driving magnet are replaced by windings which are normally connected across resistors, resulting in a moderate torque and long delay times. If a fault is sufficiently severe to reduce the system voltage appreciably, an instantaneous (attracted-armature) undervoltage relay short-circuits the shading windings; the consequent increase in torque then reduces the delay time to something more compatible with the characteristics of other protective devices in the system.

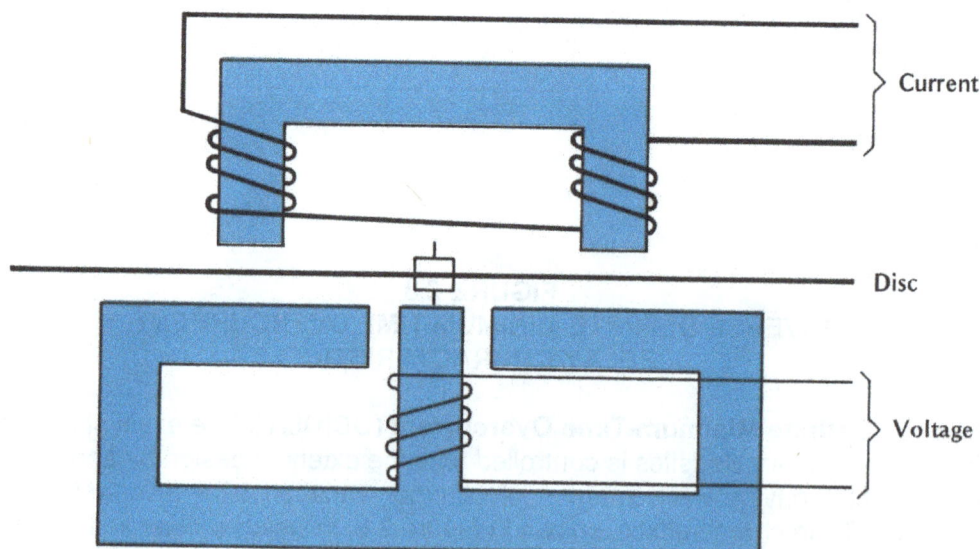

FIGURE 2.9
ELECTROMAGNETIC SYSTEM OF A WATTMETRIC INDUCTION DISC RELAY

(vii) **Power**. Power relays are 'two-quantity' relays (ie voltage and current) which incorporate the type of driving magnet structure used for induction watt-meters and kilowatt-hour meters, in which the torque exerted on the disc depends upon the product of the currents in the two energising coils, and upon the phase-angle between them. Figure 2.9 illustrates the basic structure. Because power flow is directional, power relays are also used to give a directional bias to other relays.

The sensitivity of this arrangement to the phase-angle between the inputs gives the relay directional properties; a flow of power in one direction generates torque in the direction required to close the relay contact, while the only effect of a power flow in the opposite direction is to produce a thrust against the back-stop. The principal use for such a relay is in reverse-power protection.

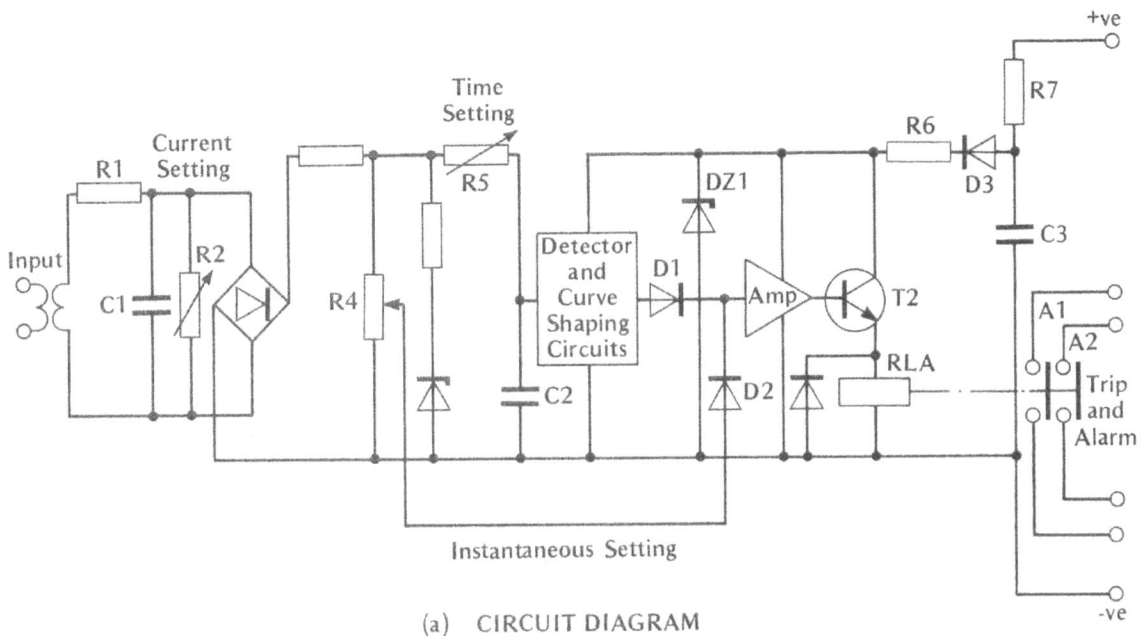

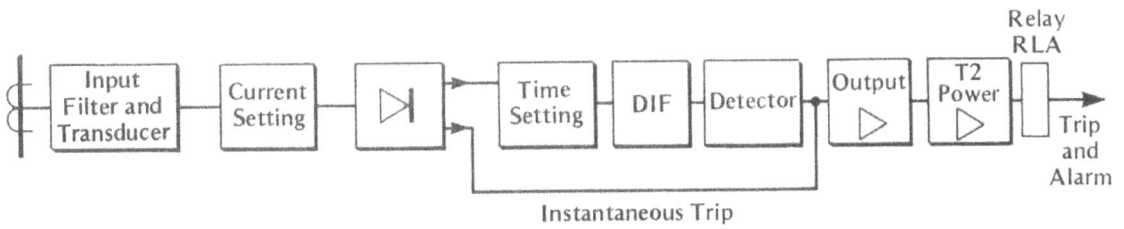

(a) CIRCUIT DIAGRAM

(b) CORRESPONDING BLOCK DIAGRAM

FIGURE 2.10
ELECTRONIC INVERSE TIME OVERCURRENT RELAY

(c) **Electronic Relays.** To a considerable extent protection relays of the electromagnetic type, in which a moving armature or disc is actuated by some kind of electromagnet, are being superseded by electronic types. In these the functions of signal detection and processing are carried out by entirely static circuits, and only the final operation of contacts is done by electromechanical relays, which can be of any suitable but simple control type. The advantages of this technique include a greater flexibility in providing virtually any desired function, however complex, better accuracy, ease of adjustment, and the usual benefits of static circuits with regard to reliability and freedom from regular servicing requirements.

The diversity of functions and principles to be found in static protection relays is such that no comprehensive discussion is possible here. The various characteristics and adjustments established in electromagnetic relay practice are readily reproduced electronically. Figure 2.10, without exactly representing any actual apparatus, illustrates as an example the application of analogue principles to inverse-time overcurrent protection. (An analogue system is one in which continuously variable internal signals are used to represent external quantities such as current and time.)

In Figure 2.10(a) the input from the line current transformer is fed through a small matching transformer to a low-pass filter R1-C1 which suppresses transient voltage surges. A voltage proportional to the input current is developed across the current-setting potentiometer R2. This voltage is applied to the bridge rectifier.

The d.c. output voltage, which is proportional to the line current, is used to charge the capacitor C2 through the potentiometer R5. The setting of this potentiometer determines the rate at which the voltage across C2 increases and hence the timing of the inverse-time operating characteristic of the relay. When the voltage across C2 reaches a predetermined value, the detector circuit operates to switch the electromechanical relay RLA through the output amplifier and power transistor T2.

'Instantaneous' operation is obtained by applying the output voltage of the bridge rectifier directly to the input of the amplifier through R4. Thus, for higher values of fault current, the inverse-time delay circuit is bypassed.

The power supply for the solid-state circuits is applied through D3 and R6. It is stabilised by zener diode DZ1, and spike protection is afforded by R7 and C3. The diode D3 guards against reversed polarity of the d.c. power supply.

Figure 2.10(b) shows the corresponding circuit in block form.

The flexibility and scope of present-day electronics enables a very wide variety of characteristics to be created with relative ease. While a simple analogue overcurrent circuit has been described above for the purpose of illustration, digital techniques have latterly been adopted very widely as a result of the availability of microprocessors and other digital integrated circuits.

2.2.3 Miscellaneous Relays

A number of other types of relay are now briefly described. They are not often encountered on installations, if at all; it is more usual for them to be used as components of larger equipment. Still other types exist but do not merit attention here.

(a) **Reed Relay**. The reed relay - normally a small, low-power device with a single contact - is an electromagnetic relay in which the magnetic structure and the contact assembly are combined. The contacts are mounted on two fingers of a springy magnetic steel sealed in a glass tube, as shown in Figure 2.11, which is placed within an elongated energising coil. Current in the winding produces a magnetic flux along the fingers (the magnetic path is completed through the surrounding air) and the contact-bearing ends are attracted together, closing the circuit. A number of contact tubes can be embraced by one coil to form a multi-contact relay, but normally-closed contacts are less easily provided than normally-open. Because of the low inertia of the reeds, the operating time can be very short - of the order of one millisecond.

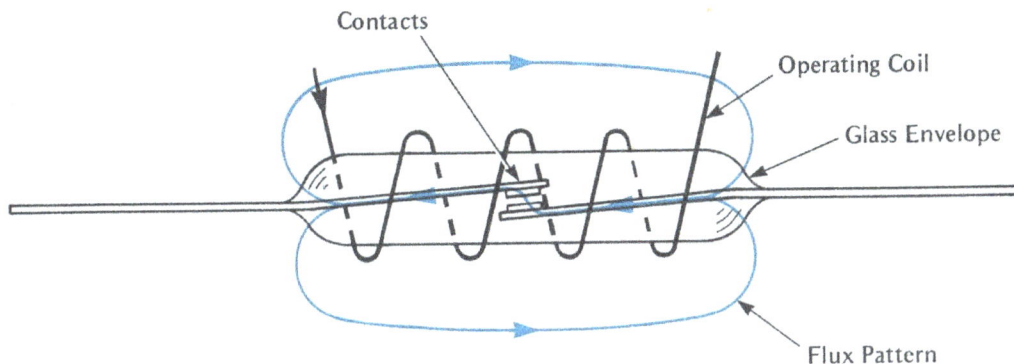

FIGURE 2.11
REED RELAY

Reed relays are not amenable to accurate calibration, and they are used purely as auxiliary relay devices. A similar contact tube, without a coil, can be actuated by a permanent magnet and used as a proximity detector.

(b) **Moving-coil Relay**. The moving-coil relay is a d.c. measuring relay capable of high sensitivity and accurate calibration. The operating coil which carries the moving contact moves in the field of a fixed permanent magnet in the same way as in a moving-coil d.c. meter, and the construction may be similar. The functions of relay and indicating instrument are sometimes combined; it is then more usually described as a 'contact instrument'.

(c) **Thermal Relay**. The contact of a thermal relay is operated by a bi-metallic strip, directly heated or surrounded by a heating coil. Its main purpose is to introduce a time delay. The principle is more generally applied to simple inverse-time protection on small circuit-breakers and contactors, particularly for motor protection where its characteristic can be arranged to match that of the motor.

(d) **Solid-state Relay**. In the solid-state a.c. power relay the switching element is a thyristor (triac) - see Chapter 3 - instead of a contact, and a triggering signal is provided in response to an input signal by a coupling device in which electrical isolation is provided by an electromagnetic or optical link. Within the limitations of the voltage and current ratings of the thyristor the arrangement has the advantages of rapid operation - about one half-period of the supply - and a long life unaffected by the frequency of operation.

(e) **Buchholz Relay**. Although not strictly a relay according to the definition here, the Buchholz relay detects gas evolved from oil in oil-immersed transformers and closes a contact when a sufficient quantity has been collected to signify a serious deterioration of the transformer insulation. It also indicates when a major fault inside the transformer produces a surge of oil. Buchholz protection is not applicable to the sealed type of transformer normally used on most installations. (See Part 10 'Electrical Protection', Chapter 10.4.)

(f) **Qualitrol Relay**. Like the Buchholz, this is not strictly a relay but is a pressure sensor with contacts. It is placed on the top of sealed transformers. If the internal pressure exceeds a certain preset level, its contacts give an alarm or trip signal. If the pressure is excessive, a relief valve opens. (See Part 10 'Electrical Protection', Chapter 10.4.)

2.3 RELAY APPLICATIONS

2.3.1 Control Relays

A relay which operates its contacts simply in response to the presence or absence of an energising input may be used for any of a number of purposes:

(a) To control a number of separate circuits from a single signal input by virtue of multiple contacts.

(b) To couple circuits which would be mutually incompatible if coupled directly - e.g. a.c. and d.c. circuits, circuits operating at different voltages, or circuits which need to be electrically isolated for any reason. Relays used for this purpose are referred to in some installations as 'interposing relays'. They provide the additional benefit that a fault in one circuit does not immediately cause failure of the other. The fact that d.c. interposing relays are not very sensitive to a.c. also greatly reduces the risk of unintended relay operation due to stray coupling of power circuits.

(c) To provide amplification so that a relatively large amount of power can be switched by a low-power control signal or a large load can be controlled by a small contact - e.g. on a protection relay, flow-switch etc.

(d) To perform operations which might loosely be grouped under the heading of 'logic' - Figure 2.12 gives two examples of such logic.

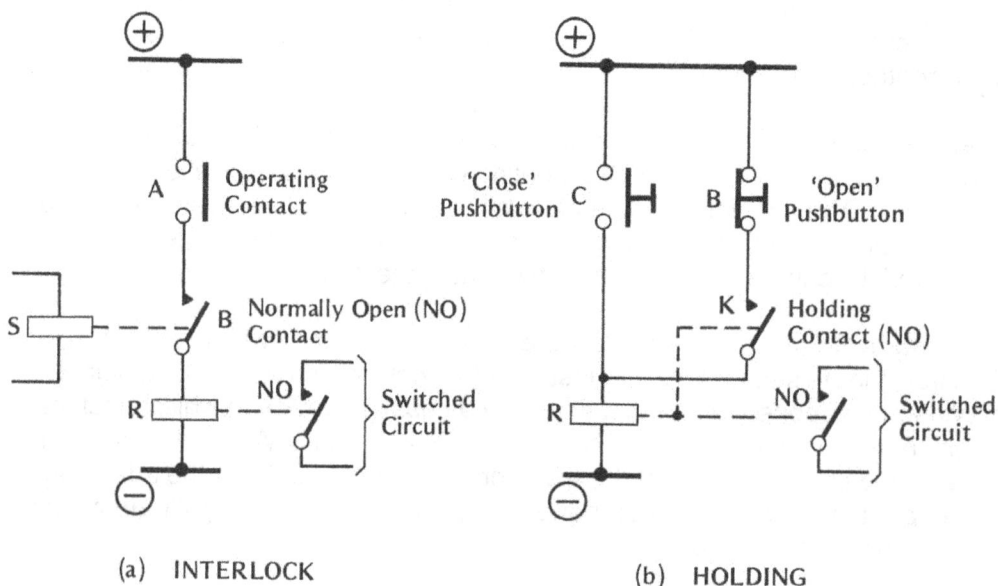

(a) INTERLOCK   (b) HOLDING

FIGURE 2.12
SIMPLE EXAMPLES OF RELAY LOGIC

Figure 2.12(a). Here relay 'S' is used as an interlock to ensure that the circuit of relay 'R' cannot be energised by the operating contact 'A' unless relay 'S' has first been energised and has closed its contacts 'B'. Relay 'R' will de-energise as soon as relay 'S' has de-energised - i.e. relay 'R' depends wholly on 'S'.

Figure 2.12(b). Relay 'R', once it has been energised by pressing the CLOSE pushbutton 'C', is kept energised by its own auxiliary retaining contacts 'K' even after the pushbutton has been released. To de-energise the relay the retaining circuit is broken by pressing the normally-closed OPEN pushbutton 'B'. This type of circuit is often called a 'retaining' or 'hold-in' circuit.

Relays that serve no purpose other than to pass on the instructions generated in one circuit to another circuit, or a number of circuits, are often referred to as 'slave' or 'auxiliary' relays. Any sequence of operations, or any system of interlocking, may be built up in a like manner from elements of relay logic.

2.3.2 Protection Relays

The applications of protection relays are described more fully in Part 10 'Electrical Protection', but, briefly, they are those applications, mostly in a.c. circuits, in which

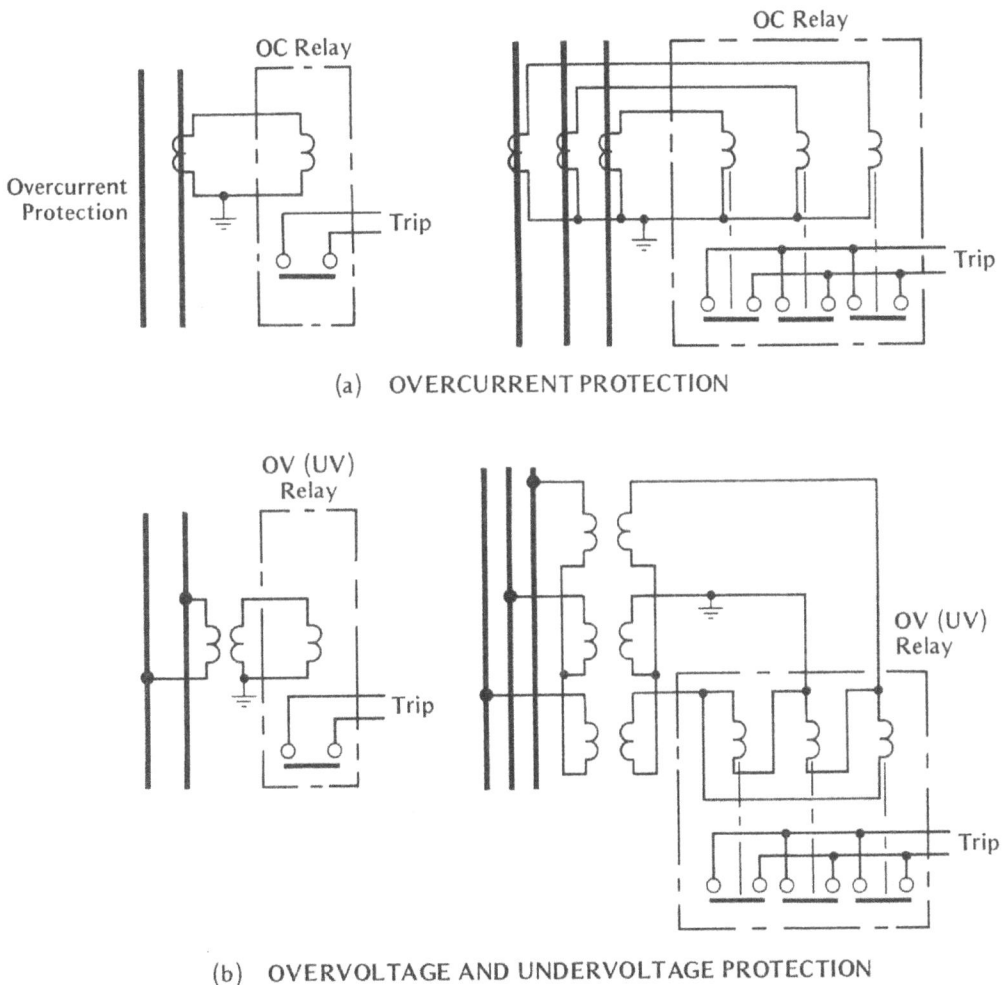

(a) OVERCURRENT PROTECTION

(b) OVERVOLTAGE AND UNDERVOLTAGE PROTECTION

FIGURE 2.13
PROTECTION RELAYS FED FROM CURRENT AND VOLTAGE TRANSFORMERS

accurately calibrated operating characteristics, in many cases in terms of time as well as current or voltage, are important, such as the following modes of protection:

- overcurrent
- earth fault
- differential
- undervoltage
- overvoltage
- neutral displacement
- reverse power
- directional fault
- negative phase-sequence
- overfrequency
- underfrequency.

Most of these can be either 'instantaneous' or time-delayed with fixed delay or inverse-time characteristics.

Current coils in a.c. relays are coupled to the supply system through current transformers with standard ratios, usually designed for a secondary current of 1A or 5A with the full rated line current. Voltage coils are usually energised at 110V through voltage transformers. Figure 2.13 shows typical connections for (a) overcurrent and (b) overvoltage or undervoltage protection. Three-phase relays for such purposes may be simply assemblies of three single-phase units in single cases, or they may be assembled in a single case with the three single-phase elements acting on a common rotor shaft.

A.C. relays normally respond to the rms values of the measured quantities. For some purposes relays are energised by rectified alternating current (not induction relays), and in this case the nature of the response depends upon details of the rectifier circuit.

CHAPTER 3 SEMICONDUCTOR DEVICES

3.1 SEMICONDUCTORS, JUNCTIONS AND DIODES

Semiconductor devices - the term 'solid-state' is often used to distinguish them from earlier valves based on vacuum, gas or mercury vapour - include switching devices capable of controlling hundreds of kilowatts, signal-processing elements operating at microwatt levels and a wide range of devices between these extremes. Their functions (disregarding some special devices) fall generally into three categories, which may be found in combination:

- rectifying
- power amplification
- switching.

The properties of semiconductor materials that make these functions possible lie between those of conductors and insulators. Good conductors, such as most metals, conduct electricity freely by virtue of the high mobility of electrons within the bulk of the material and of the consequent easy movement of electric (negative) charges. By contrast insulators, so long as they are not overstressed by high voltage, have very few mobile charge carriers, and the amount of current that can flow in them is therefore very small indeed. A semiconductor material in its pure state and at a 'normal' temperature is like a rather poor insulator, but it is possible to increase its conductivity, to the extent that it behaves more like a normal metal, by a number of methods which all have the effect of increasing the availability of free (mobile) charge carriers; the most important are:

(a) **Raising the Temperature.** Unlike metals, semiconductors (and insulators) conduct increasingly as their temperature is increased. While this has very important consequences in the design and use of semiconductor devices, it need not be considered in the present discussion.

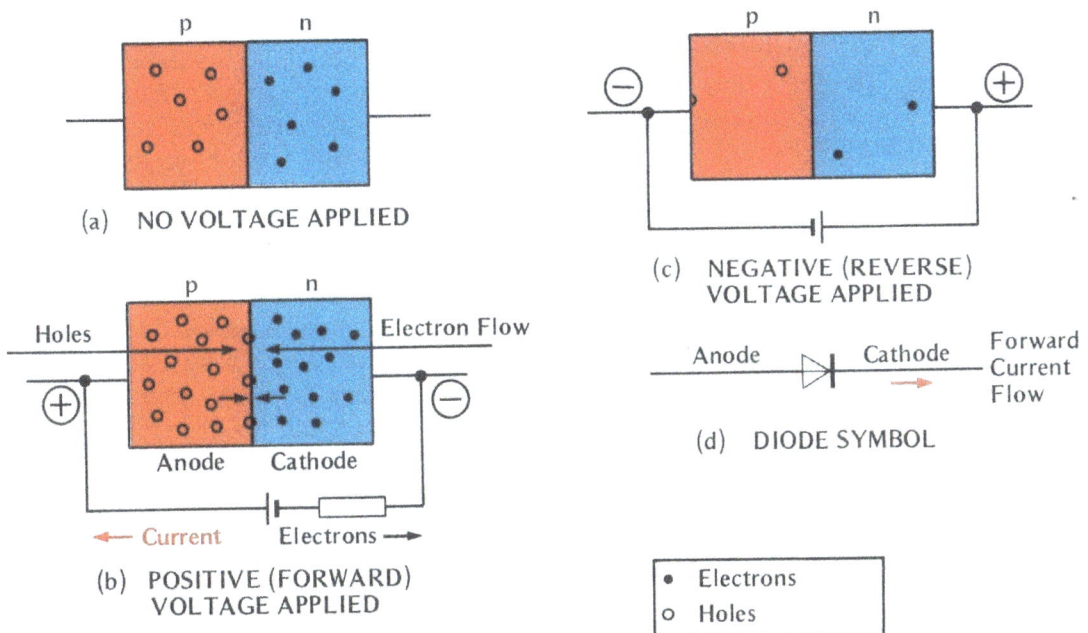

FIGURE 3.1
VARIOUS CONDITIONS IN A SEMICONDUCTOR p-n JUNCTION

(b)   **Adding Impurities**. The true behaviour of the semiconductor material is observed only when it is extremely pure. Certain elements, technically termed 'impurities', when added to the semiconductor in very small proportions, have the property of making free charge carriers available in abundance, hence increasing the conductivity.

(c)   **Injecting Current**. Since a current is essentially a flow of charge carriers, it can, in certain circumstances, be used as a means of controlling, or 'modulating', the degree of conductivity in a semiconductor material which has been rendered potentially conductive by the addition of the impurities referred to above. This provides the element of control required in the three functions listed above.

The term 'charge carrier' has been used advisedly in preference to 'electron', which signifies a negative charge. Semiconductor physics also recognises positive charge carriers, which are actually electron deficiencies - i.e. spaces vacated by departed electrons - generally known as 'positive holes', or simply 'holes'. Conductivity in a semiconductor may result from the presence of either electrons or holes, depending on the particular impurity which is added. A material that conducts through the presence of free (negative) electrons is called 'n-type' (coloured blue in the diagrams); one that conducts by virtue of (positive) holes is called 'p-type' (coloured red).

In most power semiconductor devices (all large ones), and also in most small signal devices, the desired characteristics originate from a juxtaposition of layers of dissimilar types of material, forming a so-called 'p-n junction'. Figure 3.1 illustrates in a diagrammatic and simplified form the behaviour of such a junction under various conditions of applied voltage. At (a), with no voltage applied, the two layers simply co-exist, suffused respectively with holes and electrons. When a positive voltage is applied to the 'p' layer, as at (b), the holes in the 'p' layer and the electrons in the 'n' layer are driven by electrostatic forces towards the junction (shown by black arrows), where they combine (that is, the electrons fill the holes) and their charges are neutralised; meanwhile the supply is replenished by holes from the positive side and electrons from the negative side of the electrical circuit and a current flows clockwise continuously, shown by red arrow opposite in direction to the black electron flow arrows. This current depends on the external voltage and impedance. At (c) the voltage is reversed: the holes are drawn away from the junction to the negative terminal and the electrons likewise to the positive terminal; both layers are thereupon deprived of the means of conduction, and no current flows beyond a very small leakage current.

The p-n semiconductor junction thus exhibits a one-way conduction characteristic and so provides the basis of a rectifier diode, which will pass current freely in one (forward) direction (from 'p' to 'n') but will block a high voltage in the other (reverse) direction. In a practical diode the semiconductor element is a disc of silicon, which may be anything from the size of a pinhead up to about 75mm in diameter and something less than one millimetre thick. There are semiconductors other than silicon, but they are of little significance in comparison. The 'p' terminal is called the anode of the diode, and the 'n' terminal is the cathode. By convention, forward current flows within the element from the anode to the cathode (i.e. against electron flow), in the direction of the red arrow in the symbol (Figure 3.1(d)).

3.2   THE TRANSISTOR

The transistor (strictly, the 'bipolar junction transistor') has two p-n junctions, formed between three layers, arranged more commonly n-p-n as in Figure 3.2, but sometimes as p-n-p, with an external connection to the intermediate layer.

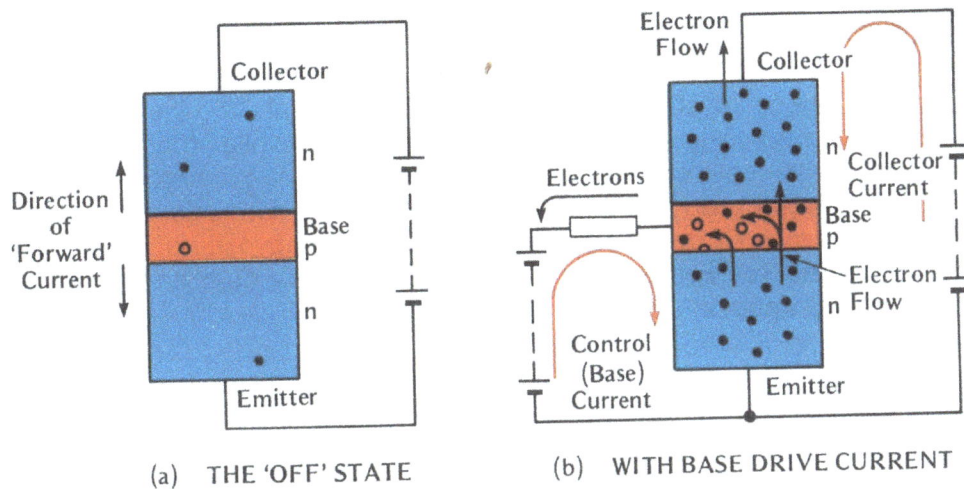

(a) THE 'OFF' STATE   (b) WITH BASE DRIVE CURRENT

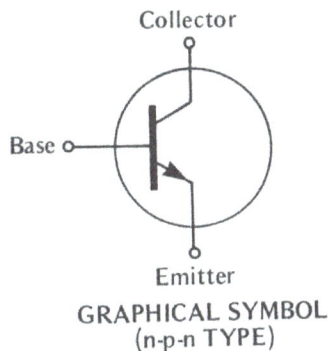

GRAPHICAL SYMBOL
(n-p-n TYPE)

FIGURE 3.2
THE TRANSISTOR

Since the two diodes so formed are back-to-back, the transistor as it stands does not conduct in either direction (Figure 3.2(a)). In Figure 3.2(b) a control current is applied to the intermediate 'p' layer, termed the 'base', and flows to the lower 'n' layer, termed the 'emitter', in the forward direction of the p-n junction. It is indicated by the lower red current arrow. This results in a supply of electrons from the emitter into the 'p' region, indicated by the black arrows, and it is arranged by careful design that most of them overshoot into the 'collector' 'n' region, where they are attracted to the collector and produce a current flowing into the collector terminal from the d.c. supply (indicated by the upper red current arrow in the opposite direction to the black electron flow).

The collector current resulting from this process varies with the control current in the base, but it can be much greater (normally from about ten to several hundred times), and only a low voltage is required at the base; the transistor is therefore capable of current gain and power amplification. It should be noted that, used in this way, the transistor is a **current** amplifier, in contrast to the old thermionic valve which was used mainly as a voltage amplifier.

While for many purposes the transistor is used in such a way that its output (collector) current or voltage is entirely under the control of its base input - that is, in a so-called 'linear' mode - a point can be reached in any circuit where the current is limited by what is available from the power supply through the resistance or impedance in the collector circuit. If the transistor is driven with a control signal so large that this condition is the normal one, it functions not as a linear amplifier but as a switch, which is either 'on' (closed) or 'off' (open), according to whether base drive is applied or not. Transistors which are intended to be used in this way are often specially designed.

3.3   THYRISTOR

For reasons beyond the scope of this book there are fundamental difficulties in making transistors capable of controlling very large amounts of power. While progress in design and manufacturing techniques has resulted in the extension of the transistor's capabilities to very substantial power levels (100kW or more), the particular problems associated with it are avoided in a more recently developed switching device, the thyristor (formerly, known as the Silicon Controlled Rectifier (SCR)). The thyristor also has a number of other advantages for some purposes.

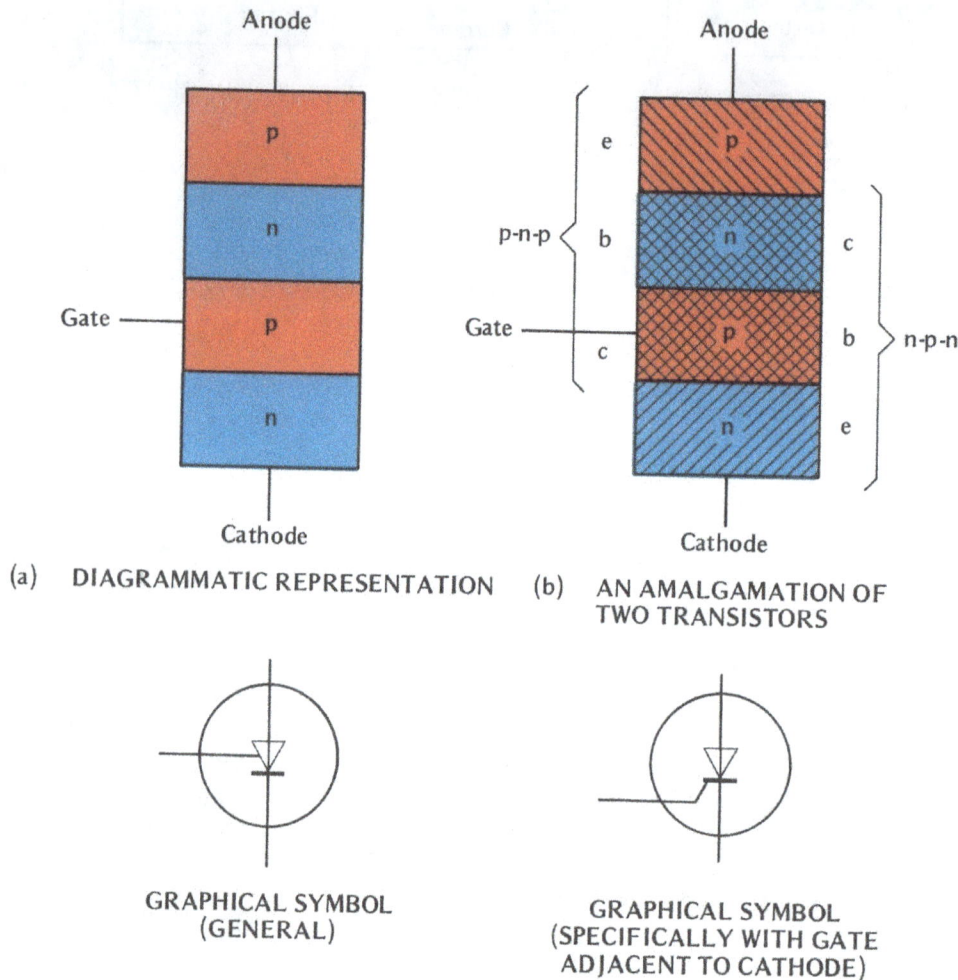

(a)   DIAGRAMMATIC REPRESENTATION

(b)   AN AMALGAMATION OF
         TWO TRANSISTORS

GRAPHICAL SYMBOL
(GENERAL)

GRAPHICAL SYMBOL
(SPECIFICALLY WITH GATE
ADJACENT TO CATHODE)

FIGURE 3.3
THE THYRISTOR

The thyristor has four layers within the silicon element, arranged p-n-p-n, and correspondingly three p-n junctions (Figure 3.3(a)). To explain its behaviour in simple terms, it may be thought of as two transistors, one p-n-p and the other n-p-n, telescoped together so that their emitters form the main terminals of the device (Figure 3.3(b)). In this imagined amalgamation the collector current of each constituent transistor is base current to the other. If collector current is flowing, therefore, there is a generous level of base current to sustain it by transistor action. Conversely, if no collector current flows, there is no base current, and hence nothing to initiate a flow of current. This gives rise to two stable states - conducting and non-conducting, or blocking. A third electrode, called the 'gates, is connected to one of the inner layers, adjacent to the cathode, and enables the thyristor to be triggered from the blocking to the conducting state by a short pulse of control current, assuming that there is a positive supply available at the anode.

Reversion to the blocking condition depends (in the great majority of thyristors) on the current in the external circuit falling to zero, commonly as a result of changing polarity in an a.c. circuit. The thyristor cannot be switched off by gate control.

Compared with the power transistor, the thyristor can handle higher currents and voltages. It also gives a higher power gain and, at least in large sizes, is cheaper to produce. In addition, unlike the transistor whose emitter junction will support only a very low reverse voltage, the thyristor will withstand a high voltage in reverse without conducting and without switching. This is because of the junction between the anode and the adjacent 'n' region, which causes it to function as a rectifier.

3.4   THE TRIAC

In many applications of thyristors in a.c. circuits (see para. 4.6) pairs of thyristors of the kind described above are connected in 'inverse parallel', as shown in Figure 3.4(a), in order to provide a path for current in either direction. A more convenient alternative for many purposes is the bidirectional thyristor, or triac, illustrated in very simplified form in Figure 3.4(b).

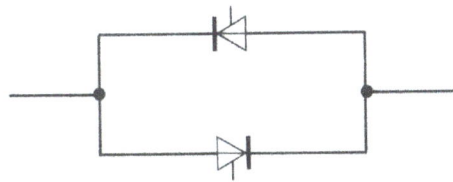

(a)   THYRISTORS IN INVERSE
      PARALLEL

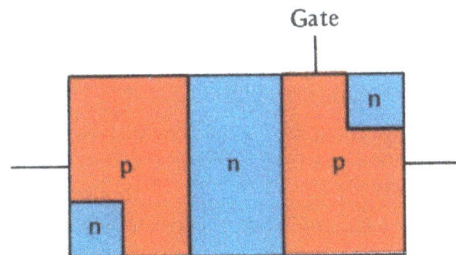

(b)   TRIAC

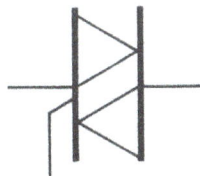

(c)   TRIAC SYMBOL

FIGURE 3.4
BIDIRECTIONAL THYRISTOR (TRIAC)

The structure of the triac is so arranged that it will function as a p-n-p-n thyristor in either direction: the apparent short-circuiting of certain regions, which might be expected from Figure 3.4(b), does not in fact occur because of the relatively high lateral resistance of the thin silicon wafer which actually constitutes the element. Control of this dual structure is made very easy by a gate triggering system more complex than that of the ordinary thyristor. This system enables the thyristor to be triggered into conduction by a gate pulse of either polarity with either polarity of voltage in the main circuit. There are thus useful benefits to be gained in terms of simplicity and economy in both the power and control circuits.

Figure 3.4(c) shows the conventional symbol for a triac.

CHAPTER 4 USE AND APPLICATION OF SEMICONDUCTOR DEVICES

4.1   POWER DEVICES

The semiconductor devices so far described have been shown in block form. In practical power devices (contrary to the impression given by the conventional diagrammatic representations) they are in the form of silicon discs or wafers. The discs or wafers are so thin and brittle that they have to be bonded to more substantial discs of molybdenum (chosen because of its similar coefficient of thermal expansion) and then supported by a relatively massive copper base and protected by a hermetically sealed housing. Plastic encapsulation is used for many small devices.

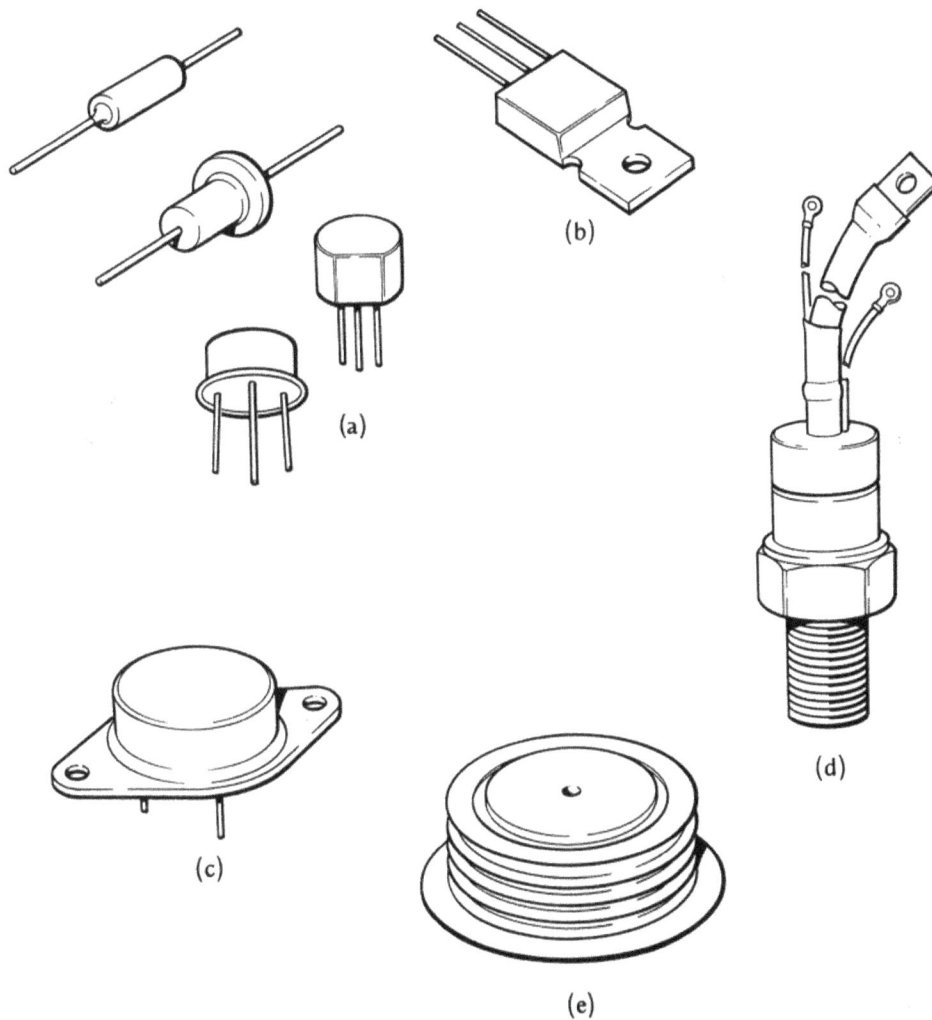

FIGURE 4.1
VARIOUS CONSTRUCTIONS OF POWER SEMICONDUCTOR DEVICES

In total there are many variations in mechanical design, but the majority of constructions are represented in Figure 4.1, which shows:

(a)   wire-ended air-cooled diodes and transistors;

(b)   a plastic-encapsulated transistor;

(c)   a transistor with a 'lozenge' base;

(d)   a thyristor of 'top-hat' construction with a threaded stud;

(e)   a 'capsule' diode.

Apart from the wire-ended devices, all the various housings are designed with a view to mounting on some kind of cooling structure. Some are liquid-cooled, but mostly they are air-cooled by natural or forced convection, with some contribution from radiation. The question of cooling is associated with two important aspects of the use of semiconductor devices - heat dissipation due to losses and the limits of permissible operating temperature.

When a transistor is used as a linear amplifier or regulator, it has to dissipate an amount of power equal to the product of the current through it and the voltage drop across it. In the case of a diode or a thyristor, or of a transistor used as a switch, there should ideally be no loss, because at any time the device is either conducting current with no voltage drop, or blocking voltage with no current flowing. In practice semiconductor devices are not perfect switching devices. In the conducting state they exhibit a significant voltage drop of approximately one volt. In the blocking or 'off' state they pass a small leakage current, which depends on the size of the device and is in any case somewhat unpredictable; it may typically be five orders of magnitude less than the rated current. In switching from one state to the other, they do not turn on or off quite instantaneously but experience current and voltage drop simultaneously for a short but finite period, so generating a switching loss. The off-state loss and, at least at normal power frequencies the switching loss, are generally very small, but the conducting loss is a limiting factor in the current rating of the device and governs the design of the associated cooling structure in any given operating conditions. It follows that anything which affects the efficiency of the cooling arrangement, such as the physical mating of the semiconductor device with its cooling fin, merits careful attention.

The limits of permissible operating temperature, which determine the cooling requirements, arise from the effects of temperature on characteristics - semiconductor devices will only work over a certain temperature range - and from physical properties of the materials used in the construction of semiconductor devices, such as melting points of solders, coefficients of expansion etc. All such considerations are taken care of by observing manufacturers' ratings, which are ultimately in terms of 'junction' temperature - i.e. the temperature of the operative part of the semiconductor element. Maximum junction temperatures in silicon are usually in the range 125 to 200°C.

4.2   RECTIFIERS, SINGLE-PHASE

Much could be, and has been, written on the subject of rectifier technology: two simple examples will suffice here to illustrate the application of rectifier diodes and the principle of controlled rectification using thyristors.

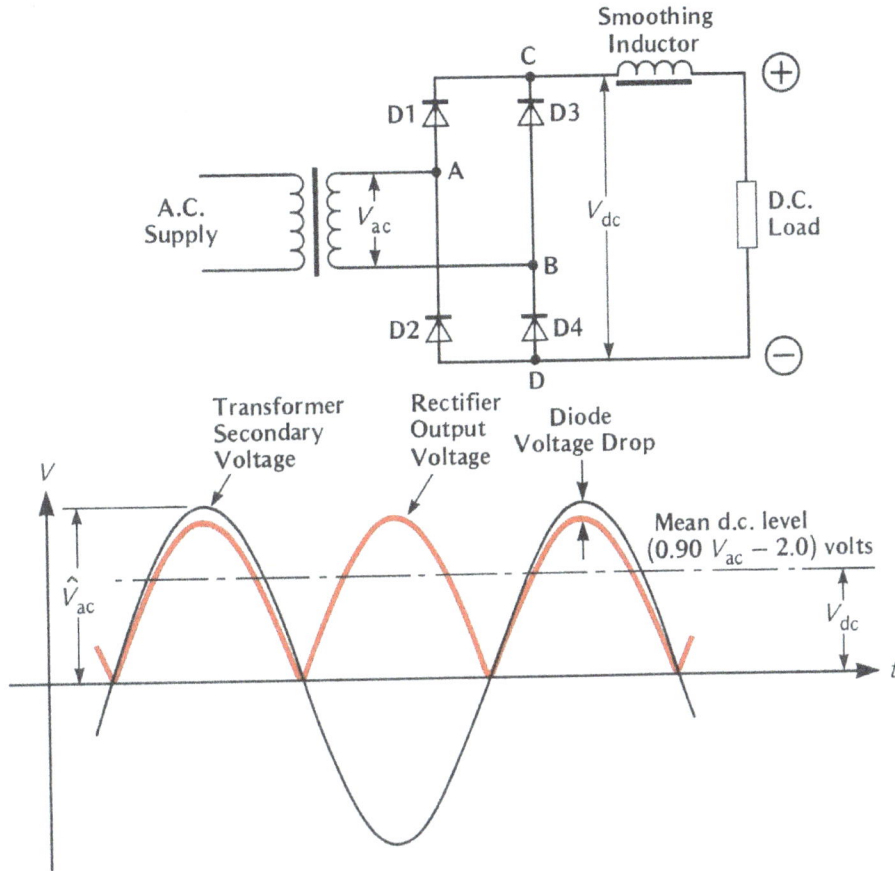

FIGURE 4.2
SINGLE-PHASE FULL-WAVE UNCONTROLLED BRIDGE RECTIFIER

Figure 4.2 shows the very common single-phase full-wave bridge rectifier, with four diodes. It takes power from the a.c. supply (usually through a transformer) and supplies the rectifier at a voltage suitable to provide the desired d.c. output voltage. During a half-cycle when terminal A of the bridge is driven positive with respect to terminal B, diodes D1 and D4 conduct current through the load, and diodes D2 and D3 block reverse voltage. In the alternate half-cycle, when B is positive with respect to A, current flows through D2 and D3, and D1 and D4 are on reverse. The smoothing inductor, ideally at least, maintains a constant flow of current to the load, so that current flows at all times through one pair of diodes or the other pair.

The voltage produced at the output terminals of the bridge, C and D, is thus a series of half-cycles all of the same polarity, and has a mean value, or d.c. component, equal to the half-cycle mean value of the transformer secondary voltage, less the voltage drop of roughly 1.0 volts across each of the two conducting diodes in series - total 2.0 volts. The mean value of the a.c. voltage is $\frac{2}{\pi}$ ($= 0.637$) times the peak value, and therefore $\sqrt{2} \times \frac{2}{\pi} (= 0.90)$ times the rms value. In terms of the rms secondary voltage V_{ac}, therefore, the d.c. output voltage is approximately:

$$V_{dc} = \frac{2\sqrt{2}}{\pi} . V_{ac} - 2.0 \text{ volts} \quad = 0.90 V_{ac} - 2.0 \text{ volts}$$

The output also contains a very large a.c. component of voltage, or 'ripple', amounting in rms terms to about 48% of the mean direct voltage and having a fundamental frequency twice that of the a.c. supply (i.e. it is a second harmonic).

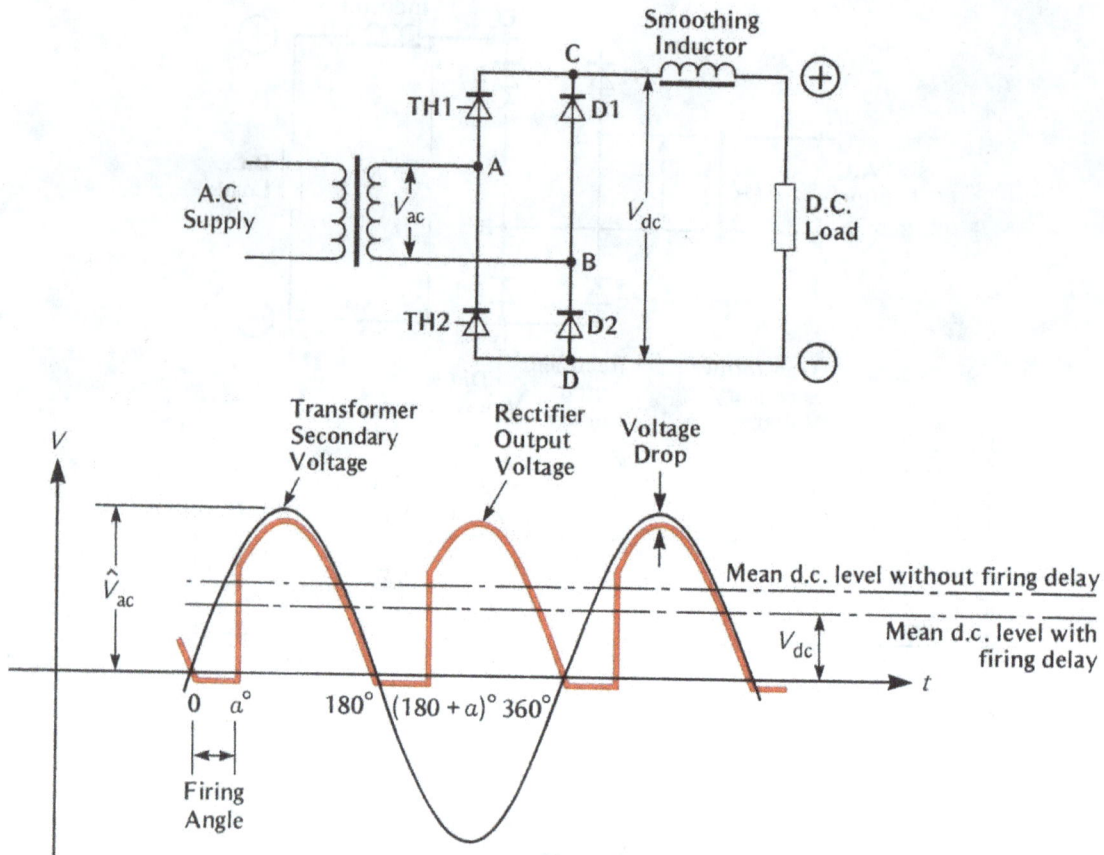

FIGURE 4.3
SINGLE-PHASE FULL-WAVE CONTROLLED BRIDGE RECTIFIER

The bridge rectifier as described produces a **fixed** d.c. voltage level, determined by the transformer's secondary voltage. It can however be modified to give a controlled and variable d.c. output voltage.

Figure 4.3 shows the same circuit with two thyristors replacing two of the diodes. If the appropriate thyristor were triggered at the beginning of each half-cycle, it would merely behave in the same way as the diode, and the output from the bridge would be the same as in the previous case (except that the voltage drop in a thyristor is a little higher (1.5 to 2.0 volts) than that of a single diode (1.0 volts)). If the thyristors were not triggered at all, in effect there would be no circuit from terminal A of the bridge, and hence no output.

In the intermediate condition the appropriate thyristor is triggered at some instant during each half-cycle, which is determined by a control circuit. Before the triggering point the current (if it is maintained by the smoothing inductor) 'free-wheels' through D1 and D2. After the triggering point, and up to the end of the half-cycle, the transformer secondary winding, voltage V_{ac} rms is connected to the load through the conducting thyristor TH1. In the next half-cycle, 180° later, the other thyristor is triggered and the transformer secondary winding is connected to the load through the conducting thyristor TH2. The output voltage of the bridge is thus a series of part half-cycles whose starts have been delayed by $\alpha°$ and which have a mean value (V_{dc}) less than its undelayed value.

If the triggering occurs at an electrical angle $\alpha°$ relative to the supply voltage zero (called the 'triggering angle', or 'firing angle', or 'angle of delay') it can be calculated that the mean d.c. output voltage V_{dc} is reduced in the ratio $\frac{1+\cos\alpha}{2}$. Therefore, if a total voltage drop of 2.5 volts is assumed (1.5 volts for the thyristor plus 1.0 volts for the diode in series), then:

$$V_{dc} = 0.90V_{ac} \times \frac{1+\cos\alpha}{2} - 2.5 \text{ volts}$$

Reduction of the rectified output voltage by delay-angle control is accompanied by a considerable increase in ripple voltage.

In calculating the open-circuit transformer secondary voltage for a practical rectifier design, allowance has to be made for various voltage drops in addition to that of the diodes, particularly those due to the resistance and leakage reactance of the transformer.

4.3 RECTIFIERS, THREE-PHASE

4.3.1 Uncontrolled Rectifier

At power levels of more than a few kilowatts it is customary to use multi-phase rectifiers. The most common form is the three-phase uncontrolled bridge rectifier using a transformer with a three-phase output connected to a bridge which has three arms, each with two diodes in it. The use of multi-phase rectifiers, and particularly the bridge connection, produces a smoother d.c. output and reduces the voltage waveform distortion on the a.c. supply system.

The circuit of a three-phase transformer bridge rectifier is shown in Figure 4.4(a) and the corresponding voltage waveforms in Figure 4.4(b). The circuit diagram shows that, although it is the a.c. line voltages that are applied to the rectifier bridge, it is the red, yellow and blue phase voltage V_{ac} waveforms that are plotted in Figure 4.4(b).

Referring to Figure 4.4(b), at the instant represented by the line AB, red phase is the most positive and diode D1 conducts. At the same instant yellow phase is the most negative and the current returns through diode D4. The a.c. voltage applied across arms D1 and D4 of the rectifier bridge at this moment is the instantaneous sum of red and yellow phase voltages and is represented by the length of the line AB. Neglecting the voltage drops in the diodes, the d.c. output voltage is then represented by the line A'B', which is equal in length to AB.

As the a.c. cycle progresses, yellow phase becomes less negative until, at the point P, blue phase becomes the more negative and diode D6 starts to conduct. It progressively takes over current from D4, which switches off as yellow voltage passes zero.

Similarly at Q, red phase ceases to be the most positive voltage, being replaced by yellow phase. If the sum of the instantaneous values of the most positive and most negative phase voltages across the bridge is plotted, taking account of the natural changeover or 'commutation' that takes place at points P, Q, R, S, etc., then the resulting d.c. voltage has a waveform as shown in black. It has six pulses per full cycle of a.c. and a mean value of V_{dc} volts, where the shaded areas above and below the V_{dc} line are equal.

Drawn above the d.c. voltage curve are coloured arrows showing for each 30 degree period those diodes which are conducting current to the load in the case of D1, D3 and D5 and from the load in the case of D2, D4 and D6.

As V_{ac} is the rms value of the a.c. **phase** voltage, then its peak value is $\sqrt{2}V_{ac}$. But it is the line-to-line voltage (V_L) that is actually applied to the rectifier bridge; therefore the effective peak line voltage is $\sqrt{3}(\sqrt{2}V_{AC})$ or $\sqrt{6}V_{ac}$.

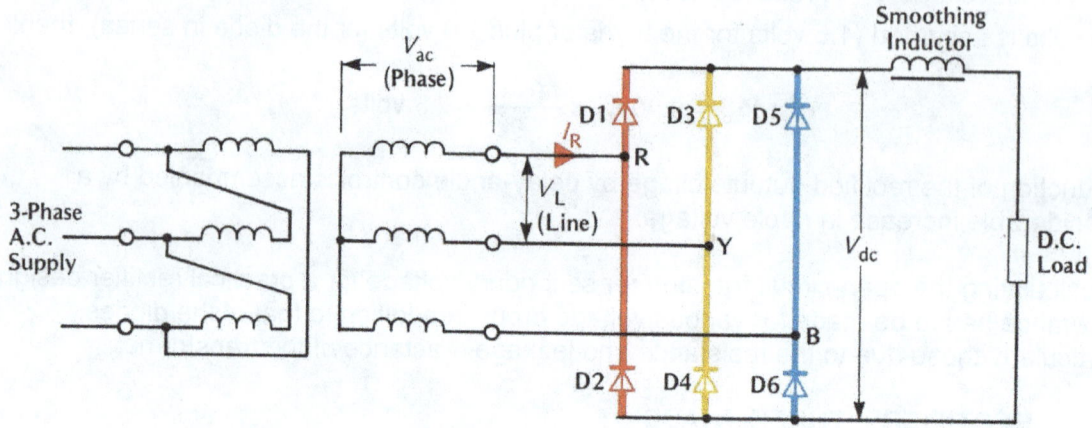

(a) TRANSFORMER RECTIFIER
     CIRCUIT

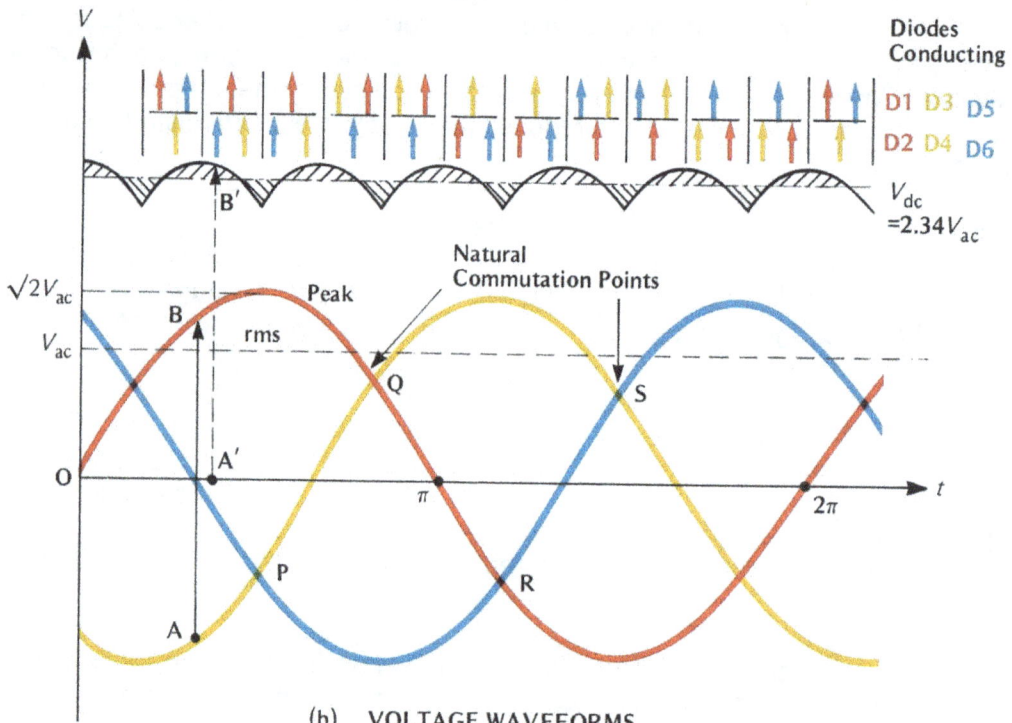

(b) VOLTAGE WAVEFORMS

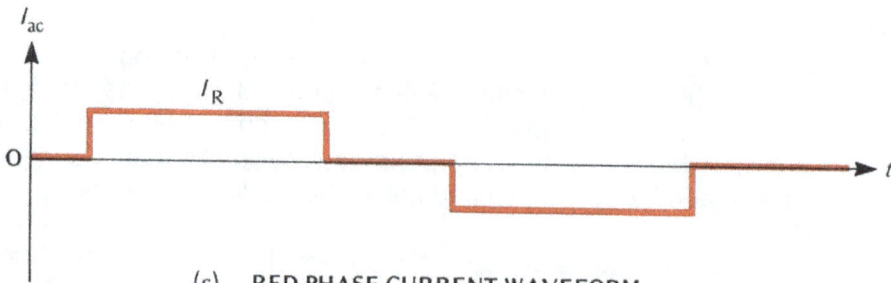

(c) RED PHASE CURRENT WAVEFORM

FIGURE 4.4
THREE-PHASE UNCONTROLLED BRIDGE RECTIFIER

For a six-pulse rectifier such as this it can be shown that the mean d.c. voltage, V_{dc}, neglecting diode voltage drops, is given by:

$$V_{dc} = \frac{3}{\pi} \times \sqrt{6} V_{ac}$$

$$= 2.34 V_{ac} \quad \text{where } V_{ac} \text{ is the secondary rms phase voltage,}$$

or, in terms of a.c. secondary rms line voltage V_L,

$$V_{dc} = 2.34 \frac{V_L}{\sqrt{3}} \quad (\text{since } V_{ac} = \frac{V_L}{\sqrt{3}})$$

$$= 1.35 V_L$$

With a six-diode bridge the d.c. voltage is thus a fixed multiple (1.35) of the a.c. rms line voltage and cannot be varied. It is determined solely by the voltage of the a.c. supply and the transformer ratio. It should be noted that in this case the d.c. voltage is actually **higher** than the rms a.c. voltage.

The multiplying factor 1.35 is strictly correct only on open circuit or with a purely resistive load. If the load contains inductance the factor will be lower.

If the d.c. circuit has sufficient inductance to sustain the current, the a.c. input current has a square-wave form as shown for red phase in Figure 4.4(c).

Each diode element is individually protected by a close-rated fuse, not shown on the diagram.

4.3.2 Half-controlled Rectifier

If one of the diodes in each phase of the bridge is replaced by a thyristor, the d.c. voltage output level of the bridge can be controlled.

The circuit for this 'half-controlled' rectifier is shown in Figure 4.5(a). The thyristors TH1, TH2 and TH3 do not conduct until triggered by a pulse from a separate control circuit. If, as shown in Figure 4.5(b), the triggering signal is delayed to a point T which is α electrical degrees later than the natural commutation point Q, then the d.c. voltage waveform shown in black in the illustration results. The action of the bridge rectifier is similar to that in the uncontrolled case, but commutation of the positive half-cycles of a.c. voltage is delayed.

In the case of the half-controlled bridge the mean d.c. output voltage V_{dc}, neglecting diode and thyristor voltage drops, is:

$$V_{dc} = 2.34 V_{ac} \times \frac{1 + \cos \alpha}{2} \quad \text{where } V_{ac} \text{ is the secondary rms phase voltage}$$

or $$V_{dc} = 1.35 V_L \times \frac{1 + \cos \alpha}{2} \quad \text{where } V_L \text{ is the secondary rms line voltage}$$

since $$V_{ac} = \frac{1}{\sqrt{3}} V_L$$

As α increases above zero, $\cos \alpha$ (which is unity when $\alpha = 0$) becomes gradually smaller, so that the multiplying term $\frac{1 + \cos \alpha}{2}$ becomes less than 1. This shows that, as the delay angle α is increased, so the d.c. voltage V_{dc} becomes less. Thus the d.c. voltage level can be controlled by varying the delay angle α.

V_{ac}
(Phase)

TH1 TH2 TH3

Smoothing
Inductance

R

V_L
(line)

Y

V_{dc}

B

D1 D2 D3

3-Phase
A.C.
Supply

D.C.
Load

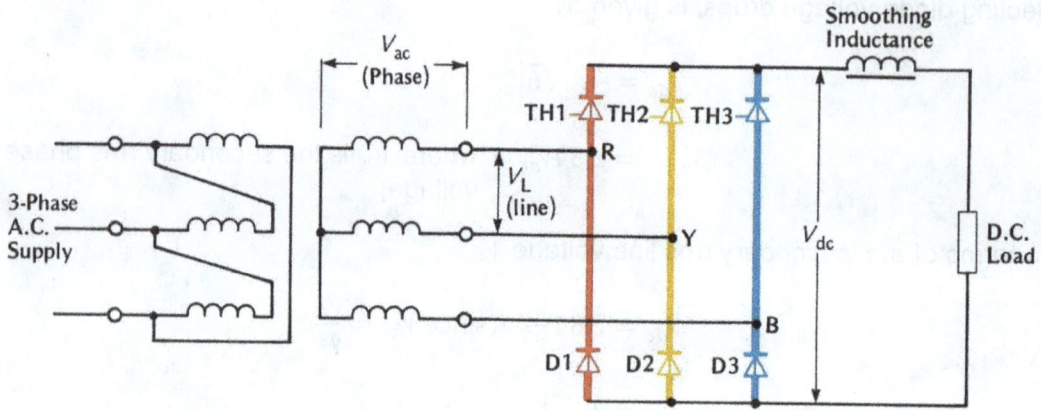

(a) TRANSFORMER RECTIFIER CIRCUIT

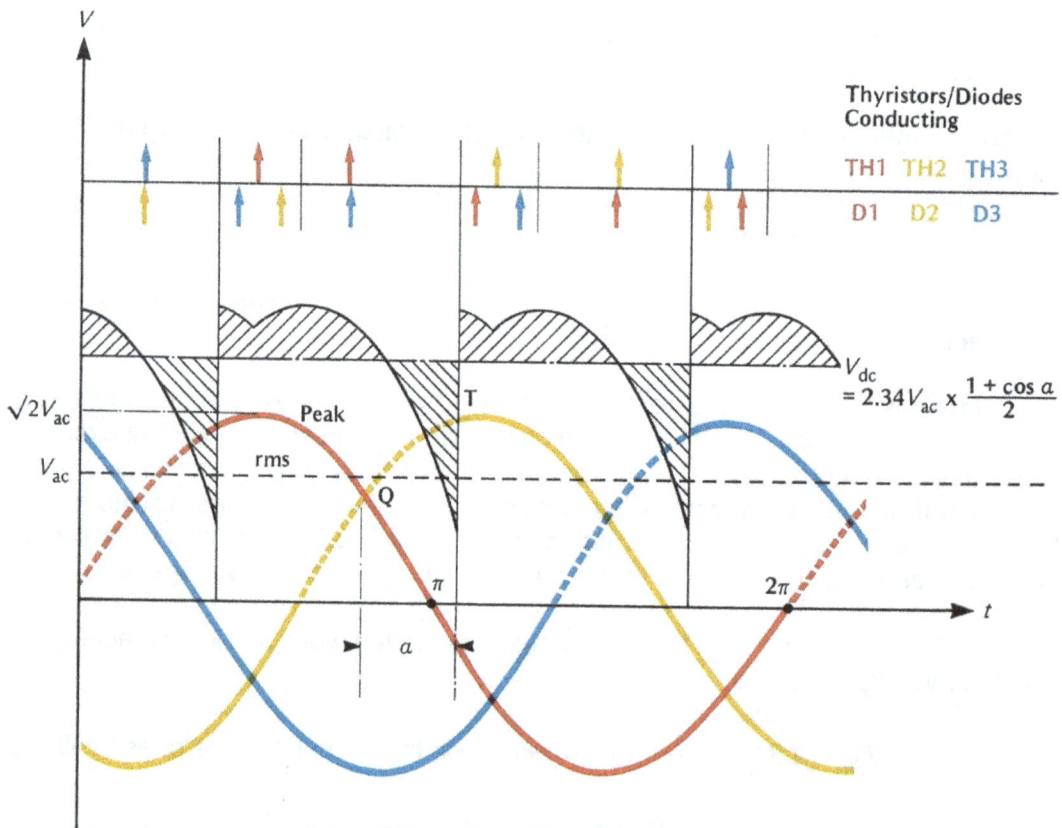

V

Thyristors/Diodes
Conducting

TH1 TH2 TH3

D1 D2 D3

$\sqrt{2}V_{ac}$

Peak

T

rms

V_{ac}

Q

V_{dc}
$= 2.34 V_{ac} \times \dfrac{1 + \cos a}{2}$

π

2π

t

a

(b) VOLTAGE WAVEFORMS

FIGURE 4.5
HALF-CONTROLLED RECTIFIER

As with the uncontrolled case, the multiplying factor 1.35 is valid only on open circuit or with a purely resistive load. When inductance is present the factor will be lower.

Each diode and thyristor element is protected by a close-rated fuse, not shown on the diagram.

4.3.3 Fully Controlled Rectifier

The rectifier circuit in Figure 4.5(a) was 'half-controlled'. By replacing all six diodes by suitably pulsed thyristors, the bridge becomes 'fully controlled'.

The circuit diagram of a fully controlled three-phase bridge rectifier and its voltage wave-forms are shown in Figures 4.6(a) and 4.6(b) respectively. In this case the commutation of **both** half-cycles of a.c. voltage is delayed by α electrical degrees after the natural commutation points. The mean d.c. output, neglecting diode and thyristor voltage drops, is then:

$$V_{dc} = 2.34 V_{ac} \times \cos \alpha \quad \text{(where } V_{ac} \text{ is the secondary rms phase voltage)}$$

or $\qquad\qquad V_{dc} = 1.35 V_L \times \cos \alpha \quad$ (where V_L is the secondary rms line voltage)

since $\qquad\qquad V_{ac} = \frac{1}{\sqrt{3}} V_L$

As in the half-controlled rectifier case, the multiplying factor depends on α, but in this case directly on $\cos \alpha$. When α is zero, $\cos \alpha$ is unity, but as α is increased it becomes smaller. Therefore the d.c. level is controlled by varying the delay angle α. It gives a wider range of control than with a half-controlled rectifier.

As in the uncontrolled and half-controlled case, the multiplying factor 1.35 is valid only on open circuit or with a purely resistive load. When inductance is present the factor will be lower.

Each thyristor element is protected by a close-rated fuse, not shown on the diagram.

By far the largest fully controlled rectifiers in offshore installations are the 'SCR Units' which provide controllable d.c. power to the drilling drives.

4.4 RIPPLE AND HARMONICS FROM A RECTIFIER

4.4.1 D.C. Ripple

Examination of the V_{dc} curves in Figures 4.4, 4.5 and 4.6 shows that the output d.c. voltage is not smooth (though it has a steady mean value) but is an uneven curve. In the single-phase full-wave cases of Figures 4.2 and 4.3 there are two d.c. peaks per full cycle of a.c., and in the three-phase cases of Figures 4.4, 4.5 and 4.6 there are six per full cycle.

These peaks, superimposed on the mean d.c. output voltage, give the effect of a d.c. 'ripple'. If such ripple cannot be accepted in the output circuit it must be smoothed out by the use of a low-pass filter of chokes and capacitors.

Figures 4.5 and 4.6 show clearly that the greater the delay angle α, the deeper the 'pits' in the rectified d.c. wave and therefore the greater the amplitude of the ripple.

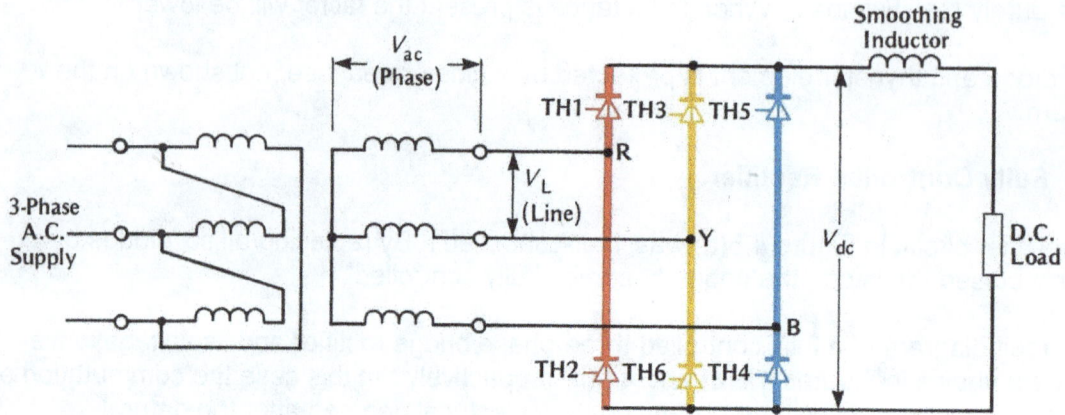

(a)   TRANSFORMER RECTIFIER CIRCUIT

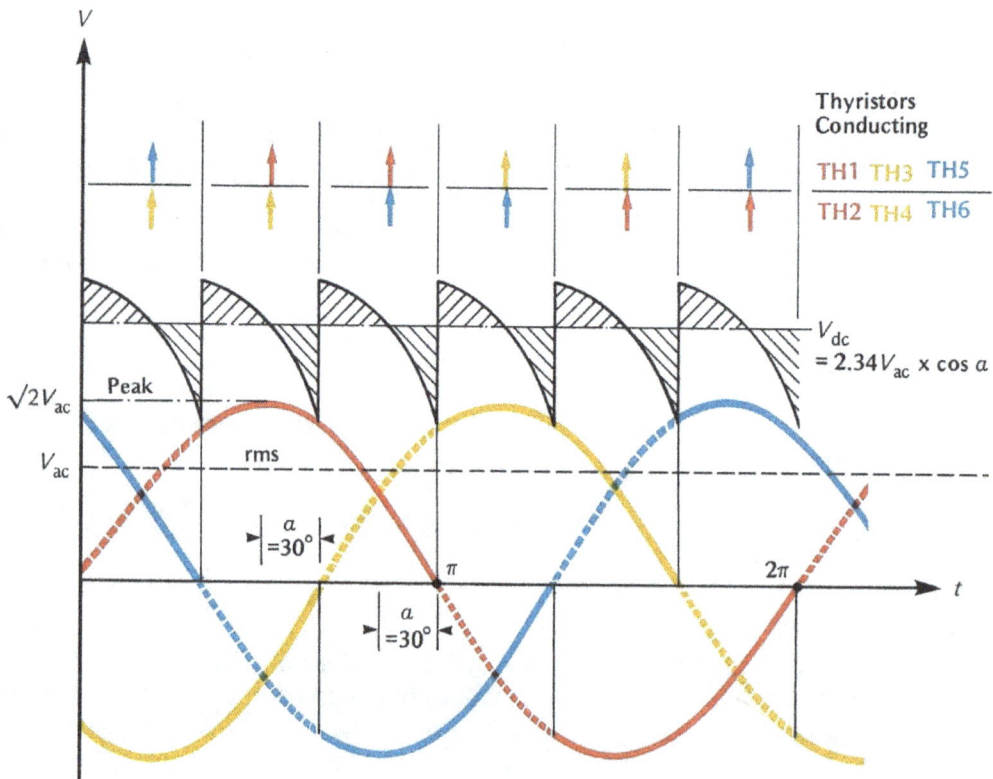

(b)   VOLTAGE WAVEFORMS

FIGURE 4.6
FULLY CONTROLLED BRIDGE RECTIFIER

The magnitude of the d.c. ripple in rms terms expressed as a percentage of the mean d.c. voltage is as follows:

Single-phase uncontrolled and controlled ($\alpha = 0$)   48%

Three-phase uncontrolled, half-controlled ($\alpha = 0$) and
fully controlled ($\alpha = 0$)   5%

For controlled values of α greater than zero the percentage ripple is considerably increased.

From these figures the advantage in the restricting of ripple by going to a three-phase rectifier is obvious.

4.4.2 Harmonics

The current drawn in each phase from the a.c. mains as each diode or thyristor starts to conduct is in a series of pulses. The current waveform is therefore far from sinusoidal, and it appears as a fundamental sine wave at system frequency together with a train of odd-numbered harmonics. The generation of harmonics is one of the major problems in the use of rectifier equipment.

Only the odd-numbered harmonic currents appear, and of these the 3rd, 9th, 15th etc. cannot flow in a balanced 3-phase circuit. This leaves the 5th, 7th, 11th, 13th, 17th, 19th harmonic currents flowing right through from generator to rectifier. Their individual amplitudes relative to the fundamental are inversely proportional to the harmonic order number. In practice these amplitudes are reduced somewhat by the supply and transformer reactance, especially those of the higher-order harmonics.

The effect of harmonic currents on a generator is fully explained in Part 2 'Electrical Power Generation'. In brief, the harmonic currents, in flowing through the reactance of the generator windings, give rise to internal voltage drops at harmonic frequencies and so distort the generator's output voltage waveform. This is often not acceptable to other consumers on the same system.

The harmonic currents due to rectifiers, expressed in rms terms as a percentage of a generator's full-load rated current, is as follows for a typical offshore main generator where the rectified load forms 50% of the generator's rated load:

	5th	7th	11th	13th	17th	19th harmonic
Three-phase, uncontrolled, half-controlled and fully-controlled	11%	8%	5%	4%	3%	3%

Increasing the firing delay angle α has no effect on the magnitude of the harmonics, although the supply power factor is reduced.

With a view to reducing harmonics and their distorting effect on the supply voltage, and to meet the stringent limitations imposed by Area Board supply authorities, large rectifier systems onshore are often designed as 6-phase, 12-phase or even 24-phase units.

4.5   INVERTERS

4.5.1 General

Inverters are of two types: the 'Synchronous Inverter' and the 'Free Inverter'.

A synchronous inverter converts d.c. to a.c. power, which it returns to the existing a.c. system. The inverted a.c. is at the same voltage and frequency as the main a.c. power and is in synchronism with it. The inversion is carried out by the same rectifier equipment as provides the d.c. power, but it operates in reverse by controlling the firing angle α nearer to 180°. The equipment is thus a combined rectifier or inverter unit according to how it is used (sometimes also called a 'convertor'). This form of inversion is used mainly in large d.c. systems such as railway traction, pit winding gear and rolling mill drives, where a large amount of braking power is recovered from the d.c. system and returned to the a.c. power source. Synchronous inverters are not used offshore or in many onshore installations and are not further discussed here.

A free inverter converts d.c. to supply a limited passive a.c. consumer system and does not return power to the mains. It cannot be used in reverse as a rectifier. It can be designed to run at any desired voltage and frequency. It is widely used offshore to provide battery-supported a.c. supplies for essential equipment. The following paragraphs deal exclusively with the free static inverter.

4.5.2 The Free Inverter

Using thyristors it is possible to provide an a.c. supply from a d.c. source with purely static equipment.

The principle of a simple static inverter giving a single-phase a.c. output is illustrated in Figure 4.7. Only the thyristor circuits themselves are shown; separate control circuits for triggering the thyristors in turn at the desired frequency are also required.

When thyristors TH1 and TH4 are triggered together they both conduct, and a current I_T flows through the transformer primary winding in the direction of the red arrow in Figure 4.7(a). At the next triggering TH3 and TH2 conduct and TH1 and TH4 are turned off, effectively reversing the transformer primary current, shown by the blue arrow. This is repeated continuously, so that the transformer primary is made to carry forward and reverse currents alternately.

Although the voltage applied to the transformer primary approximates to a square wave as Figure 4.7(b), the current waveform is modified by inductance in the circuit as in Figure 4.8(c). The output transformer itself and further waveshaping circuits (not shown in the diagram) give the output from the transformer secondary a voltage waveshape that approaches sinusoidal. It will nevertheless not be a pure sine-wave and will contain harmonics.

The frequency of the a.c. output of the static inverter is controlled by the rate at which the thyristors are fired.

A suppression diode across each thyristor reduces transient voltages due to the switching action of the thyristor.

Adding a third arm to the thyristor bridge and providing a three-phase output transformer makes it into a three-phase inverter. The six thyristors are fired by the control circuit at the appropriate intervals.

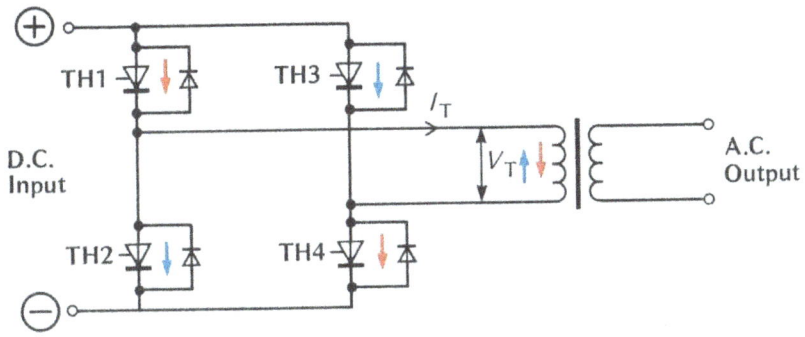

(a)   INVERTER CIRCUIT

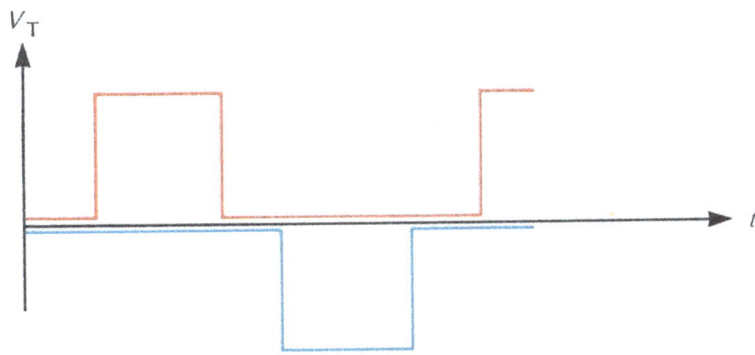

(b)   VOLTAGE WAVEFORM
ACROSS TRANSFORMER PRIMARY

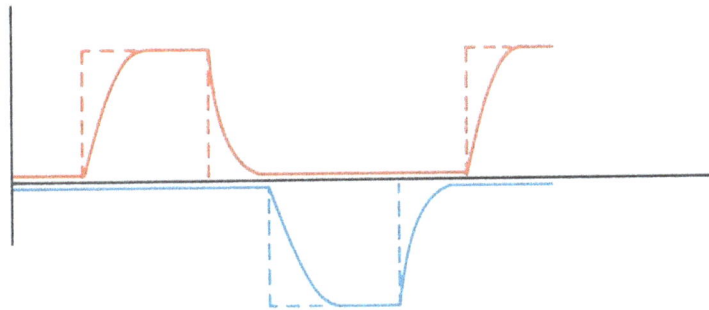

(c)   CURRENT WAVEFORM
IN TRANSFORMER PRIMARY

FIGURE 4.7
PRINCIPLE OF THE STATIC INVERTER

As with rectifier bridges, each thyristor element is protected by a close-rated fuse, not shown on the diagram.

4.6   A.C. THYRISTOR CIRCUITS

Reference has already been made above to the use of thyristors in inverse parallel pairs in a.c. circuits, and to the alternative bidirectional triac (Figure 3.4). Two common applications are:

(a)   **Power Regulators**. Triggering the thyristors at an intermediate point in each supply half-cycle, as in the controlled rectifier, produces the kind of wave-form shown in Figure 4.8: varying the angle of delay α from 180° to zero causes a variation in rms output voltage from zero to the full supply voltage, less the small voltage drop across the thyristors. Regulators of this type are mainly applicable to heating and lighting loads and to small a.c. motors.

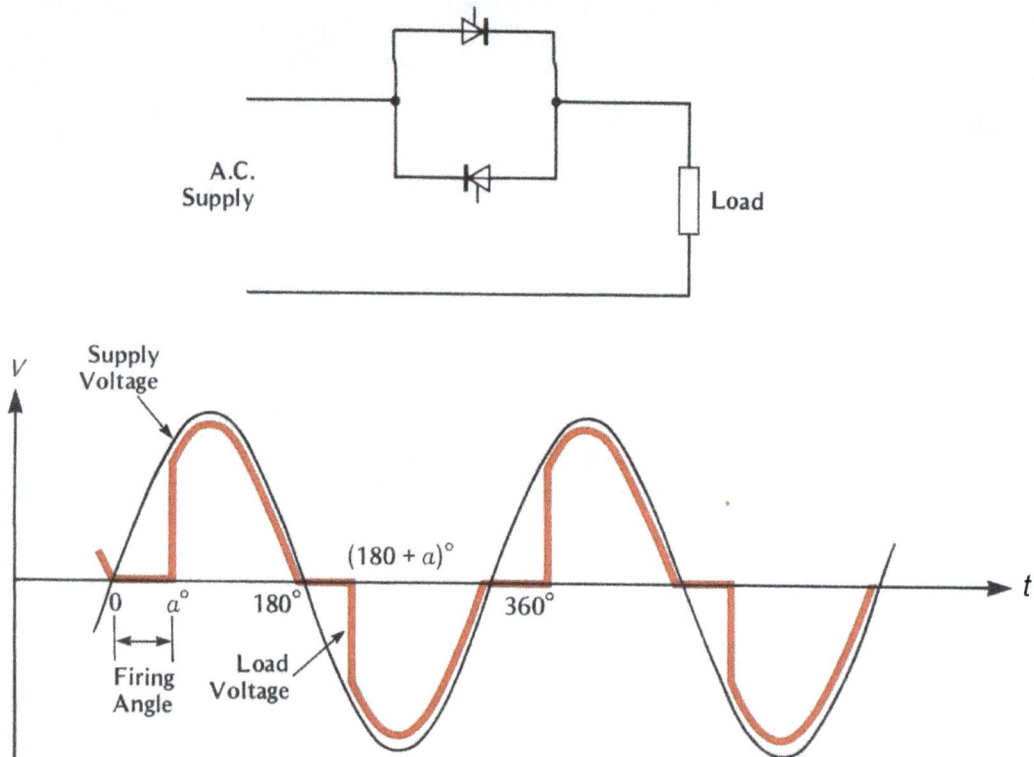

FIGURE 4.8
THYRISTOR A.C. REGULATOR

(b)   **Static Switches**. If delay-angle control is not used, the thyristors are either conducting continuously or blocking, according to whether triggering signals are applied or not; the thyristors then function as a.c. switches rather than as regulators. For switching in power circuits they have a number of attractions in comparison to contactors: they have an indefinite life, they can operate without generating radio interference or mechanical disturbance, and they can switch with a time delay of less than one half-cycle of the supply. The switching speed, in particular, makes them specially useful for switching in a.c. no-break systems (see Part 3 'Electrical Power Distribution').

4.7   HARMONICS, RADIO INTERFERENCE AND SUPPLY LOADING

Because of the rapid periodic switching in an a.c. regulator on which their operation depends, thyristor circuits, and to some extent also diode circuits, tend to generate harmonics which may, depending on circumstances, extend in frequency up to the radio frequency range. At the lower frequencies the harmonics may reach amplitudes comparable with that of the fundamental. This often necessitates considerable care in introducing thyristor equipment into a power system, paying particular attention to such factors as:

- radio frequency interference (RFI) (more recently referred to as 'electromagnetic contamination'), possibly requiring RFI suppression to prevent interference with communications;

- supply loading, which may be higher than expected because of the high current form-factor and low power-factor;

- supply voltage waveform distortion, possibly leading to instrument errors, harmonic currents in capacitors and other problems. Also the voltage wave-form offered to other consumers may be outside the permitted limits;

- with distributed single-phase loads, harmonic currents in neutral conductors, sometimes causing spurious 'earth fault' tripping.

Various control techniques have been devised to reduce harmonic problems, an example being the a.c. regulator widely used in industry for heating loads, in which the load is switched on and off periodically at a frequency of about 1Hz or less, the switching being effected entirely at voltage zeros ('on') and current zeros ('off').

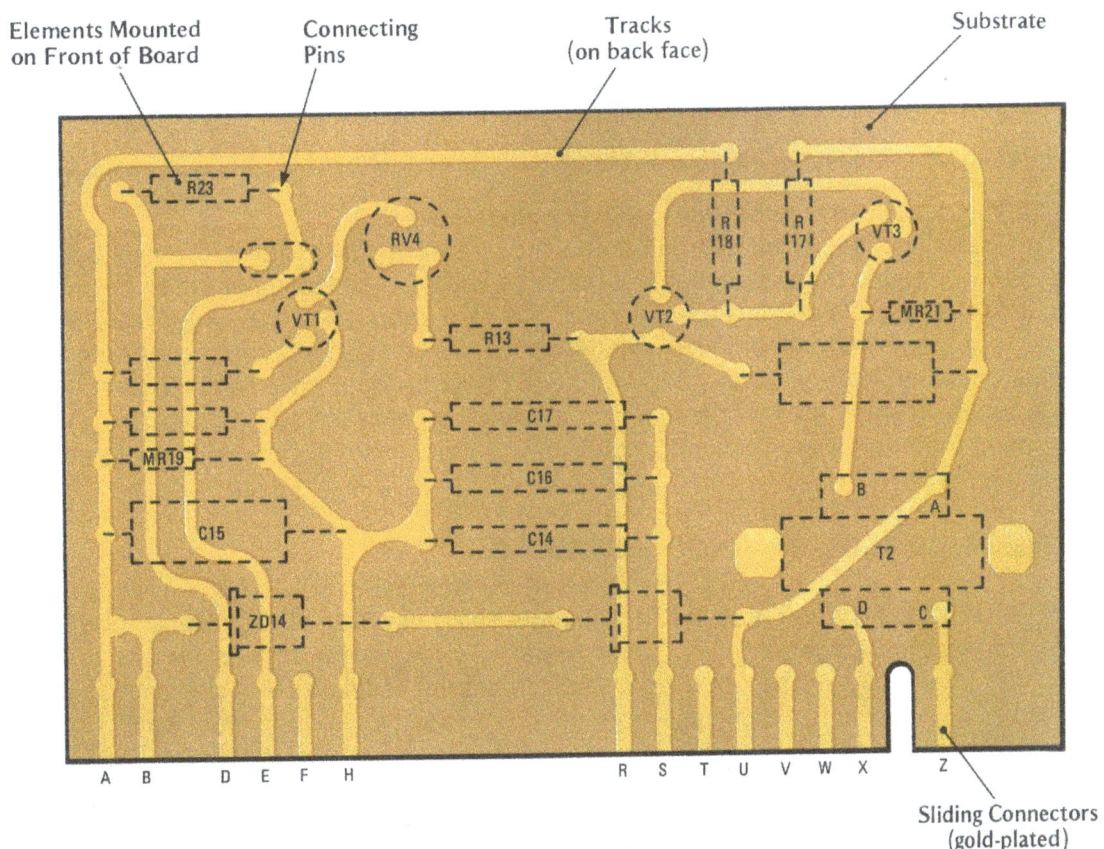

FIGURE 4.9
PRINTED CIRCUIT BOARD

4.8   PRINTED CIRCUIT BOARDS

The great majority of electronic assemblies at the present time are on printed-wiring boards, more often referred to (particularly when the components are mounted on them) as Printed Circuit Boards (PCB). They may be rigidly mounted in an equipment or accommodated in a rack of uniformly sized boards.

Printed circuit boards replace the former wiring between individual elements by 'tracks' of conducting metal, usually copper, arranged so as to avoid crossovers. The tracks are made by an etching process on the board's insulating substrate. Connections are made between the tracks and the individual elements mounted on the board by soldered pins. In Figure 4.9 the tracks, though shown uppermost, are in fact on the back of the board and the elements on the front.

External connections to the board are usually made by sliding contacts on one edge, allowing the board to be pushed in to corresponding contacts on the fixed assembly. This allows rapid replacement of a board. In some applications these contacts are gold-plated to resist corrosion and consequent poor contact.

In large assemblies where there may be many printed circuit boards of identical construction, the outer edges may be coloured to distinguish one type of board from another. Spare boards of each type would be carried.

Printed circuit boards are used, in preference to assembling circuits with connecting wire, for a number of reasons, including the following:

- they offer a cheap and satisfactory method of combined mechanical and electrical assembly;

- in repeated production the connections are guaranteed to be correct and consistent;

- they lend themselves to automatic assembly and soldering;

- inspection and testing is relatively easy.

There is an increasing tendency to treat printed circuit boards as entities in themselves, to be replaced when found to be defective but not to be repaired, unless later in the manufacturer's works or a suitably equipped service workshop. This makes for rapid repair of even complicated electronic or electronically controlled apparatus, which could be out of commission for inconveniently long periods if it were necessary to trace faults to their ultimate sources. 'Repair by replacement' is facilitated by using the plug-and-socket connections to the boards as seen in the figure, and by incorporating checking systems into the equipment.

4.9   SIGNAL DEVICES, CONTROL SYSTEM

A signal device is by definition a device for the processing of electrical signals rather than the control of electrical power, but it is not fundamentally different from a power device. Rather, it falls into a different part of a continuous spectrum of ratings and characteristics, is designed with a view to the particular properties of value in low-power electronic systems, whether in linear or switching modes, and is suited mechanically to the requirements of electronic assemblies.

The circuit applications of semiconductor signal devices are too many and diverse to be considered here in detail. However, a useful division can be made between two kinds of application which are broadly speaking of equal importance and may be found in close association within a particular circuit - namely, linear and digital. These two modes of use have already been mentioned in connection with power devices, with slightly different significance.

In control electronics linear operation is concerned with analogue systems and with continuously varying quantities. A simple example is an amplifier which produces an output proportional to its input signal; a slightly more complicated one is the electronic overcurrent relay illustrated in Figure 2.10. The term 'linear' is not always to be taken literally in this context; in general use it would embrace, for example, a logarithmic amplifier or a multiplier.

Digital circuits on the other hand operate in a switching mode. Any element of the circuit is considered to have only two levels of output and two levels of input - 'on' and 'off', or 'high' and 'low' or, more often than not, labelled '1' and '0'. Operations in this mode are referred to as 'binary', and numerical quantities are processed in binary arithmetic - that is, in terms of the number base 2 instead of the base 10 more familiar in ordinary decimal arithmetic. The principal operations and functions which feature in digital control electronics are as follows:

- counting;

- memory;

- arithmetical operations;

- logic.

The last item might be extended to include any kind of manipulation of information that could be carried out in a microcomputer, and in fact it is quite common in some circumstances to include a standard computer as part of a complex control system.

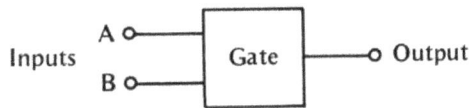

INPUTS		OUTPUT			
A	B	OR	AND	NOR	NAND (= NOT AND)
0	0	0	0	1	1
1	0	1	0	0	1
0	1	1	0	0	1
1	1	1	1	0	0

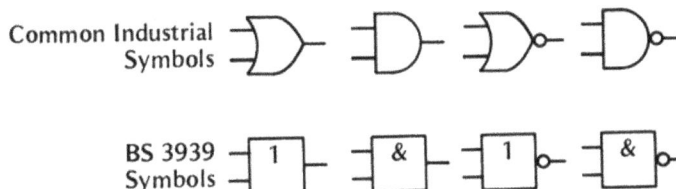

FIGURE 4.10
COMMON LOGIC GATES

To illustrate the basis of digital logic in its restricted sense, Figure 4.10 shows the reactions of four common types of logic gate - OR, NOR, AND and NAND (= 'NOT AND') - to two inputs, each of which may be high (i) or low (0). ('Gate' is the term used to denote a single stage of a logic system such as one of the four gates just mentioned.) For each gate the three columns indicating the two inputs and the resulting output constitute a 'truth table'; in the OR gate, for example, the output is '1' when input A is '1', or input B is '1', or both inputs are '1', but it is '0' when neither input is '1 '.

For comparison, Figure 4.11 shows how the same four basic logic functions are performed by relays.

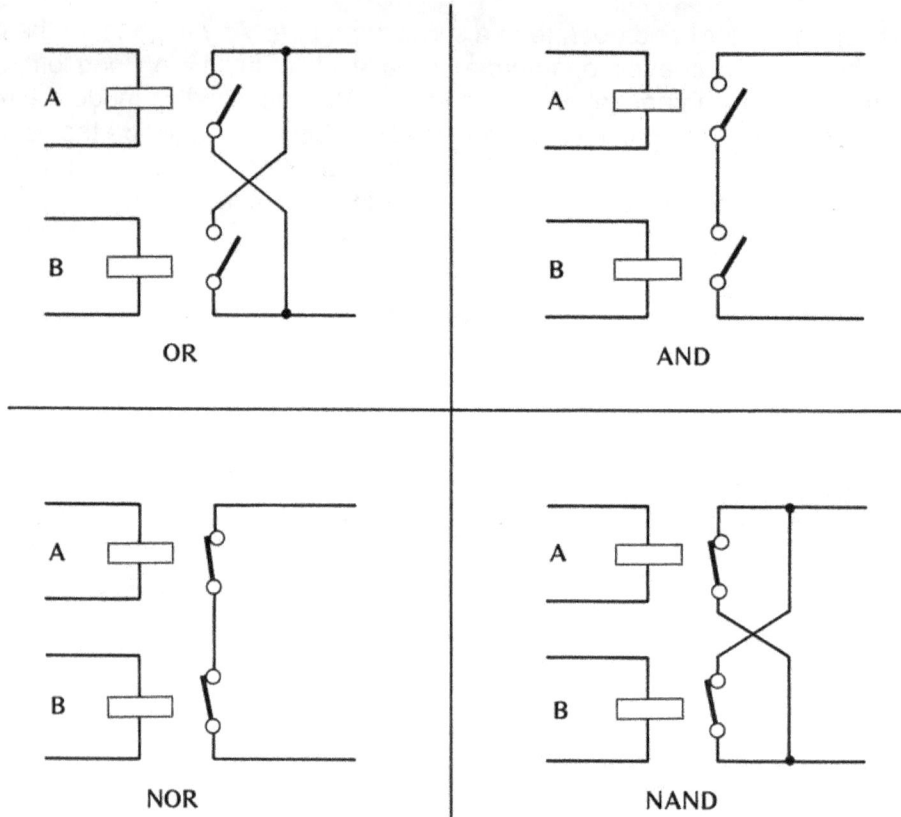

FIGURE 4.11
RELAY LOGIC CIRCUITS CORRESPONDING TO FIGURE 4.10

The fact that a control system is digital does not mean that it cannot deal with varying quantities. They can be processed very effectively by sampling - that is, taking their values at very short intervals - and by numerical representation, which can be made as accurate as desired by using the necessary number of digits. The only limitation in performance in this respect is, ultimately, the time required for the computations entailed, and consequently in the speed of response.

Digital systems are increasingly preferred to analogue for a number of reasons - principally because they are more accurate, more reliable and less susceptible to interference. The main disadvantage of digital systems (apart from limited speed of operation) is their complexity and hence their cost; this has, however, been overcome to a very large extent by the development of integrated circuits (see below) and by the very economical large-scale production of components. A control system does not have to be wholly digital or wholly analogue: 'digital' systems include, more often than not, analogue sections, particularly where it is necessary to deal with analogue inputs and outputs, coupled by means of analogue-to-digital or digital-to-analogue converters.

4.10   INTEGRATED CIRCUITS

The development of more complex and more accurate ways of fabricating semiconductor components has made it possible to combine a number of components, not necessarily similar, into a single very small element called a 'chip'. It is provided with interconnections, so forming a complete functional circuit or group of circuits; it is called an 'integrated circuit'.

The process is achieved by a sequence of photo-etchings of the basic chip. For mass-production purposes the process is carried out on thousands of identical circuits on a large silicon sheet, later broken up into individual integrated circuits.

Such circuits may be analogue (amplifiers, multipliers etc.) or digital (memory modules, logic gates etc.), or in some cases both, and may combine power and signal functions as in the case of a d.c. voltage regulator. The scale of integration can be anything from a very small grouping to the fabrication of large functional blocks (large-scale integration, or LSI) or even complete systems of great complexity (very large-scale integration, or VLSI).

The circuit **diagram** of an integrated circuit is no different from that of a similar circuit formed from separate components, but the integrated circuit itself occupies only a minute amount of space. Indeed it is the connections to its elements which determine the size of the unit, not the elements themselves.

Figure 4.12(a) shows a typical assembly. All the elements such as resistors, capacitors diodes, thyristors, etc. are embodied in the very small central chip itself (coloured blue on Figure 4.12(a) and only a few millimetres across). The chip is sometimes housed in a metal can, but the many internal connections to the various parts of the circuit must be brought out so that external connection may be made to them, either directly or on a printed circuit board. These internal connections can be seen in Figure 4.12(a).

Another form of construction is to embody the chip in a flat plastic or ceramic package, as shown typically in Figure 4.12(b). In this design the internal connections are brought out to two rows of connecting pins; this form is consequently known as 'dual-in-line package' (DIP).

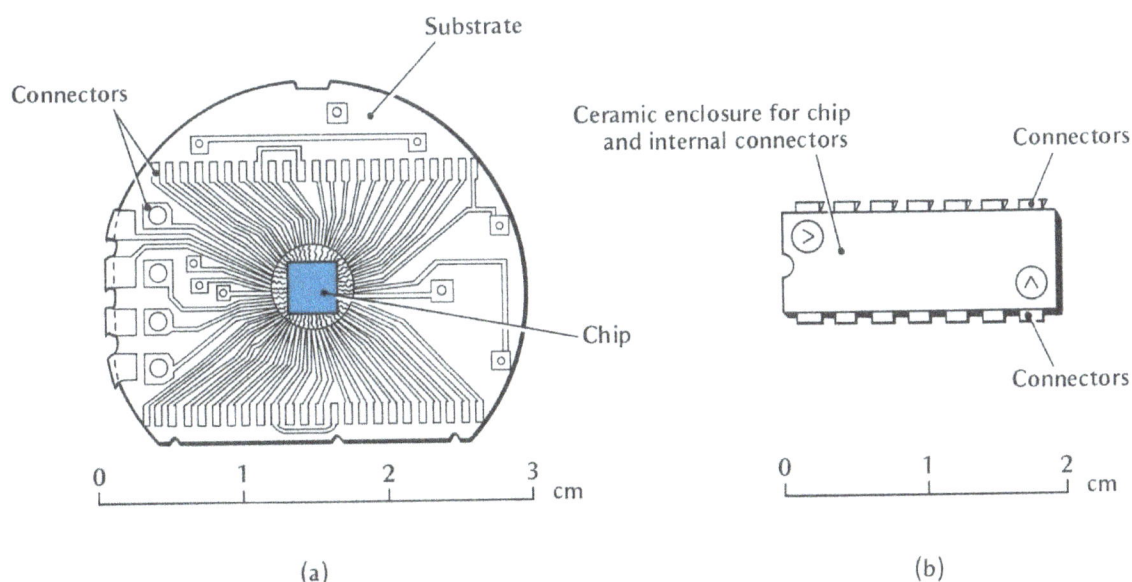

(a)   (b)

FIGURE 4.12
TYPICAL INTEGRATED CIRCUIT

The currents handled by integrated circuits are usually minute, being measured in 'nano-amperes' (nA), equal to 10^{-9} amperes.

Integrated circuits do not incorporate any devices that are not known, at least in principle, as discrete components, but their introduction has made it possible in very many cases to employ techniques that would be impracticable or uneconomic with discrete components. Specifically, they offer the following advantages:

- **Low Cost**. In principle, given a sufficiently large scale of production, quite a complex integrated circuit costs little more to produce than a discrete semi-conductor device, and there are big savings in assembly costs.

- **Reliability**. Whereas the probability of failure in a system increases in proportion to the number of components, an integrated circuit provides tens, hundreds or even thousands of components with a reliability of a similar order to that of a single discrete component. It can thus facilitate a very great improvement in the reliability of a complex system or, conversely, a great increase in complexity without sacrificing reliability.

- **Performance**. Several aspects of performance are improved by integration, in particular because of the uniformity of temperature within the very small silicon chip. Also the uniformity of manufacture of all the components of the circuit, and the compactness of the circuit reduce its sensitivity to external electric and magnetic fields and minimises unwanted inductances and capacitances.

- **Size**. Many equipment designs would be unacceptable simply from the point of view of overall size without the benefit of integrated circuits.

There are other methods for reducing the size and cost of control systems, notably thick-film and thin-film circuits, whereby resistors and interconnections are produced with the aid of small-scale printing techniques. These can be used in conjunction with integrated circuits in so-called 'hybrid circuits'.

One of the most significant developments which has been made possible by integrated-circuit techniques is that of the microprocessor, which has made true computing power available at relatively very low cost.

CHAPTER 5 OTHER SEMICONDUCTOR DEVICES

There are in total a large number of components which exploit various properties of semi-conductors. Detailed discussion of all these is not warranted here, but the following are some of the more important.

5.1 FIELD EFFECT TRANSISTORS (FET)

Unlike the bipolar devices mentioned above, whose operation depends on the properties of junctions between p-type and n-type silicon, a field-effect device has an element of one type of silicon only - either p or n - and its conduction is modulated by an electric field produced by a control electrode which in most cases is insulated from it. Corresponding to the emitter, base and collector of the bipolar transistor, the electrodes of the FET are respectively the source, the gate and the drain; the fact that the gate is insulated results in a very high input resistance, which is one of the main distinguishing features of the FET (see Figure 5.1).

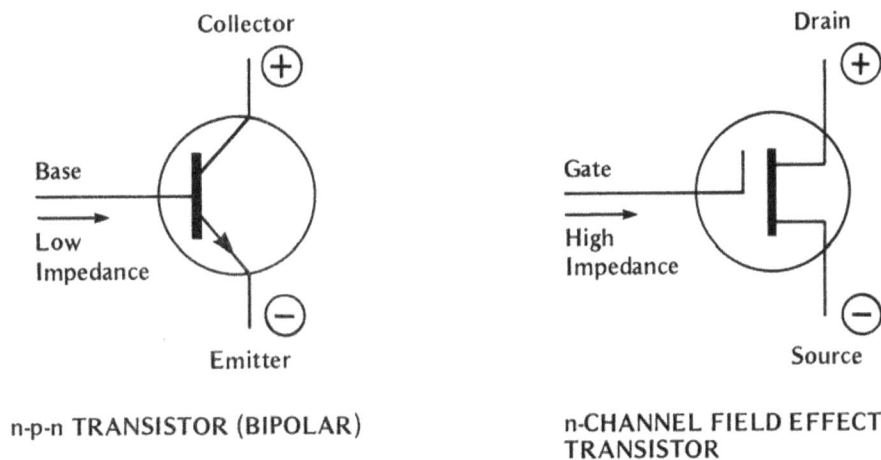

n-p-n TRANSISTOR (BIPOLAR)   n-CHANNEL FIELD EFFECT TRANSISTOR

FIGURE 5.1
COMPARISON OF BIPOLAR AND FIELD EFFECT TRANSISTORS

Insulated-gate field effect transistors (IGFET) are also referred to as 'metal-oxide-silicon' (MOS), and they are further sub-divided according to the method of incorporating the gate insulation. 'NMOS' and 'PMOS' indicate the use of n-type and p-type material respectively, and 'CMOS' integrated circuits incorporate both types in 'complementary MOS' stages. These terms are used as prefixes to FET, such as NMOSFET.

CMOS circuits are particularly advantageous in digital systems because of their very low power consumption and high noise-immunity - that is, low susceptibility to electrical interference - and have assumed great importance in computers and low-power control systems.

Field-effect power transistors ('Power MOSFET') also have some useful advantages, notably their very fast switching characteristics, which make them attractive for many small power supplies and inverters.

Because of their very-high-resistance gates, MOS devices tend to be sensitive to electrostatic charge and can easily be damaged by handling without appropriate precautions.

5.2 VOLTAGE REGULATOR AND VOLTAGE REFERENCE DIODES

For the purposes of discussing rectifier diodes above it has been assumed simply that a p-n junction subjected to a reverse voltage passes virtually no current. Its reverse characteristic however, if extended to its limits, is actually of the form shown in Figure 5.2(a).

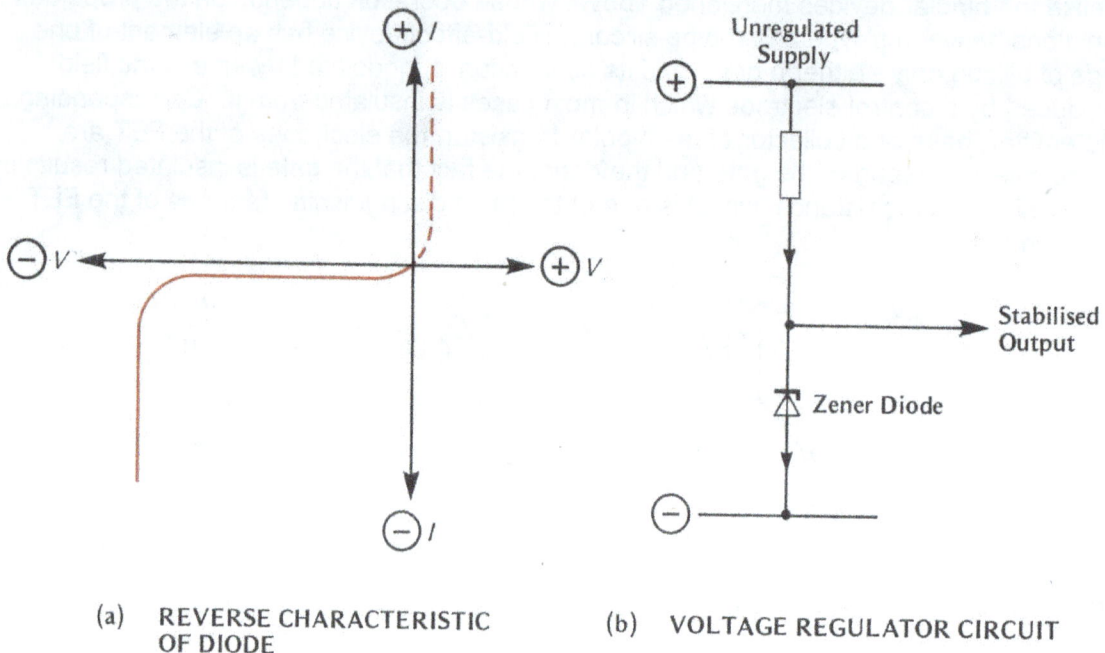

(a)   REVERSE CHARACTERISTIC OF DIODE

(b)   VOLTAGE REGULATOR CIRCUIT

FIGURE 5.2
VOLTAGE REGULATOR CHARACTERISTIC AND CIRCUIT

This shows a very small leakage current at low reverse voltages, changing sharply to a region where a small increase in reverse voltage results in a very large increase in current.

The part of the characteristic which is used is that in the third quadrant, as shown in the figure, because it is the **reverse** characteristic in terms of the convention applied to rectifier diodes. In virtually all cases it is the only part of the complete characteristic that is used in a voltage-regulated diode. In fact, if the full characteristic were drawn, the forward part (in the first quadrant) would be as shown dotted in the figure and would be similar to that of an ordinary rectifier diode.

In a rectifier application the working voltage is normally kept below the level where the sharp change occurs. The effect, known as the 'avalanche' or 'Zener' effect depending on the level of voltage at which the change occurs, is widely used in special diodes ('Zener diodes') for the purposes of stabilising d.c. supplies to electronic circuits or of providing low reference voltages for voltage or current regulators.

Figure 5.2(b) shows a straightforward voltage-regulator circuit, capable of an accuracy of the order of ±1%. With a slightly more elaborate construction, voltage-reference diodes can achieve even better stability.

5.3   OPTO-ELECTRONIC DEVICES

Since light is a form of electromagnetic energy, it should not be surprising that semi-conductors (not only silicon) provide the means to generate electrical signals in response to light (in photodiodes and phototransistors) and also to produce light from an electrical power source (in light-emitting diodes). Common uses in modern electronic equipment are in photocouplers, which provide signal transmission without electrical connection, and also in visual displays. Photocouplers, for instance, are used in telephone circuits in hazardous areas, where the intrinsically safe speech and call-up circuit is coupled to, but isolated from, the power ringing circuit which is not intrinsically safe.

5.4   HALL EFFECT AND MAGNETORESISTOR DEVICES

The 'Hall-effect' is a three-dimensional phenomenon in which the interaction of an electric current and a magnetic flux within a plate of conducting material generates a small emf at right angles to both the flux and the current. For practical purposes the effect is significant only in semiconductors. It is appreciable in germanium and silicon, but the most useful results are obtained with certain compound semiconductors such as indium antimonide.

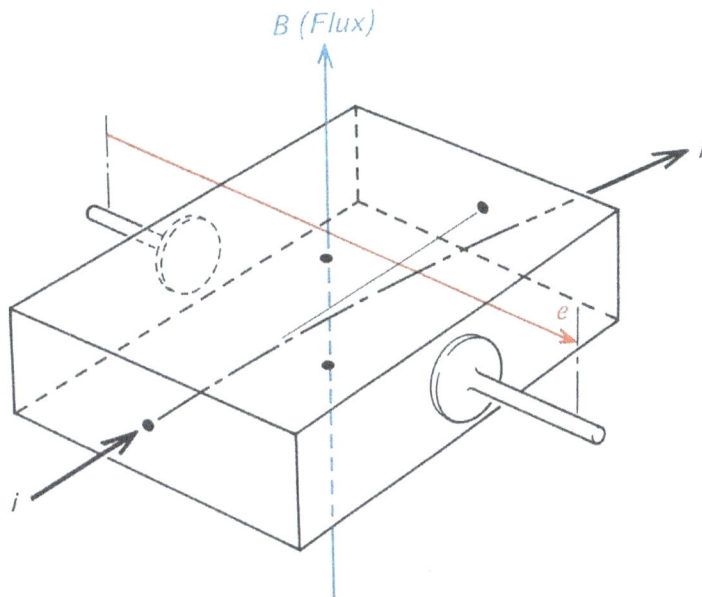

FIGURE 5.3
HALL EFFECT ELEMENT

The basic arrangement of a Hall element is illustrated in Figure 5.3. A magnetic flux B (shown blue) passes through the plate which is carrying a current i and cuts its surface at right angles. Electrons and holes flowing along the length of the plate, representing the current i, are deflected laterally by the flux in much the same way that a free conductor carrying a current would be deflected (in accordance with Fleming's left-hand rule for motors). The resulting distortion of the distribution of charge carriers gives rise to an emf e (shown red) - the 'Hall voltage' - which may be detected externally by electrodes on the surfaces. The Hall voltage is proportional to the current i and to the flux density B, and also to the 'Hall coefficient' R_h for that semiconductor material - that is to say,

$$e \propto R_h.i.B$$

If the voltage output electrodes are short-circuited, the Hall voltage drives a current across the plate. This in turn generates a Hall voltage in a direction opposite to that of the original current i, so that the apparent resistance of the plate is increased. A short-circuited Hall element thus constitutes a so-called 'magnetoresistor', where the resistance can be varied by controlling the magnetic flux B.

Hall elements and magnetoresistors are used, to a large extent interchangeably, mainly for the following purposes:

- in magnetometers, for measuring magnetic field strength - for example the earth's magnetic field and the air-gap flux density in electrical machines;

- in current measuring devices, by measuring the flux density produced by that current flowing in the winding of an electromagnet;

- as multipliers of analogue quantities, making use of the fact that the output depends on the product of current and flux;

- in transducers which detect mechanical displacement, for example by responding to a consequent change in magnetic flux due to that displacement.

5.5   PIEZO-ELECTRIC DEVICES

A further physical phenomenon is displayed by certain semiconductor materials when subjected to mechanical stress. When this occurs an emf is generated along an axis at right angles to the direction of stress and proportional to that stress. Such devices are called 'piezo-electric' after the Greek piezein, 'to press', and are in effect stress/emf transducers.

Only semiconductors of a crystalline nature exhibit this property. There are a large number of them, but natural quartz is by far the most common because of its ease of working and low cost.

Like all crystals, quartz has three 'axes', termed x, y and z. The z-axis is the optical axis running along the crystal; no piezo-electric effect is observed along this axis. A thin plate is cut across the crystal at right angles to the z-axis, and from this a further small chip is cut parallel to one pair of faces. This contains the x-axis and y-axis mutually at right angles, as shown in Figure 5.4(a).

Consider one group of ions, three positive and three negative which form one molecular group of the quartz crystal material. The three positive ions arrange themselves at the corners of an equilateral triangle, and the three negative ions at the corners of another equilateral triangle displaced 60° from the first, as shown in Figure 5.4(b). The 'centre of gravity' of the three positive ions is P, and that of the three negative ions is N. Under normal conditions point P coincides with N. If now a mechanical stress is applied along the y-axis, the resulting strain will distort the triangles into more elongated shapes, causing the centres of gravity of the ions to move outwards - N (negative) to the left and P (positive) to the right, as shown in Figure 5.4(c). Repeated throughout every group in the crystal, this will result in a net emf along the x-axis, positive to the right. That is to say, a mechanical stress applied along the y-axis causes an internal strain which produces an emf along the other axis of the plate.

Piezo-electric materials will also behave in the reverse mode. If a **voltage** is applied along one axis, mechanical strain will result along the other. In Figure 5.4(d), if an electric field is applied along the x-axis from right to left, the positive ions and their centre of gravity P will be attracted to the left and the negative ions, with N, to the right.

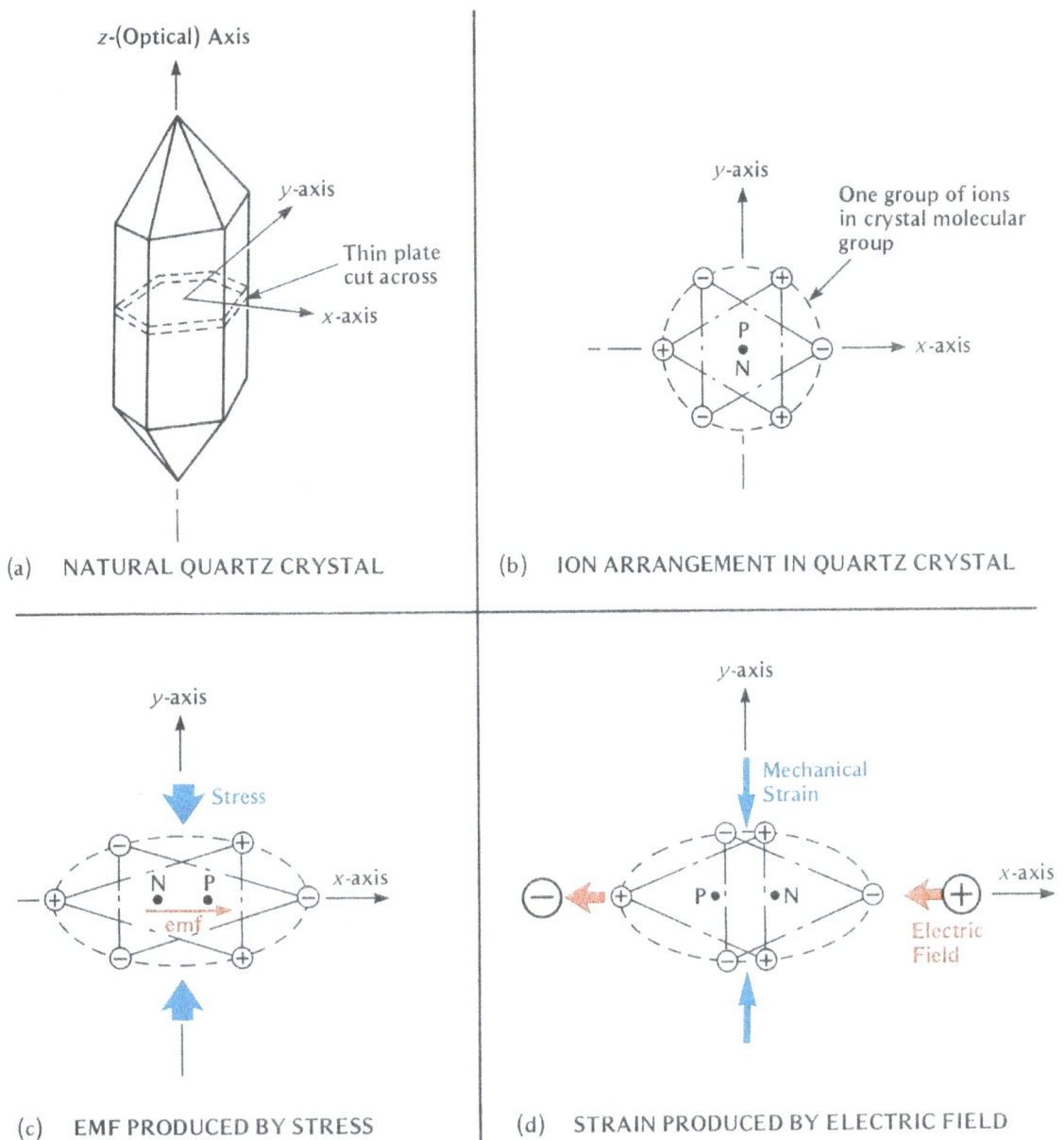

FIGURE 5.4
PIEZO-ELECTRIC CRYSTAL

The two triangles will move horizontally relative to each other but are constrained to remain in their group, which thus takes on an elliptical shape, as shown in Figure 5.4(d). The vertical height of the group is reduced, causing some shrinkage - or mechanical strain - along the y-axis. Thus an electric field applied along the x-axis will cause mechanical strain along the other axis of the plate. If the direction of the electric field is reversed, the two triangles are moved **towards** each other and the ellipse is distorted vertically and therefore stretched, causing the direction of mechanical strain also to be reversed. In particular, if an alternating field is applied, the mechanical strain alternates with it.

One particular application of this property is to arrange that the frequency of the applied electric field resonates with the natural mechanical frequency of the crystal itself. This is determined when the crystal is cut and cannot be altered. With such an arrangement an extremely accurate and stable frequency source is provided which has a great number of uses, particularly in radio tuning and instrumentation.

Applications of piezo-electric devices include:

(a)   As a stress/emf transducer:

- strain gauge
- pressure gauge
- hydrophone, microphone or record player pick-up.

(b)   As an emf/stress transducer:

- loudspeaker/hydrophone
- vibrator (fixed frequency source).

(c)   As a combined transducer:

- quartz clock
- frequency control of radio transmitters or receivers
- calibration.

CHAPTER 6 ADVANTAGES OF SEMICONDUCTOR DEVICES

In comparison with electromechanical devices, for which they may often be regarded as substitutes, semiconductor devices offer a number of advantages which are realised to a greater or lesser extent in practice, depending upon the circumstances. The principal features are as follows:

- **Longer Life**. Semiconductor devices, generally speaking, do not have a 'wear-out' mechanism such as mechanical devices have, so that they have an indefinite life without the need for (or the possibility of) periodic servicing. This is a particular advantage when a solid-state device is compared with a mechanical one which has to operate frequently, since the life of contacts, bearings etc., unlike that of semiconductor devices, is directly related to the number of operations performed.

- **Greater Reliability**. The reliability of mechanical devices is not always easy to assess, since it depends very much on the circumstances of use and the way in which reliability is defined. The reliability of semiconductor components and systems, based on the absence of any moving parts, can be much better in some applications, particularly in cases of great complexity and adverse environments.

- **Robustness**. Semiconductor devices are mechanically robust and are normally encapsulated in sealed housings which protect them against hostile environments and make them suitable for incorporation into any form of equipment construction.

- **Smaller Size** . This depends on the kind of function in question. Control systems, in particular, can be made much more compact by using appropriate semiconductor techniques, especially with the benefit of integrated circuits. On the other hand a contact rated at a given current is much smaller than a similarly rated thyristor switch.

- **Flexibility**. Semiconductor components and functional circuits can easily be arranged in different combinations and configurations to perform any number of desired operations - a process often facilitated by the availability of electronic modules with matching inputs and outputs. Indeed, a wide variety of control functions can be provided by completely standard hardware, using computer techniques, with all variations encompassed in the associated software; functions can even be changed extensively after installation without any modifications to the hardware.

Other benefits may become apparent in particular cases. Some particular disadvantages or limitations may also appear in some applications. For example, the static thyristor switch discussed above, while it has definite advantages from certain points of view, does not provide the same degree of isolation in the 'off' state as do physically separated contacts, and it dissipates heat in the 'on' state.

CHAPTER 7 TEMPERATURE CONTROL DEVICES

7.1 GENERAL

The sensing of temperature for both measurement and control purposes has many applications in power plant, especially in the windings of generators, motors and transformers.

Apart from the simple thermometer, three main methods are used:

- Thermocouple
- Resistance Temperature Device (RTD)
- Thermistor.

7.2 THE THERMOCOUPLE

The thermocouple makes use of the physical phenomenon known as the 'Thompson Effect'. This states that, when a conductor is subject to a temperature gradient along its length, the heat flow resulting will also cause a movement of electrons resulting in an emf along the conductor. This shows as a voltage appearing between its ends. The effect differs in degree with different conductor materials and depends on the magnitude of the heat gradient.

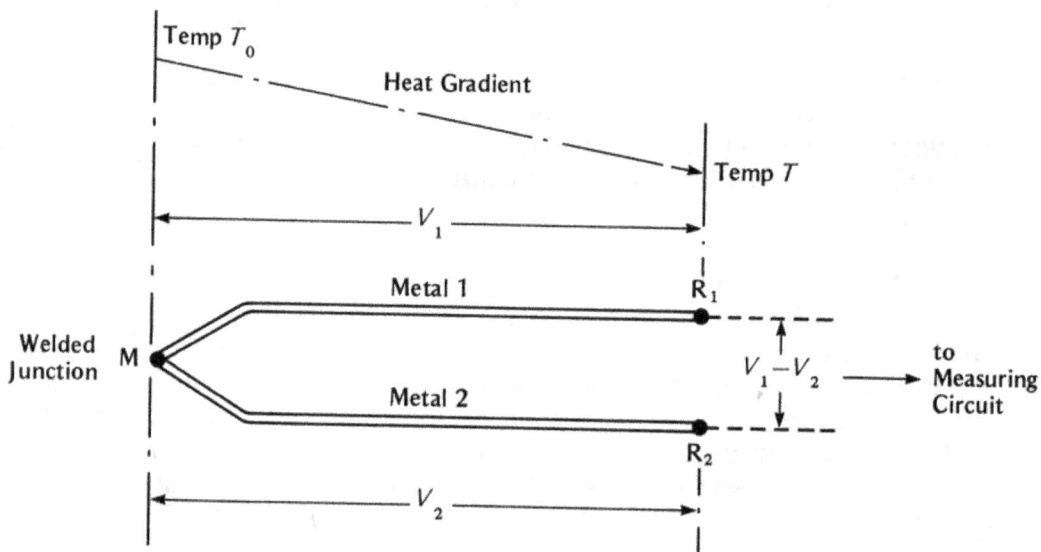

FIGURE 7.1
THERMOCOUPLE

Figure 7.1 depicts two wires joined at one end 'M' and with the other two ends R_1 and R_2 open. Both wires are subjected to a temperature gradient, the temperature being T_0 at the M-end and T at the two R ends. If both wires were made of the same material, the emf generated in each would be the same and there would be no net voltage difference between the ends R_1 and R_2. As however they are of different materials as shown in Figure 7.1, the emfs V_1 and V_2 generated by the same heat gradient in the two conductors will differ, and there will be a net voltage $(V_1 - V_2)$ between R_1 and R_2.

Since the magnitudes of the emfs depend on the heat gradient - and therefore on the difference of temperature $(T_0 - T_1)$ between the M and R junctions - the voltage is a measure of that temperature difference. If the temperature of the R ends is known, the voltage is a direct measure of the M-end temperature, though not necessarily linear. This is the principle of the thermocouple. The M end is referred to as the 'Measuring' junction and the R ends as the 'Reference' junction.

The voltage, amplified if necessary, gives a direct indication of temperature on a calibrated scale, or it can be used to operate an alarm or protective system if it reaches a certain pre-set value.

The thermocouple element is usually mounted in a tube with the M junction as nearly as possible in contact with the point whose temperature is to be measured - for example embedded in the windings of a machine. The R junction terminals are connected by normal or extension cable to the measuring unit.

The operation of a thermocouple is entirely passive and of itself requires no external power supply.

7.3   RESISTANCE TEMPERATURE DEVICE (RTD)

The RTD makes use of the general property of all conductors that their resistance increases with temperature. Some materials show this effect more strongly than others: copper, gold, nickel, platinum and silver have the smoothest and most stable property. Of these, platinum is generally preferred, as its resistance is six times that of copper and it has a wide temperature range.

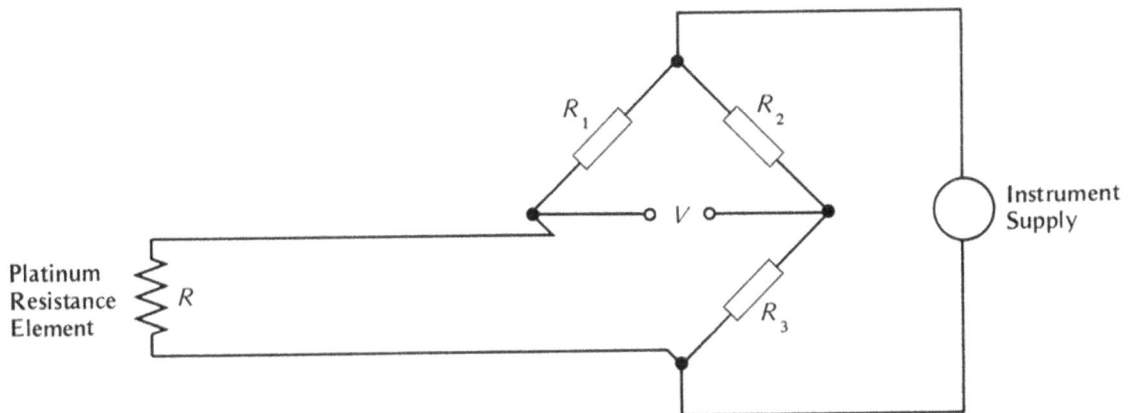

FIGURE 7.2
RESISTANCE TEMPERATURE DEVICE

In the RTD a small sensing element consisting of a fine coil of platinum wire is enclosed in a protective tube and placed near the point to be measured. Its resistance R is measured externally by a Wheatstone Bridge or similar method as shown in Figure 7.2.

Instead of balancing the bridge as is usually done, the unbalance voltage V is measured and operates an indicating instrument or an alarm or trip circuit. This voltage is then a measurement of the temperature of the sensor. In general it is not linear and must be calibrated.

One important difference between an RTD and thermocouple is that the RTD is not passive and requires a small external supply.

7.4   THE THERMISTOR

All solid-state devices are temperature-sensitive to varying degrees, and their resistance varies with temperature. A thermistor is such a device made with certain selected materials which exhibit this property to a marked degree.

There are two kinds of thermistor: the 'Negative Temperature Coefficient' (NTC) type whose resistance falls with increase of temperature. It is made from metallic oxides. The other is the 'Positive Temperature Coefficient' (PTC) type whose resistance rises sharply at the centre of its characteristic, although it exhibits a negative coefficient at the lower and upper ends. One common PTC material is barium titinate.

Figure 7.3 shows typical characteristics of both types, the NTC in blue and the PTC in red. It will be seen that the resistance of the NTC type falls smoothly and steadily, though not

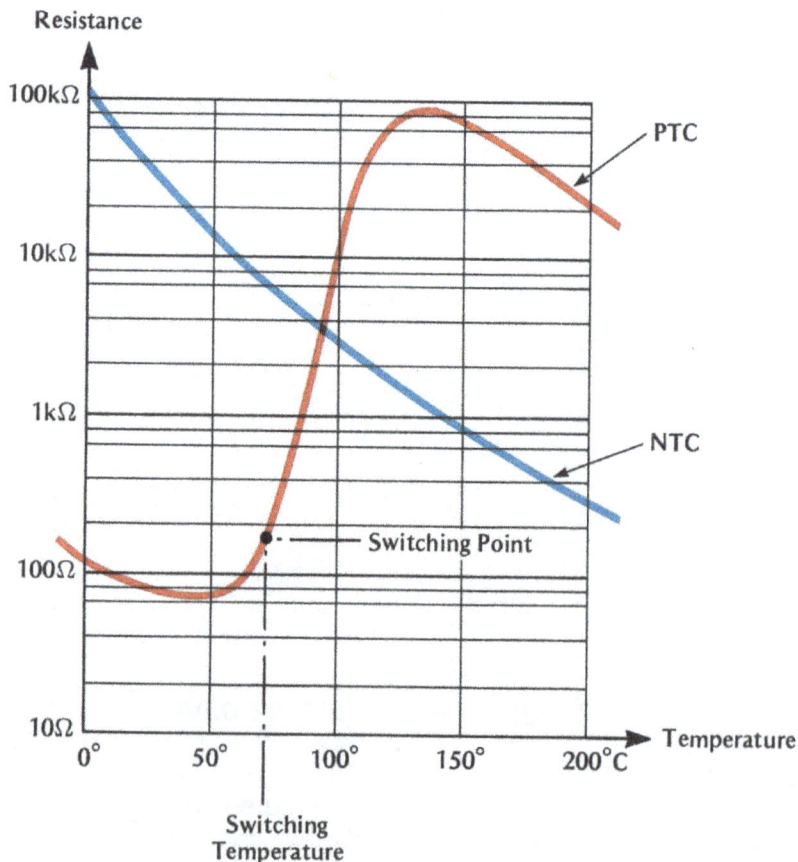

FIGURE 7.3
THERMISTORS

linearly, with increase of temperature. It lends itself therefore to continuous measurement of temperature by measuring the resistance with a bridge or similar means and calibrating. The resistance varies from 6 to 1% per °C, giving a range of resistance of some 300:1 between 0°C and 200°C in the example shown.

Use is made of the PTC for switching purposes - indeed the PTC is sometimes referred to as a 'switching thermistor'. In the example of Figure 7.3 the characteristic changes from negative to positive at about 50°C and begins to rise steeply. At its maximum gradient the resistance is changing some 70% per °C. This rapid increase of resistance over a few degrees of temperature is used to provide a switching signal for alarms or protective over-temperature circuits. It is customary to regard the point where the resistance has risen to twice its minimum value as the 'switching point', and the corresponding temperature as the 'switching temperature'. The purpose of choosing this point low on the rising characteristic is to allow for thermal lag. If the machine windings are heating up rapidly due, say, to a short-circuit, a finite time exists before the thermistor will heat up to the switching temperature. It is therefore prudent to make this point early on the characteristic.

Different solid-state materials have different characteristics, and they can be chosen so that the switching point is arranged to occur at different temperatures. For convenience three classes of PTC are recognised:

Class B   145°C switching temperature
Class F   165°C switching temperature
Class H   190°C switching temperature.

Like other temperature-sensing devices the thermistor can be embedded in the insulation of a machine winding. Since it employs a bridge resistance-measuring circuit it requires an external supply to operate.

They can be tested by disconnecting the thermistor leads at their outer ends and measuring their resistance with a low-voltage instrument. It should be approximately within the range 30 to 250 ohms.

CHAPTER 8 QUESTIONS AND ANSWERS

8.1   QUESTIONS

1.   What is the essential function of a relay, and how is a sensor different?

2.   What is the purpose of a transducer?

3.   What are the most common types of electromagnetic relay? How are they divided broadly in terms of characteristics?

4.   How should a relay with normally-open contacts be shown in a circuit diagram when it is (a) usually de-energised, (b) usually energised?

5.   What is the significance of an inverse-time characteristic in a protection relay?

6.   How are the operating current and time characteristics of an OCIT induction disc relay adjusted?

7.   How is the OCIT relay principle modified to produce a voltage-restrained over-current relay?

8.   What determines the torque in a wattmetric induction disc relay? What is its principal use in protection?

9.   What is the principal application of thermal relays?

10.   What are the fundamental purposes of control (auxiliary) relays?

11.   What are the three general categories of semiconductor device applications?

12.   How is a p-n junction formed, and what are the material properties that enable it to rectify?

13.   What are the two modes of operation of a transistor?

14.   How does a thyristor differ functionally from a transistor? What are its advantages?

15.   What factors necessitate the cooling of power semiconductor devices?

16.   How is the output voltage of a thyristor rectifier varied?

17.   If a single-phase bridge rectifier with two thyristors and two diodes is supplied at 40V a.c. rms and delivers a direct voltage (on load) of 24V, what, approximately, is its angle of delay?

18.   What are the uses of thyristors in a.c. circuits?

19.   What are the attractions of printed circuit boards in electronic equipment?

20.   Construct a truth table for four logic gates.

21.   In what circumstances is the long life of semiconductor devices particularly an advantage?

22. What are the particular merits of integrated circuits?

23. Name three types of temperature-sensitive device.

24. Explain briefly how each operates.

25. Sketch the characteristics of typical NTC and PTC thermistors. Why is the PTC type suited to protection of machine windings?

8.2   ANSWERS

(Figures in brackets after each answer refer to the relevant chapter and paragraphs in text.)

1.   A relay operates contacts in response to an electrical input. A sensor responds to a non-electrical input.   (1.2)

2.   It converts an input in one form into an output in another form - e.g. temperature to voltage.   (1.2.3)

3.   Attracted-armature, induction disc and reed. Those which have no precisely defined characteristics and those which are calibrated in terms of operating level or operating time.   (2.1)

4.   In both cases, with the contacts open - i.e. as in the de-energised state.   (2.2.1)

5.   The operating time delay decreases as the input signal level increases, so that protection becomes more rapid as the severity of a fault increases.   (2.2.2)

6.   The operating current is adjusted by moving a setting plug at the front of the relay to select different tappings on the coil; to adjust the time multiplier setting a dial is moved to alter the disc travel required to close the contacts.   (2.2.2)

7.   A second driving element is fitted, responsive to the system voltage and producing a torque on the disc in opposition to that produced by the current.   (2.2.2)

8.   The product of the currents in the two coils and the phase-angle between them. The principal use is for reverse-power protection.   (2.2.2)

9.   Inverse-time overcurrent protection, especially for motors.   (2.2.3)

10.   (a)   To control a number of circuits from one signal input.
       (b)   To couple mutually incompatible circuits.
       (c)   To provide power amplification.
       (d)   To perform logic control functions.   (2.3.1)

11.   (a)   Rectifying.
       (b)   Power amplification.
       (c)   Switching.   (3.1)

12.   By adjacent layers of semiconductor material with different impurities added to produce p- and n-type characteristics, n-type material conducts by virtue of the availability of free electrons, p-type by 'holes' (notional positive charge carriers).(3.1)

13.   (a)   Linear - in which the collector current is directly controlled by the base current.

       (b)   Switching - in which the transistor is always either 'on', conducting a current limited by the external circuit, or 'off', when it is virtually non-conducting. (3.2)

14. It is purely a switching device, triggered to its conducting state by a pulse of gate current and turning off only when the current in the anode circuit is reduced to zero by external factors. It can handle higher currents and voltages, has a higher power gain and will function as a rectifier. (3.3)

15. The generation of heat by electrical losses and the need to keep the device within the permissible limit of operating temperature. Losses arise mainly (at power frequencies) from the product of current and voltage drop. (4.1)

16. By triggering the thyristors after a variable delay relative to the beginnings of their operative half-cycles. (4.2)

17. Assuming a voltage drop of 2.5V in the diodes and thyristors, the d.c. output voltage is:

$$0.90 \times 40 \times \frac{1+\cos\alpha}{2} - 2.5 = 24 \text{ volts}$$

from which $\cos\alpha = \frac{2\times(24+2.5)}{(0.9\times40)} - 1 = 0.472$

or $\alpha = 61.8°$ (4.2)

18. In power regulators, mainly for heating and lighting loads and small motors, and in static switches in a.c. no-break power supplies. (4.1)

19. (a) Cheap mechanical and electrical assembly.
(b) Accuracy and consistency of connections.
(c) Suitability for automatic assembly.
(d) Ease of testing and inspection. (4.7)

20.

INPUTS		OUTPUT			
A	B	'OR'	'AND'	'NOR'	'NAND'
0	0	0	0	1	1
1	0	1	0	0	1
0	1	1	0	0	1
1	1	1	1	0	0

(Fig. 4.10)

21. When operation is frequent, since semiconductor devices do not have an inherent wear-out mechanism related to the number of operations. (6)

22. Low cost and high reliability for a given complexity, high performance and small size. (4.9)

23. Thermocouples
Resistance Temperature Devices (RTD)
Thermistors. (7.1)

24.   A thermocouple depends on the different emfs in two dissimilar conductors when the point where they are electrically joined (the M-junctions) is hotter than the remote ends (the R-junction). The resulting voltage difference is measured and is an indication of the temperature of the hot end.

    An RTD depends on the large increase in the resistance of a platinum wire when heated. The resistance is measured by a bridge and is a direct indication of the temperature.

    A thermistor is a solid-state device whose resistance is sensitive to temperature. In the NTC type the resistance has a negative coefficient, and in the PTC type it has a steep positive coefficient.   (7.2-7.4)

25.   In the PTC type the sudden change to a large positive coefficient within a few degrees is used to activate switching for alarm or tripping. A suitable class of PTC thermistor is chosen to provide the switching at the desired temperature.   (7.3)

PART 8 UNINTERRUPTED POWER SUPPLIES

CHAPTER 1 SYSTEM CONFIGURATIONS

1.1   INTRODUCTION

Uninterruptible power systems (UPS), also called uninterruptible power supplies, are machines that are designed to take over the supply of electrical power to a load whenever the normal supply fails. Energy is either supplied by a storage device (battery) or conversion of fuel (generator) or both.

In this section we concentrate only on solid-state equipment (chargers, inverters and static switches) and the associated battery systems.

1.2   DC LOADS

An uninterruptible power system can be as simple as a battery and charger connected in parallel with a load When the mains supply is healthy the charger supplies current to the load and keeps the battery on a float charge.   When the mains fails the battery takes over the supply of load current. This is a true 'no-break' supply.

Generally, in applications where a high degree of reliability is needed (most offshore systems) the chargers and batteries are duplicated, see Figure 1.1.

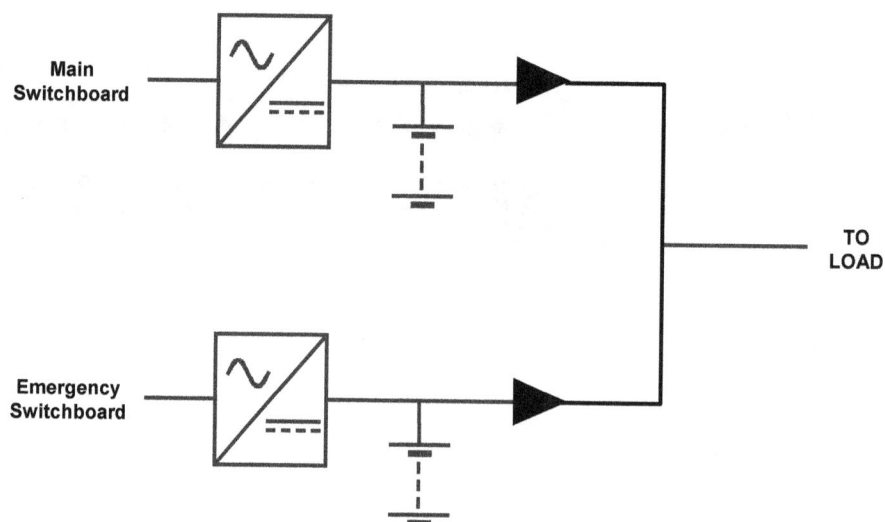

FIGURE 1 .1
DUAL BATTERY CHARGER SYSTEM

1.3   AC LOADS

To provide uninterruptible power to an ac load a battery and charger is still used (ignoring the mechanical alternatives), the difference being that the battery output is converted to ac.

The conversion of power from dc to ac is called 'inversion' and these days inverters are machines using solid state devices like silicon controlled rectifiers (SCRs) or transistors.

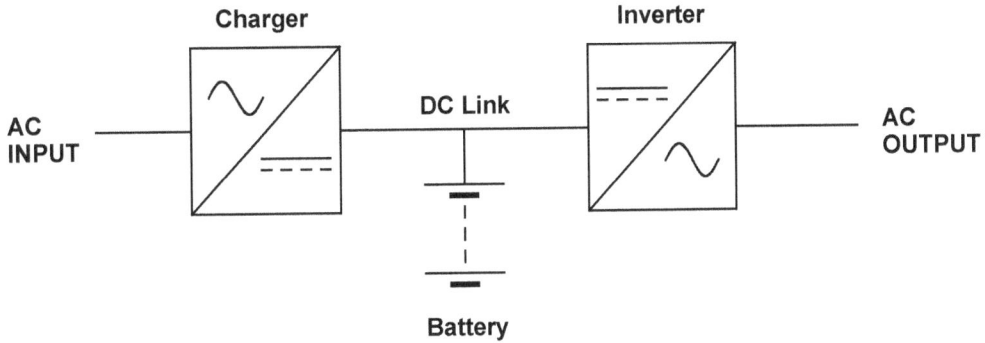

FIGURE 1.2
SIMPLE A.C. ON-LINE SYSTEM

The system in Figure 1.2 is a simple 'on-line' system, that is to say, the load is powered by the inverter all the time, whether the mains is available or not.

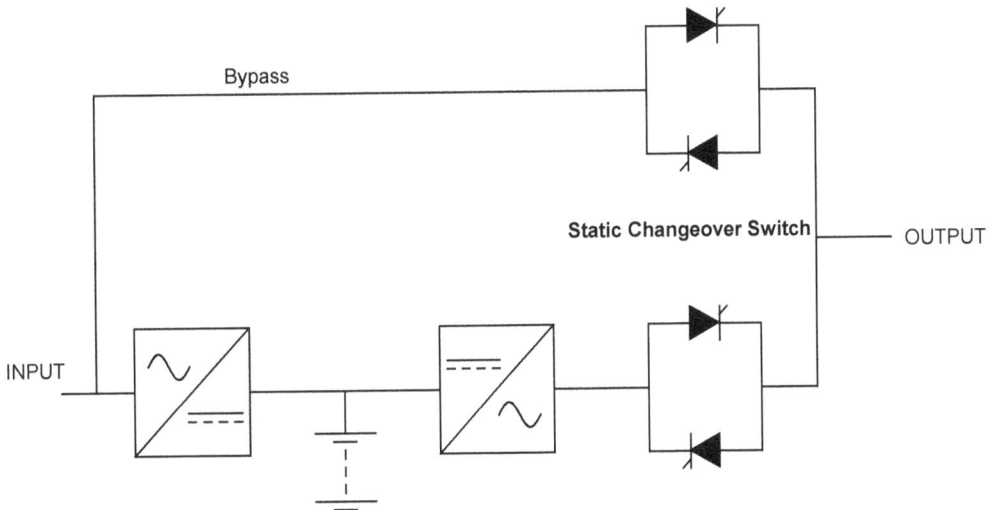

FIGURE 1.3
A.C. ON-LINE SYSTEM WITH BYPASS

A more practical on-line system is the type shown in Figure 1.3. Here a static switch is used to switch to bypass in the event of inverter failure.

1.4   ECONOMICS

Referring again to Figure 1.3, a system with a static switch can be made less expensive by running it normally in the bypass condition, switching to inverter only when the mains fails. This is because the battery charger never has to supply the load and is consequently smaller. An **off-line system** of this type is not suitable for all types of load as there is a slight delay after mains power failure while the inverter starts up. The changeover delay can be reduced somewhat if the inverter is kept running off load during mains healthy periods.

Another less expensive option is the use of contactors instead of a static switch in the bypass circuit. Again there is a penalty in changeover time, but this is often acceptable for loads like emergency lighting, where a break of less than half a second will not cause danger or damage.

1.5   MAINS DISTURBANCES

Often, UPS equipment is purchased to provide isolation from mains borne surges (spikes) and voltage sags, which can damage computers, bring them to a halt or simply corrupt data. In any installation where the load is sensitive to disturbances other than simple power cuts the bypass line has to be protected too. Common solutions to the problem involve the fitting of filters and ferroresonant isolating transformers in the bypass line.

In the following paragraphs we will cover the major components of an uninterruptible power system, namely chargers, batteries, inverters, static switches and controls.

CHAPTER 2 BATTERY CHARGERS

2.1   INTRODUCTION

A battery charger is a power converter providing the appropriate current and voltage to ensure the proper charging and charge maintenance of a secondary battery. In most cases the battery is connected to its charger and load in parallel. The charger then supplies the load current as well.

2.2   CONTROLLED RECTIFICATION

In battery chargers the conversion of ac to dc is usually achieved using silicon controlled rectifiers (SCRs) in a bridge circuit. Using phase control usually regulates the power delivered to the load. This is the commonest method of controlling the output of an SCR bridge. Altering the amount of time the SCR spends in the ON condition varies the amount of power supplied to the load.

SCRs are, as the name implies, rectifiers; therefore they will only conduct in one direction. The difference between SCRs and plain rectifiers occurs when a forward voltage is applied. Whilst plain rectifiers will conduct when a forward voltage is applied, SCRs will not switch on (**trigger** or **fire**) unless they receive a firing signal or pulse, at their gate terminal. (Terminology: when a forward voltage is applied to any rectifier device it can be said to be **forward biased**.)

Delaying the moment of firing in every half cycle will therefore reduce the average output of the rectifier, and varying the delay will vary the power transmitted. This is called phase control because only the phase of the firing pulses is altered, not their frequency (See Figure 2.1). Although SCRs need a firing pulse to switch them ON they switch themselves OFF whenever the current through them falls to almost zero. Since the ac input to a rectifier is reversing twice per cycle the SCRs switch off automatically at the end of each half cycle.

Rectifier Output

Firing Pulses: Heavy load, small delay angle

Rectifier Output

Firing Pulses: Medium load and delay angle

Rectifier Output

Firing Pulses: Light load, large delay angle

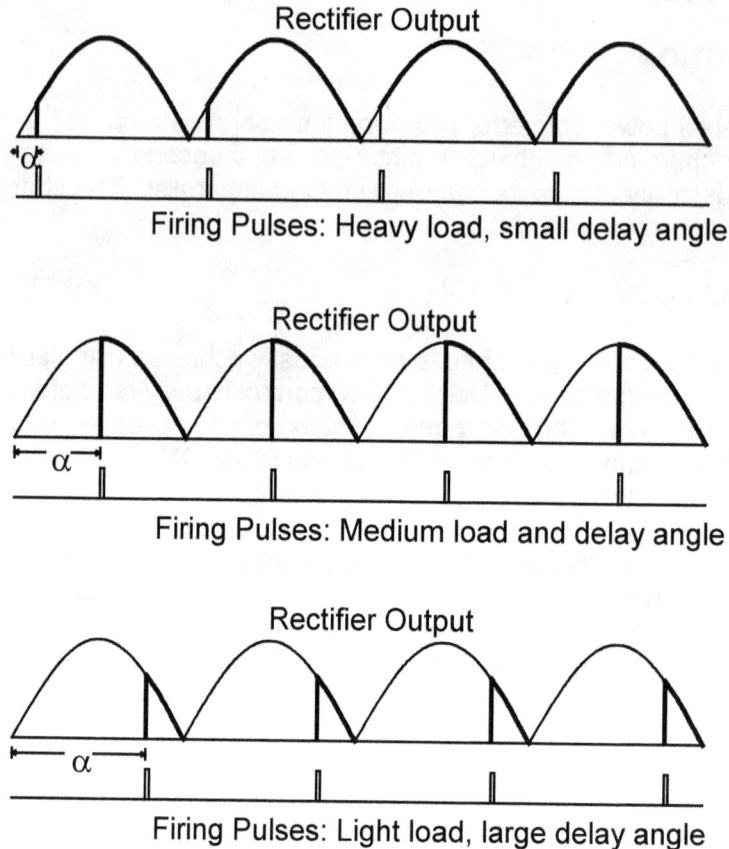

FIGURE 2.1
SCR PHASE CONTROL

Because of the ability of the SCR to switch very rapidly the output (and input) of a phase-controlled rectifier is rich in harmonics that extend into the radio frequency spectrum and precautions have to be taken by manufacturers to keep radio frequency interference (RFI) or electromagnetic interference (EMI) to acceptable and legal levels.

2.3   CONTROL CIRCUITS

The control circuit of a battery charger will usually be a voltage controller. Therefore there will be voltage feedback, where a fixed proportion of the output voltage is fed back to the control card. The voltage that is fed back will be compared with a reference voltage and the difference between the two will be an error signal that the controller will act upon.

Most chargers have some current limiting capability, preventing dangerous overloads without the need for fuses etc. (Important when operating for long periods unattended). Again, if current is to be controlled there must be some current feedback. This is usually achieved by the use of shunts (low value resistors) placed in the load and battery lines.

2.4   CHARGER OPERATION

Chargers usually have two voltage settings. Known by various names, we will refer to them as 'float' and 'boost'. A fully charged battery, left for any length of time, will begin to discharge itself due to internal leakage. The purpose of float charging can be thought of as maintaining sufficient voltage at the battery terminals to force a trickle charge into the battery, compensating for the internal discharge current.

Float charging will restore a discharged battery in most cases to more or less full charge, but it will take typically three days to restore 90% of full charge. The discharged battery will take a relatively large current initially, but as its internal emf rises the current taken from the charger will taper off, resulting in extended time to fully charge.

Boost charging places a higher voltage on the system. This forces more current in during the latter part of the charge cycle, shortening the charge period required after a battery has been discharged. Boost charging a fully charged battery will cause it to gas, therefore boost charging is never continued indefinitely. The ideal situation is to maintain a battery on float charge while the mains is healthy, and to use a period of boost charging after an outage. Some machines take care of this automatically.

Since most chargers are operated from the ac mains they are essentially rectifiers. The majority of machines are voltage regulated because the requirements of batteries on float charge are very precise. Also, to protect the battery, most machines have some form of current limiting.

2.5   CHARGER CHARACTERISTICS

We have already mentioned that the control of a typical charger is by reference to voltage, but with current limiting. This means that under normal operating conditions a charger may be regarded as a constant voltage machine. Should the load current exceed a predetermined level the machine will go into current limiting mode, preventing any further increase in current. Current limiting is achieved simply by reducing the output of the SCR bridge (increasing the firing delay).

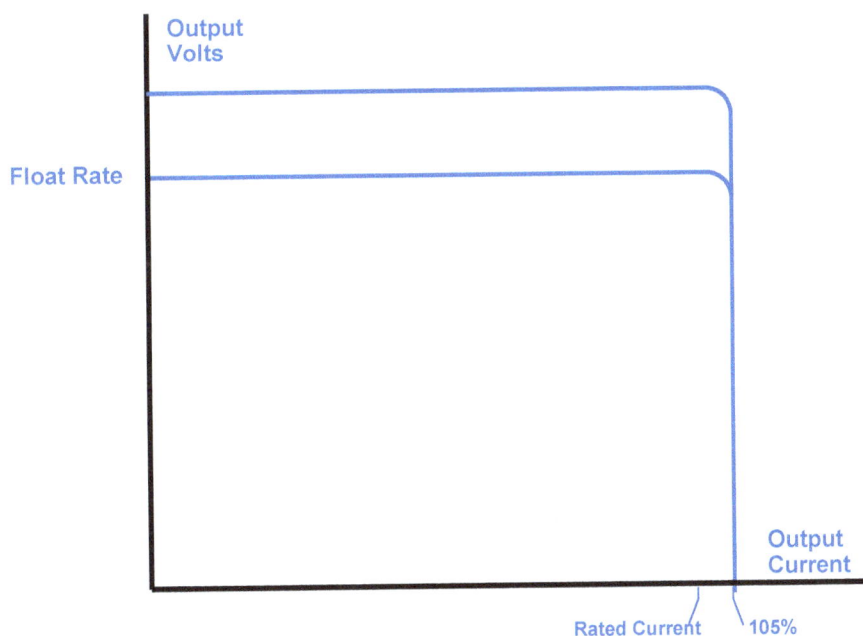

FIGURE 2.2
TYPICAL CHARGER OUTPUT CURVE

The horizontal part of the curve (Figure 2.2) represents the machine in constant voltage mode and the vertical part represents the machine in current limiting mode. You can only adjust a voltage whilst the machine is in constant voltage mode, i.e. the operating point is on the horizontal part of the curve, and you can only adjust the current setting whilst the machine is in current limiting mode (operating point on the vertical part of the curve).

2.6 CHARGER CIRCUITS

Most chargers encountered on offshore installations are of the SCR phase control type. Single and three phase types are in use, and where we cover a single phase circuit here for simplicity, it can be assumed that the same principles will apply to the equivalent three phase machine.

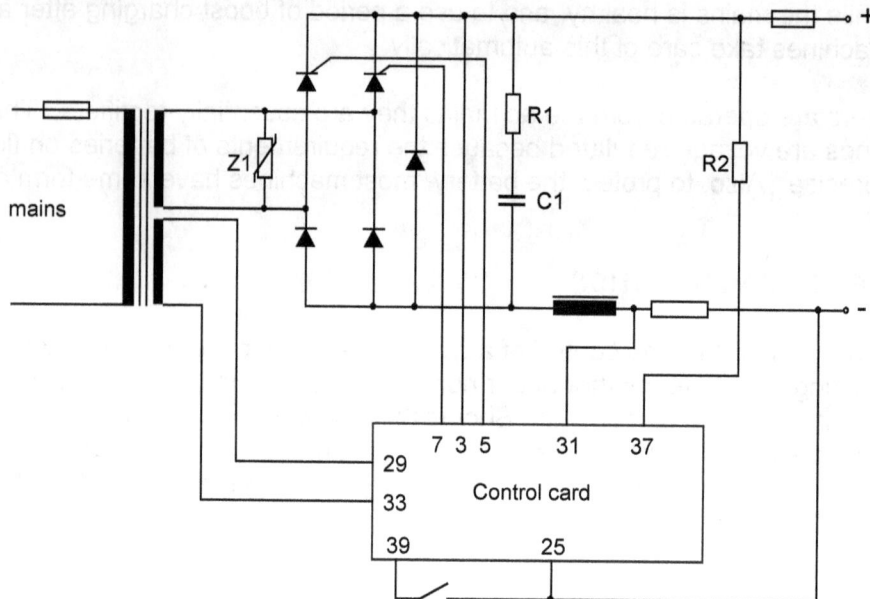

FIGURE 2.3
TYPICAL CHARGER CONTROL CIRCUIT

The circuit in Figure 2.3 is typical. A transformer is used to obtain the correct voltage for the system, and then the ac is fed into a single phase half controlled bridge, so called because it consists of two controlled rectifiers and two plain rectifiers. Fully controlled bridges (all SCRs, no plain rectifiers) are gaining in popularity because of their reduced ripple voltage.

The output is smoothed by the action of the battery connected to the terminals of the charger (some machines have capacitors too), and by the in-line choke that carries the output current. A consequence of having the choke is that it tends, together with any other inductance in the external circuit, to produce high forward voltages when the SCRs are trying to switch off. The diode connected across the charger output provides a path for current when the SCRs are switched off, quenching any inductive voltage. This diode is the freewheeling or flywheel diode and may be incorporated in the same package as the main bridge rectifier.

Note the essential connections to the control card. The function of the card is to regulate output voltage with current limiting by providing firing pulses at constant frequency but varying phase to the SCRs. Therefore there must be a mains frequency reference, which is provided by a separate transformer winding in our example, this winding also provides the power for the electronics on the card. There is output voltage and current sensing, and of course connections to both the SCRs.

The switch in Figure 2.3 is for boost charging. It works by reducing the output voltage feedback signal, causing the control card to reduce the firing delay on the SCRs, thus increasing the average output voltage.

As stated previously, most battery chargers are capable of operating with two voltage settings, one for float charging, the other for boost charging. When sealed lead-acid type batteries are used the boost voltage setting will be inhibited as the higher voltage may damage these cells.

CHAPTER 3 BATTERIES

3.1   INTRODUCTION

The operation of an uninterruptible power system relies on energy storage and that usually means batteries. Batteries that can be recharged after being discharged are called secondary batteries. Batteries are made up of cells, the cells being connected electrically in series to provide the required battery voltage (parallel connection is possible, but is rare within a given battery). Batteries work by chemical reaction and it is the type of chemical reaction that determines cell voltage. The battery types most commonly encountered on offshore installations are Nickel-Cadmium (NiCad), which give about 1.2V/cell and Lead-Acid, which give about 2V/cell, both are secondary types.

3.2   ELECTROLYTE

The electrolyte used in lead-acid cells is dilute sulphuric acid and the electrolyte used in NiCads is potassium hydroxide solution, an alkali. Although chemically opposite, both types of electrolyte have some properties in common: they can burn body tissue (skin and eyes) and are poisonous. Therefore the handling of electrolytes is subject to safety procedures and should only be carried out by trained persons.

The two types of electrolyte are mutually damaging to each other and must never meet. Even a minor contamination by the wrong electrolyte can ruin a cell. Good practice is to completely segregate the battery types, with only one battery type allowed in any battery room, together with its dedicated maintenance equipment.

3.3   BATTERY CHARACTERISTICS

In service, parallel connection with a charger and a load is usual. In this configuration the terminal voltage of a battery can vary depending on charger and load factors.

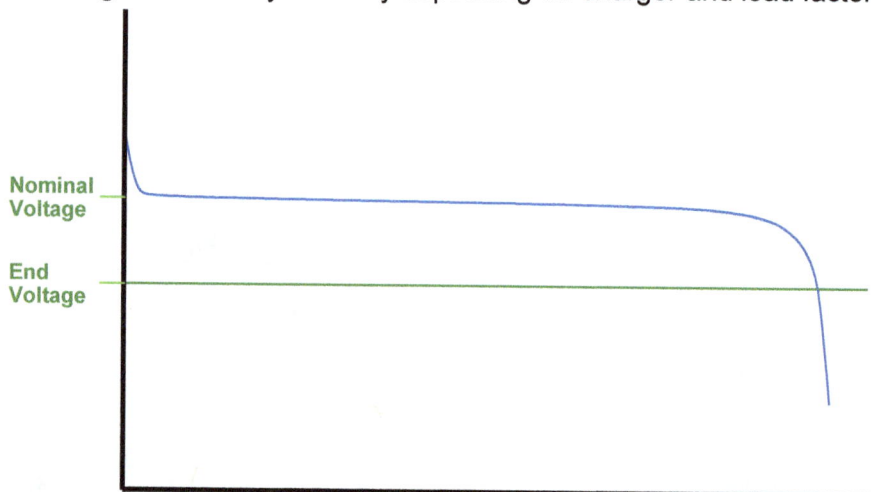

FIGURE 3.1
BATTERY DISCHARGE CHARACTERISTIC (CONSTANT CURRENT)

During discharge into a normal load the terminal voltage of a battery is called <u>nominal voltage</u>. It will hold this voltage more or less constant until it has given up most of its charge, then terminal voltage will fall quite rapidly. Usually, manufacturers as marking the end of useful discharge quote an end voltage. Deep discharging and most inverters employ low dc voltage tripping damage lead acid batteries.

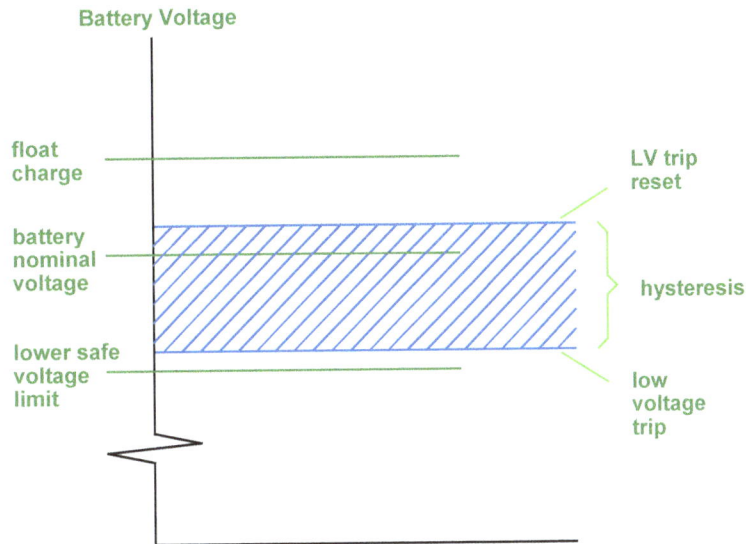

FIGURE 3.2
LOAD SHUTDOWN ON LOW BATTERY VOLTAGE

When a charger is working there are two further voltage conditions possible - float voltage and boost voltage. The charger to maintain a battery in the fully charged condition provides float voltage. A discharged battery that is connected to a float voltage will get charged, but only slowly (i.e. 3 days). Systems which use sealed lead-acid batteries have this limitation because boost voltages are not permitted.

If the charger float voltage is set too low a battery will become partially discharged over a period of time. The loss of charge is unlikely to be equal in all cells, so if the battery is then required to deliver a high current a cell with a lower emf than the others may suffer a reversal of polarity (it becomes a load instead of a source). This can damage the cell and cause electrolyte to be expelled violently. To reduce the risk of unequal charge, batteries are usually given a boost charge once or twice a year. This procedure is known as **equalising.**

When float voltage is set too high water will be lost, (even sealed cells can lose water) resulting in more frequent maintenance for open cells and rapid deterioration for sealed cells. Float voltage is therefore critical and has to be set using an accurate instrument.

To charge an open cell type battery from a discharged condition most machines can provide a **boost voltage** that will enable the battery to receive its charge in a shorter period. Excessive boost charging is not good for batteries, the main problem in the short term being loss of water. Therefore some machines have automatic timers controlling their boost voltage.

TABLE 3.1
TYPICAL BATTERY VOLTAGES

	Volts per Cell	
	NICAD	**LEAD-ACID**
Nominal	1.2	2.0
End	0.9 - 1.1	1.63 - 1.85
Float	1.4 - 1.42	2.21 – 2.35
Boost	1.55 - 1.7	2.4 max

3.4   BATTERY CHARGING

Some manufacturers recommend that an initial or commissioning charge be given at a constant current. Following the manufacturer's recommendations on initial charging will ensure full battery life and capacity. Most SCR chargers can be temporarily converted to work as constant current chargers for commissioning. If a completely discharged battery is given a constant current charge the battery voltage will look similar to the curve in Figure 3.3:

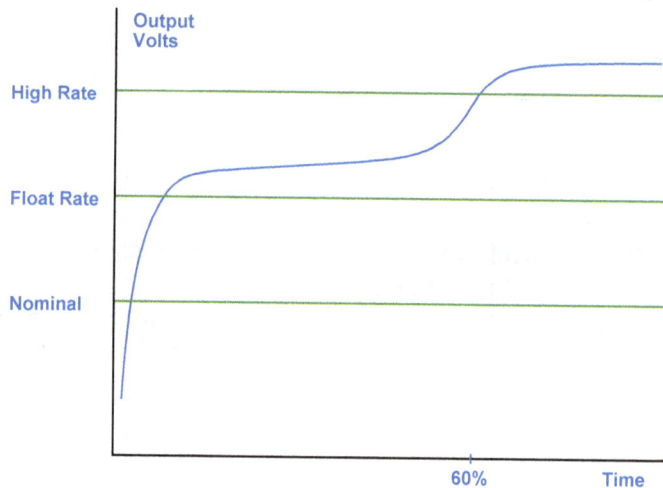

FIGURE 3.3
CONSTANT CURRENT CHARGING (COMPRESSED CURVE)

Notice the 'knee' in the constant current charging curve. This significant change in voltage always occurs when the charge has reached a certain point (about 60%). Most chargers are configured for constant voltage operation, but in the early stages of charge (into a discharged battery) will operate in current limit. Advantage can be taken of the knee in the charging curve to provide automatic control of boost charging after a power outage.

An example: Refer to Figure 3.4. Suppose an outage occurs and a battery becomes partially discharged. When power is restored the charger is likely to go into current limit. A timer starts and if the machine is still in current limit after, say, 30 seconds it will be automatically switched to boost rate. (This makes no immediate difference as the machine is in current limit anyway.)

The purpose of the 30 second timer is to ensure that if the outage was just for a short time the machine does not go into high rate charge unnecessarily. If the battery has not lost any significant charge during the outage the charger will come out of current limit before the 30 seconds is up.

When the knee voltage is reached another timer is started which allows the battery to reach full capacity before automatically switching the machine back to float charging.

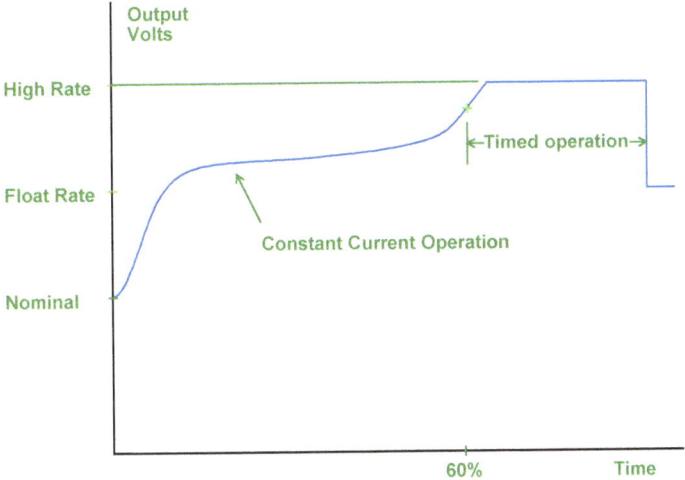

FIGURE 3.4
RECOVERY AFTER OUTAGE (COMPRESSED CURVE)

CHAPTER 4 INVERTERS

4.1 INTRODUCTION

Often referred to as a "static inverter" to distinguish the solid-state type from the older rotary machine type, the purpose of the inverter is to:

- Convert unregulated dc to regulated ac of fixed frequency
- Produce sinusoidal ac with limited distortion
- Incorporate protection against continuous overload and have good fuse blowing capability in the event of fault

To achieve the conversion of dc to ac the inverter needs a power or switching stage, in which the raw dc is switched to produce bi-directional current flow (ac). The output from the power stage may be passed through a transformer to provide the required voltage level. After conversion the waveform will need filtering to make it closer to the ideal sine wave. Different conversion processes cause differing amounts of distortion (deviation from the ideal sine wave). Filtering components are expensive and bulky so trade-offs are made with less than perfect sine wave outputs. In some machines the transformer may form part of the filter.

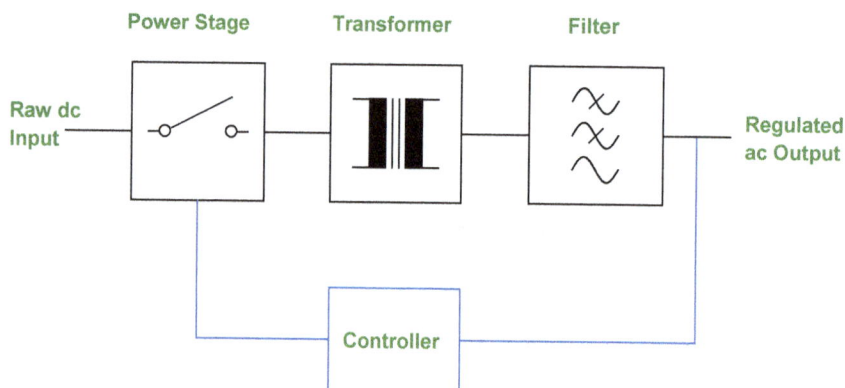

FIGURE 4.1
INVERTER CONTROL

The control circuit ensures that the output voltage and frequency remain within required limits. Control circuits for inverters will be sensing output voltage, frequency and current, and will usually have facilities for automatic synchronising. DC link voltage is also monitored on most machines and when the battery becomes fully discharged this signal is used to trip the inverter off to prevent damage to batteries.

4.2 EVOLUTION OF THE INVERTER

The development of the static inverter power stage has depended on developments in the power semiconductor industry. In the early days the only component big enough to handle the currents required was the SCR. The SCR itself has improved, both in its current carrying capacity and in switching speed, allowing larger and more efficient machines to be built. More recent developments have meant that transistors are now available with the switching speed and current carrying capacity to make them useful in power applications.

4.3   FIRST GENERATION INVERTERS

The so-called first generation inverters had a power stage that produced a square wave, simply switching the dc input direction at the required frequency. To achieve a measure of voltage regulation and filtering the **ferroresonant constant voltage transformer** (CVT) was developed. These transformers have a fairly constant output voltage over the entire load range and there is no output feedback or active voltage regulation necessary. Under overload conditions though, the output voltage collapses quite significantly. This may be thought of as a form of inherent current limiting but it poses problems under fault conditions when the inverter may fail to blow a fuse. All inverters tend to have a high output impedance (which is what causes the problem) but the CVT types are the worst.

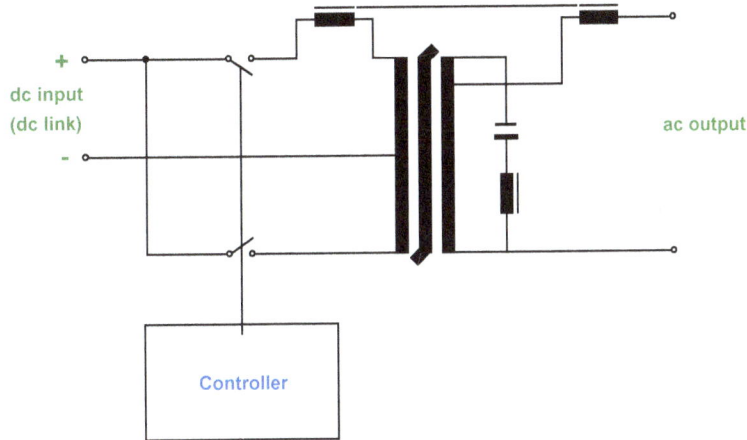

FIGURE 4.2
INVERTER WITH CONSTANT VOLTAGE TRANSFORMER

Figure 4.2 shows a form of first generation (square wave) power stage feeding into a CVT. The switch symbols could represent transistors or SCRs with appropriate means of switching them off. CVTs are characterised by their high and harsh noise level. Most of them sound as if their laminations are loose. This is fine if they are sited in unoccupied rooms.

4.4   BRIDGE CIRCUITS

The bridge circuit forms the basis of many power stages in common use. Consider the circuit in Figure 4.3 and assume that the SCRs T1 and T4 have just been fired (by the control and firing circuit).

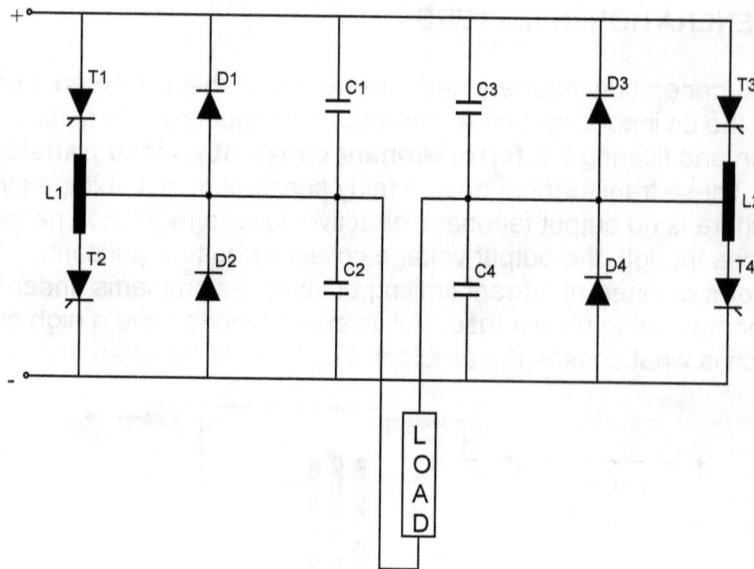

FIGURE 4.3
TYPICAL INVERTER BRIDGE

Current passes through the load (in the upwards direction in our picture) and two of the capacitors will be charged.

So, with T1 and T4 conducting C2 and C3 will be charged at about link voltage. If T2 and T3 are now switched on they will provide short circuit paths for the charges in C2 and C3 respectively.

Consider now what happens on the left hand side of Figure 4.3. (A similar thing will be happening on the right hand side too). When T2 fires, a large pulse of current will flow in the circuit formed by C2, the lower half of L1 and T2, causing a back-emf to be generated in the lower half of L1. A similar voltage will be generated in the upper half of L1 at the same time and this will oppose (and stop) the current flowing in T1. By this time T1 will not be receiving firing pulses and the cessation of current will switch it off. This process is known as **commutation**.

At the same time T3 will fire and, together with C3 and T2, will commutate T4. Now the current will be passing through SCRs T3 and T2 and the current in the load will be reversed. By repeating this process the current in the load is continually reversed, creating square wave ac. The "load" in Figure 4.3 would usually be the primary winding of a transformer.

4.5   QUASI SQUARE WAVE (SECOND GENERATION) INVERTERS

Quasi square wave (QSW) is achieved by using the same basic bridge as a power stage but with a different method of control.

Consider Figure 4.3 again. T1 and T4 are conducting and current is flowing in the load. In the previous example we fired T2 and T3 simultaneously to commutate T1 and T4 and reverse the current in the load. What we will do this time is fire only one of them, say T. This will have the effect of commutating T1 and current in the load will fall to zero. When that happens, T4 will be switched off. Then, after a suitable interval T2 will be fired again, this time together with T3. With both T2 and T3 coming on together there will be a circuit for the load and current will flow again, in the reverse direction.

The same thing will happen in the second half cycle: When the time comes to stop the current in the load only one SCR will be fired, let's say T1. This will commutate T2 and current will stop flowing in the load. T1 will have to be fired again in conjunction with T4 to get the current flowing again when the time comes for the start of the next cycle. The diodes D1 - D4 provide a current path for inductive loads as the bridge switches off.

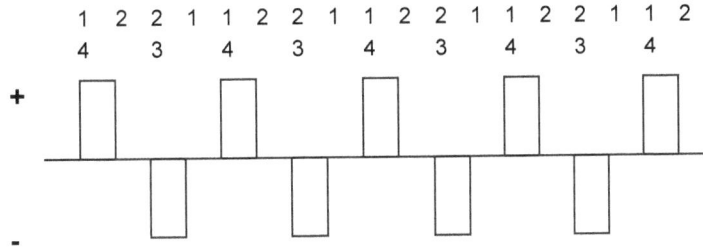

FIGURE 4.4
SCR FIRING SEQUENCE & BRIDGE OUTPUT

QSW has two advantages: First, the output waveform is a little bit more like the shape required. The closer the power stage gets its output to a sine wave the less filtering equipment will be required (this means cheaper and smaller machines). Secondly, the control circuit can vary the pulse widths of the output, making them longer to increase average output voltage or shorter to reduce it. This provides a small amount of voltage regulation, enough to compensate for the fluctuations in input (link) voltage which normally occur as the charger switches from outage to boost etc. Once we have a power stage that can be controlled by a voltage regulator we can dispense with the noisy, high impedance CVT. Second and subsequent generations of inverter use more normal power transformers and filtering components (although the functions of transformer and filter can still be interwoven, with resonant circuits using transformer windings for their inductive parts).

4.5   THIRD GENERATION MULTIPLE BRIDGE INVERTERS

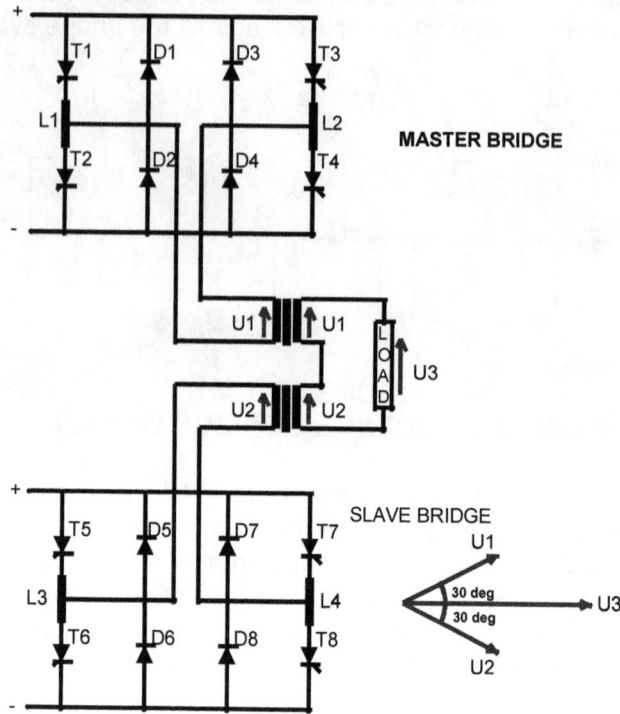

FIGURE 4.5
TWIN BRIDGE INVERTER

The heading says it all. Two QSW inverter bridges are connected in series. The "series" connection will not work if the bridges are connected directly to each other when the same dc source is used, so the two bridges feed separate primary windings on two transformers. The secondary windings are connected in series and the outputs of each bridge are thus added.

The two bridges of Figure 4.5 are operated out of phase with each other by 60 degrees (electrical) and a waveform similar to that in Figure 4.6 is obtained.

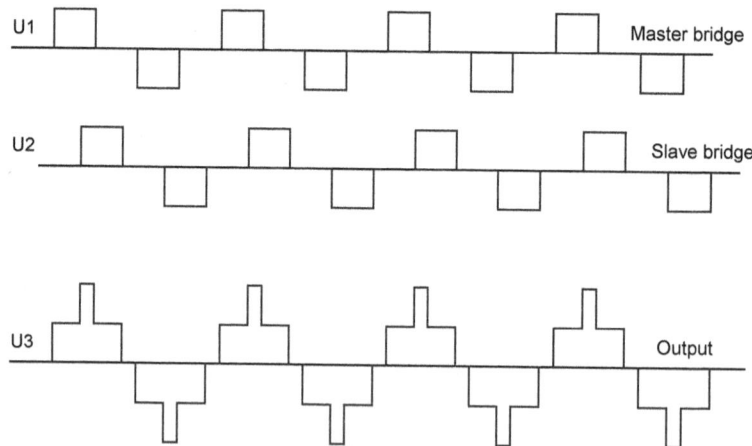

FIGURE 4.6
TWIN BRIDGE WAVEFORM

This waveform is even closer to a sine wave and therefore requires even less filtering. In practice triple bridges seem to represent the limit. The series connection of bridges can be magnetic instead of electrical, i.e. each bridge feeds a separate primary winding on the same transformer. The filtering requirements of a twin bridge are less than those of a single bridge and often will consist of a simple parallel resonant circuit:

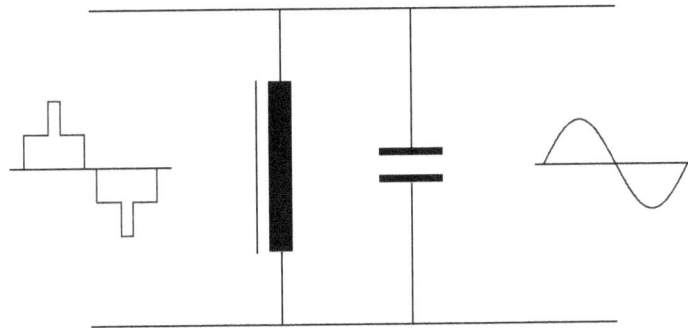

FIGURE 4.7
PARALLEL RESONANT CIRCUIT FILTER (TUNED TO FUNDAMENTAL FREQUENCY)

4.7 PWM INVERTERS - THE FOURTH GENERATION

PWM stands for pulse width modulation. Look at the bridge circuit of Figure 4.8 below and the associated waveform. If transistors B and C are switched on together current flows in the load (via a transformer and filter, usually). The way the current, and hence the voltage across the load, is regulated is by rapidly switching B and C on and off, the ratio of time spent in the ON condition to that in the OFF condition determines the average instantaneous voltage. In other words the bridge output is a series of pulses, the average voltage being determined by the pulse width.

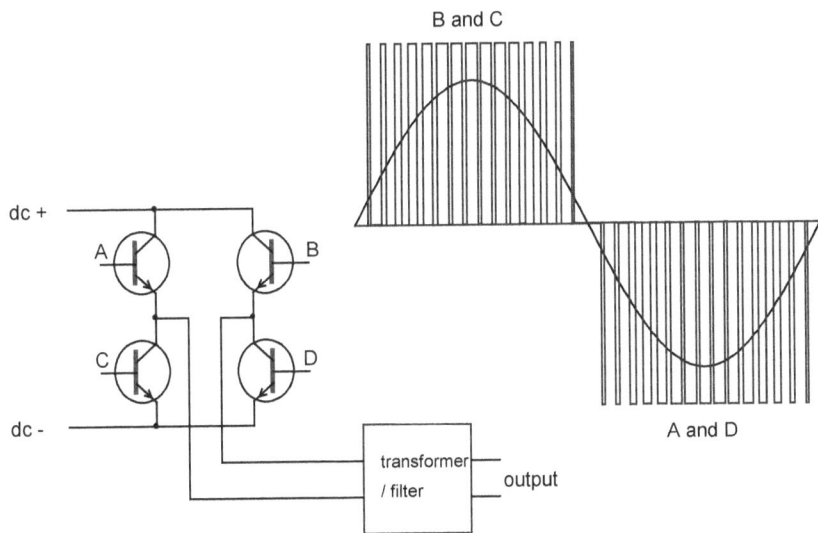

FIGURE 4.8
PWM BRIDGE AND WAVEFORMS

The control circuit produces the required output by switching B and C on and off several times during a half cycle period, then it leaves them switched off and operates A and D for a similar period, pulsing them on and off. From previous treatments of bridge circuits in these notes it will be obvious that transistors A and D together will produce negative half cycles in the load (assuming B and C produced positive ones).

The output waveform of a PWM bridge contains only the desired output frequency and the pulse or carrier frequency that is typically 16 or more times the output frequency. The carrier is a pulse train so it will have harmonics of its own, all higher than the fundamental carrier frequency. The output filter has only the carrier to get rid of, and the higher the frequency the smaller (and cheaper) the filter needs to be.

CHAPTER 5 STATIC SWITCHES

5.1   INTRODUCTION

The normal arrangement of a UPS involves the provision of a bypass, allowing power to be fed to the load straight from the mains under certain circumstances. The hardware involved is fairly simple. A pair of electronic switches formed by back-to-back SCRs provides the necessary changeover switch (Figure 5.1). SCRs are ideal components for a static changeover switch because of their inherent ability to remain in the ON condition, once triggered, until the end of the current half cycle. Provided the SCRs are triggered at the beginning of half cycles the supply to the load will normally be free of spikes or surges.

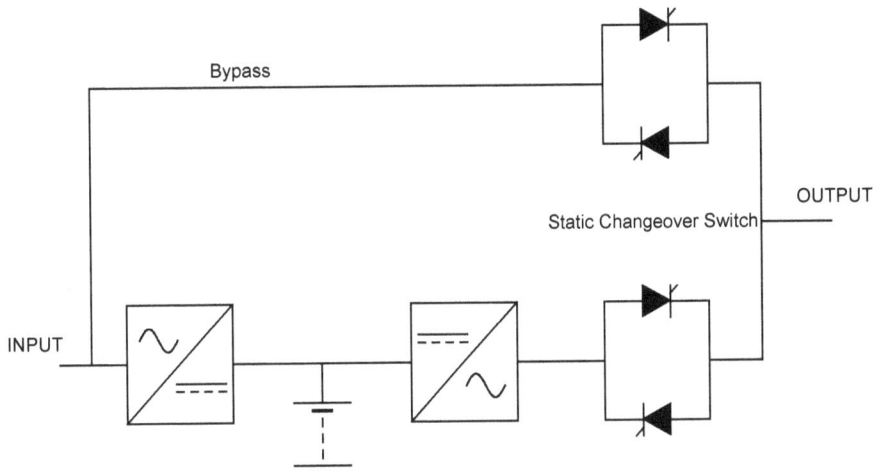

FIGURE 5.1
STATIC SWITCH

The control of the switches varies from machine to machine, so we will look at some of the control possibilities. First, the two switch units must be electrically interlocked so that they cannot both switch on simultaneously, ie the device functions as a true changeover switch. Any synchronisation on the part of the inverter will need to monitor both inverter output and mains, and this is usually done at the changeover switch position.

5.2   UPS CONFIGURATION

Using the standard hardware of Figure 1.3 it is possible to have two different modes of operation:

No Break or **On Line** operation. In this configuration the inverter supplies the load all the time. In the event of a mains failure the battery takes over and there is no interruption in the supply to the load. The purpose of the bypass in this configuration is to ensure continuity of supply in the event of inverter failure. Inverter performance is monitored and if it shuts down or the output goes outside defined voltage or frequency limits the bypass automatically takes over. If the inverter was synchronised with the mains immediately before the failure then there may not be any significant disruption to the supply.

Short Break or **Off Line** operation. This means that the inverter is not normally used to supply the load and power is normally fed through the bypass. If a mains failure (or mains out of limits condition) occurs the inverter takes over. There is a short delay from the time the outage is detected until the inverter takes up the load, and this delay depends on whether the inverter was running or not at the time of the outage. Keeping the inverter running gives the shortest break because there is a finite start-up time for these machines. The break is minimised if the inverter was running synchronised with the mains at the time of the outage.

Contactor Operation. Using a contactor instead of a static switch gives a finite changeover time and is only suitable where such delays are acceptable (emergency lighting is an example).

5.3   COST

As may be expected, the existence of options usually implies that performance is related to cost, and inverters are no exception. The short break option is less expensive to buy because, as the inverter only supplies the load during an outage, the battery charger never has to supply the load. Consequently, off line machines can have smaller chargers.

A contactor is an even cheaper option. A contactor may cost about the same as a single SCR, but a static changeover switch needs four SCRs (for a single phase switch) and a sophisticated control circuit.

5.4   BYPASS ARRANGEMENTS

As well as full automatic operation most on line UPS equipments allow manual changeover between inverter and static bypass. Additionally, many machines have a full manual changeover switch to allow machines to be taken out of service for maintenance without disrupting the supply to the load (although the load is then subject to the same risk of outage as the mains supply).

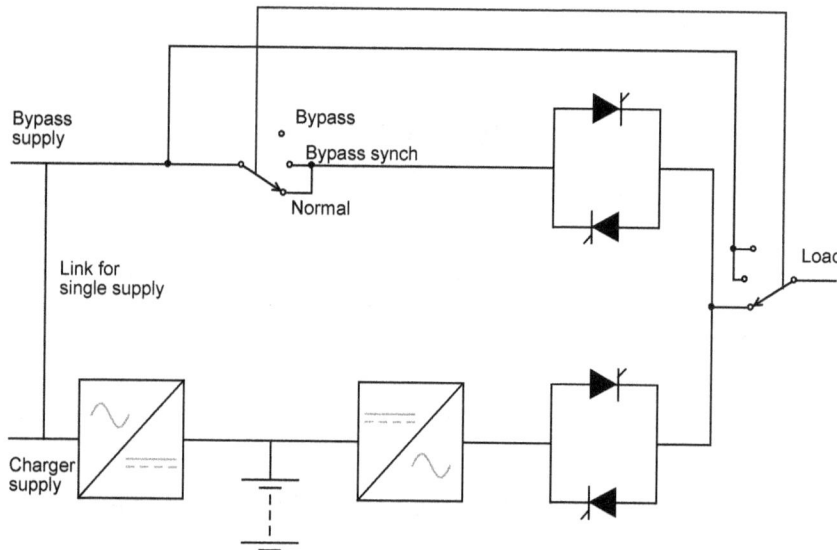

FIGURE 5.2
UPS WITH MANUAL BYPASS

The manual bypass system shown in Figure 5.2 will enable transfers both to and from full bypass without disruption to the load. A full bypass means bypassing the static switch as well as the inverter. This inverter has automatic synchronising facilities and mains monitoring is at the static switch, which is normal. When the equipment has to be taken out of service the first thing that is done is to check that the "In Synch" lamp is on. If so, the transfer to full bypass can go ahead.

To return the machine to normal service the inverter must be running. The transfer switch is moved one step to the "Bypass Synch" position and this provides the mains connection to the static switch, allowing the inverter to synchronise with mains. When the inverter achieves synchronism the "In Synch" lamp will come on and the transfer switch may then be moved one more step to the "Normal" position, connecting the UPS to the load.

5.5   UPS CONTROL LOGIC

Any discussion on UPS control logic must be of a general nature since every manufacturer will have different ways of providing control, and within that framework each of his customers will have differing requirements.

Some aspects like synchronising have already been covered. To summarise, most modern machines fitted with a bypass link will have the facility to automatically synchronise the inverter with the bypass supply. For synchronism to take place the bypass supply must come within certain parameters relating to voltage and frequency. If the bypass supply moves outside those parameters at a time when the inverter is synchronised with it the inverter should break away rather than follow the bypass supply.

The greatest risk of an outage is immediately after the restoration of supplies following an outage, so if there is to be any automatic return to mains supply after an outage then the usual method is to delay the changeover with a timer. This practice is common with automatic standby generators for the same reason.

For no-break systems the inverter normally supplies the load whether the mains is healthy or not, and the main purpose for the static switch and bypass is to provide supply if the inverter ever fails. To achieve this the inverter performance must be continuously monitored and parameters set which define a failure. Once the system has transferred to bypass there will be some indication or alarm of fault and the system will have to be transferred to inverter operation manually after the trouble has been corrected.

Most inverters are protected against overloading by some form of current limiting. Since the effects of overloading are thermal the current limiting may have a thermal (inverse time) characteristic, allowing relatively large overcurrents for short durations with the time decreasing for larger currents. A curve similar to that in Figure 5.3 is common.

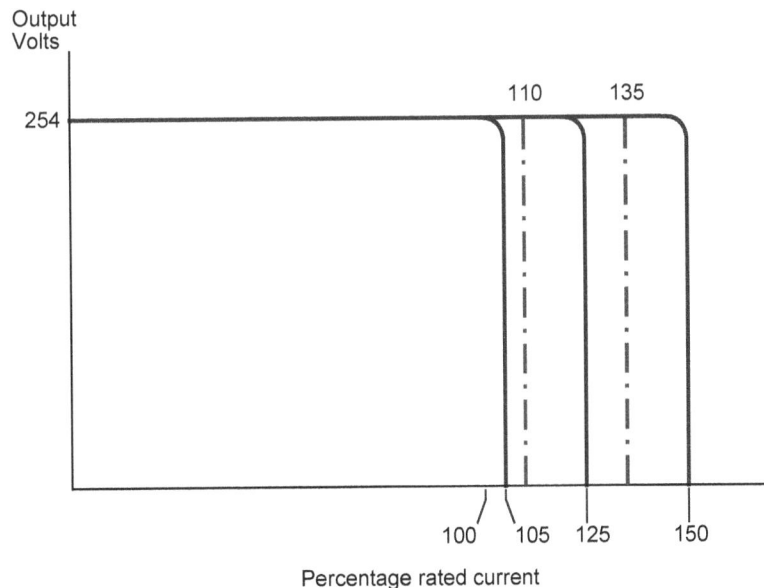

FIGURE 5.3
TYPICAL INVERTER CURRENT LIMIT THRESHOLDS

Referring to Figure 5.3, we can see that the machine can tolerate a mild overload of 105% indefinitely. If the current exceeds 110% at any time a 10 minute timer starts and if the overload still persists at the end of 10 minutes then a current limit of 105% will be imposed.

If the current exceeds 135% a 1 minute timer starts. If the overload still persists at the end of 1 minute then a current limit of 125% will be imposed. (The other timer will have started too, so after a further 9 minutes the 105% limit will cut in). The timers and limits do not reset until the machine comes out of current limit, ie the overload is removed.

In addition, for short duration overloads there is a blanket current limit of 150% for machines fitted with static switches (200% for those without).

One of the consequences of having a good current limiting system, together with the generally high output impedances encountered with inverters, is that they are not very good at clearing faults by blowing fuses. This problem has to be addressed at the design and specification stages.

The problem is tackled in two ways: First, the manufacturer can introduce a short delay in the current limiting facility, effectively bypassing it for a few seconds, long enough to blow a fuse. Secondly, the application must be carefully vetted before a machine is specified. The ideal is for a UPS to feed a distribution fuseboard where the load is split up. This implies smaller fuses and these require less energy to blow them. In instances where the UPS is designed to supply only one load (most cases) it may be necessary to purchase a bigger machine than the load rating would suggest, to ensure reliable fuse blowing. The specifier must satisfy himself that the machine he plans to buy will meet the required fuse blowing criteria and manufacturers should be consulted.

Finally, most inverters have a low voltage shutdown feature. This feature protects the battery as discussed under Batteries above. As the inverter output is regulated, a falling link voltage means the inverter will tend to draw more current from the battery to meet its power requirements, causing unstable operation and possible overheating. The low voltage shutdown feature avoids this possibility, so it will be found even on machines where the battery cannot be directly damaged by running flat (i.e. NiCads).

With recent developments in battery technology the low voltage trip has become more important. Sealed gas recombination lead-acid cells are terminally damaged (no pun intended) if their voltage is allowed to fall below prescribed levels for more than a few minutes. Anything less than 1.6 volts per cell and the cell is at immediate risk of being destroyed.

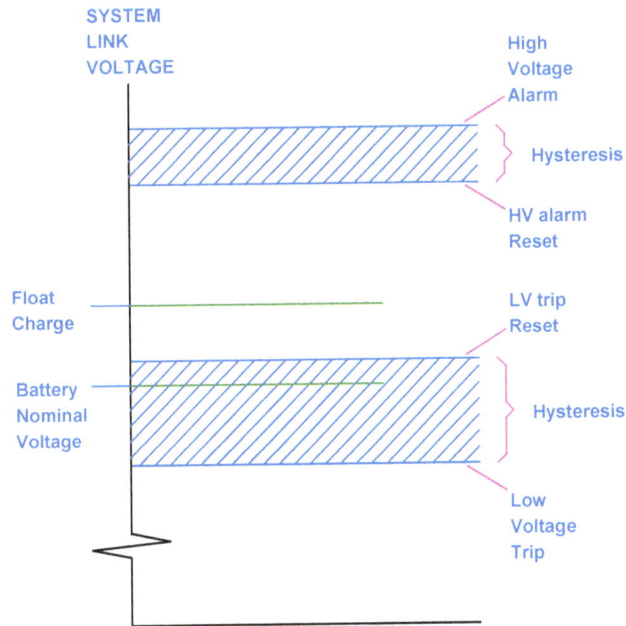

FIGURE 5.4
DC LINK ALARM AND TRIP LEVELS

Any battery which is loaded and at the end of its capacity will have a low, falling terminal voltage, but if that battery is taken off load its terminal voltage will rise again fairly quickly, perhaps even as high as nominal voltage. Any low voltage trip must therefore have a degree of hysteresis built in or the inverter will cycle on and off every few seconds.

CHAPTER 6 QUESTIONS AND ANSWERS

6.1   QUESTIONS

1   Why is an 'off-line' UPS less expensive to buy than an 'on-line' system?

    a The charger never has to supply the load and can be smaller
    b The inverter never has to supply the load and can be smaller
    c The charger runs cooler therefore needs less heatsinks or fans
    d The inverter runs cooler therefore needs less heatsinks or fans

2   An inverter will trip off when its (lead-acid) battery voltage falls to about 1.6V/cell. Hysteresis is introduced in the trip circuit to

    a   prevent the battery from gassing
    b   prevent the battery from being damaged by sulphation
    c   prevent the current in the inverter from becoming excessive
    d   prevent the inverter cycling on and off

3   What is the purpose of the 30 second timer in the battery charging circuit?

4   What are the advantages of a quasi square wave inverter over earlier types?

6.2   ANSWERS

1 a

2 d

3 The purpose of the 30 second timer is to ensure that if the outage was just for a short time the machine does not go into high rate charge unnecessarily.

4 QSW has two advantages: First, the output waveform is a little bit more like the shape required. The closer the power stage gets its output to a sine wave the less filtering equipment will be required

PART 9 ELECTRICAL SAFETY

CHAPTER 1 ELECTRICAL SAFETY AND WORK PERMITS

1.1 SAFETY DOCUMENTATION

Safety information will be contained in many company specific publications, typically named 'Standing Instructions Electrical'. This document will contain instructions, safety rules and procedures to ensure that the electrical systems and electrical equipment in company installations, both offshore and onshore, are operated, controlled and maintained in a safe manner. They are provided for the protection and safety of all personnel who operate and work on the electrical equipment of these installations. All operating companies have their own electrical safety rules and associated permit to work documentation.

It is the duty of all persons who have to operate, work on or otherwise be concerned with the Company's electrical systems and plant to make themselves thoroughly conversant with all relevant Company Safety Rules. Safety also demands that each person is thoroughly conversant with the system or equipment on which he is working. Ignorance of Safety Rules is not accepted as an excuse for neglect of duty.

It is not the purpose of this book to duplicate the information contained in specific Company safety manuals, but only to summarise the principal points. An outline of Government legislation, a description of the Permit to Work System and associated procedures for a typical offshore installation are included.

1.2 STATUTORY LEGISLATION

Offshore work is governed by Statutory Instruments, and some electrical aspects which are dealt with are:

- the overall responsibility of the OIM,

- issue of Permits to Work,

- accompanied working,

- use of written instructions,

- use of portable insulation, screens and protective clothing.

- special precautions when working near bare conductors,

- use of portable electric equipment,

- treatment of electric shock.

In addition the Electricity Regulations lay down that 'no person except an AUTHORISED PERSON .... shall undertake any work where technical knowledge or experience is required in order adequately to avoid danger'.

1.3   ELECTRICAL PERSONS

The Company's Standing Instructions Electrical will define the person appointed specifically as a Competent Electrical Person (CEP), Authorised Electrical Person (AEP), or Senior Authorised Electrical Person (SAEP) to carry out a defined duty in a defined installation. Authorisation does not follow automatically with a higher grade. The Instructions deal with the certification of such persons.

1.4   THE PERMIT TO WORK SYSTEM

No work whatever may be undertaken on electrical equipment without strictly following the formalised procedures of such work.

The Permit to Work System is designed to ensure that:

- all potentially hazardous activities have written authority,

- spheres of influence and line responsibilities are clearly established,

- line of responsibility is maintained throughout the execution of the work,

- all plant and process systems associated with the work to be done have been made safe,

- the work environment is free from dangerous concentrations of gas, vapour or fumes.

The Permit to Work System generally makes use of two basic work permits, one complementary permit, and three special electrical permits. A typical set of permits may include:

Work Permits

Cold Work Permit

For all work not involving equipment or material producing flames or heat or classified as a source of ignition.

Hot Work Permit

For all work involving the use of flames, spark or heat-producing equipment or materials.

Complimentary Permit

Preparation/Reinstatement Permit

This permit is raised and completed separately for such operations as:

Gas Test Requirements
Safety/Emergency System Isolation/De-isolation
Electrical Isolation/De-isolation
Mechanical Preparation/Reinstatement.

Electrical Permits

Electrical permits are issued to named individuals and are not transferable. They may be issued only by a Senior Authorised Person.

Applications and conditions for use are defined in Standing Instructions Electrical. Typical electrical permits include:

Permit to Work on Electrical Equipment

Sanction for Test - High Voltage Electrical Equipment

Limitation of Access Permit - issued to cover work in close proximity to live conductors

Where only electrical work is involved and where Cold or Hot Work Permits are not required for other reasons, then the appropriate Electrical Permit may be used on its own. The use of an Electrical Permit in this way must first receive the agreement of senior staff such as the Asset Holder, the Sphere Supervisor and the Offshore Installation Manager (OIM) in writing; there is provision for this on the permit itself.

1.5   SAFETY RULES

The Company's Standing Instructions Electrical will also give detailed safety rules for work on high-voltage apparatus. It will cover such subjects as:

- Ensuring that apparatus is dead,

- Isolation,

- Earthing-down,

- Screening, CAUTION and DANGER notices,

- Permit to Work - arrangements for issue,

- Accompanied working,

- Switching.

Rules are also given for work on low-voltage apparatus - that is, apparatus operating below 1 000V a.c. Although these rules are not so rigid as those for high voltage, a Permit to Work may be issued if deemed necessary.

1.6   ELECTRICAL ISOLATION AND SECURITY PROCEDURE

The Company's Standing Instructions Electrical define the isolation and security procedures for safe-guarding against access to, or operation of, the various electrical systems. A typical method is the 'padlocking' procedure which covers the following requirements:

Switch rooms	General use for preventing access to switchrooms and similar areas containing electrical equipment.
High Voltage	To prevent unauthorised interference with high-voltage equipment.
Low Voltage	To prevent unauthorised interference with low-voltage equipment.
Isolation	An Electrical Safety padlock to prevent unauthorised interference with electrical equipment which is not available for operational use and which is under cover of an appropriate permit.

Each type of lock, except the Safety padlock, has a set of common keys which are held by those persons who have a need to use them. Each Safety padlock is unique, and the lock and its key are clearly identified.

The Company's Standing Instructions Electrical also define the procedure for accounting for padlocks and keys and the use of a Padlock Cupboard, Lock-out Box and Key Safe.

CHAPTER 2 ELECTRIC SHOCK HAZARD

2.1   GENERAL

The effect of an electric shock on an individual varies widely and depends on a number of factors:

- The voltage level and the time the casualty is subjected to it

- The physical characteristics of the casualty

- The degree of contact the casualty has with the conductors or earth

- The state of perspiration or dampness of the casualty's skin

- Environmental conditions.

All the above factors have an effect on the electrical resistance of the human body.

It is generally accepted that any current between the two hands or between one hand and the feet which exceeds 15mA a.c. or d.c. can, under certain circumstances, cause death even to a fit person. The more prolonged the time of contact, the greater the risk.

The resistance of the body drops with increasing voltage. The following table gives typical body resistances with dry contact at different a.c. or d.c. voltages:

A.C. or D.C. Volts	Resistance (ohms) (Full Dry Contact)	Resulting Current (mA)
25	5 000	5
42	4 200	10
50	4 000	12.5
100	3 000	33
250	2 000	125

As stated, these figures are valid for normal, dry conditions. Under humid conditions, possibly with perspiration, and on a steel-decked installation, the resistances could be far lower and the currents correspondingly higher.

It will be seen that, as the body resistance falls rapidly with rise of voltage, the consequent current increases more rapidly still (under a 'square law' effect). The body is therefore particularly sensitive to increasing voltage, and steps must be taken to keep to a minimum any voltages with which an operator may possibly come into contact.

For this reason a working voltage of 42V is chosen for hand-held Class III portable tools (see Chapter 3). For the same reason an upper limit of approximately 0.1 ohm is set for the continuity resistance of earthed apparatus (see Chapter 4) so that the voltage drop across the earthing strap, even with full earth-fault current, will not rise to a dangerous level before the supply is broken.

Since electric shock may be accompanied by falling from a height and possible serious physical injury, normal first aid methods may be necessary in addition to resuscitation. Resuscitation however takes priority.

Notwithstanding the formal procedure of the Work Permit system and the various safety devices built into electrical equipments, it is at all times the responsibility of the individual to ensure the safety of himself and of other personnel.

2.2   TREATMENT FOR ELECTRIC SHOCK

Procedure for action in the event of electric shock is published in the Standing Instructions Electrical for the operation and control of the electrical systems and plant on offshore and onshore installations. The following notes are for guidance only and do not supersede the official published material.

2.2.1 Order of Action

Immediate and speedy action is necessary. Do not panic.

Switch off current	Do this immediately; if this is not possible, do not waste time searching for a switch.
Remove casualty from contact	Safeguard yourself when removing the casualty from contact. Stand on non-conducting material (rubber mat, dry wood or other material) bearing in mind the level of voltage involved. Use rubber gloves, dry clothing, a length of dry hemp or manilla rope (not nylon) or a length of dry wood to pull or push the casualty away from contact.

2.2.2 Procedure

If casualty is unconscious then proceed as follows:

1.   Ensure that the airway is open (see Figure 2.1).

2.   Check if casualty is breathing (see Figure 2.2).

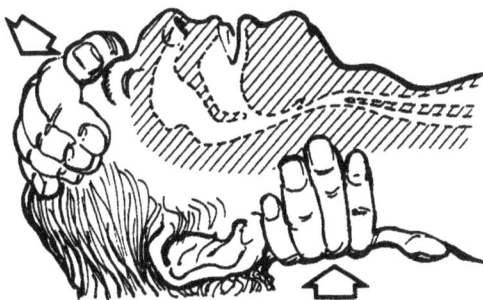

FIGURE 2.1

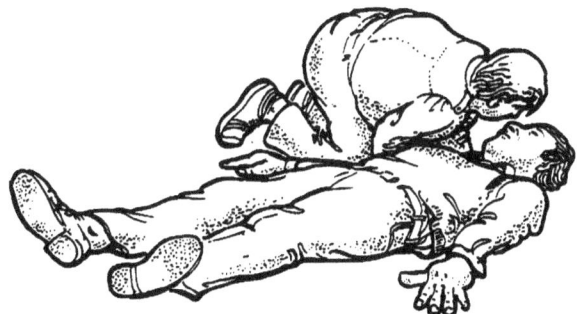

FIGURE 2.2

If breathing

Place casualty in recovery position (Figure 2.3). Give oxygen if available. Do not leave unattended. Send for medical assistance.

If not breathing

Commence expired air resuscitation, giving 4 quick breaths initially. (Figure 2.4).

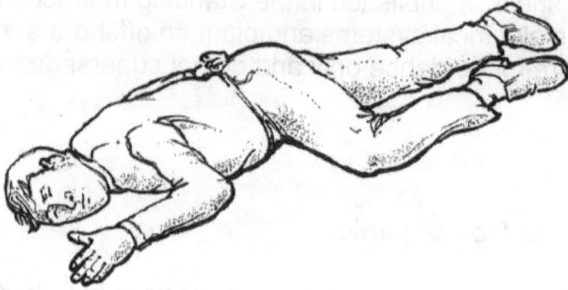

FIGURE 2.3

FIGURE 2.4

Check if heart is beating by feeling carotid pulse (Figure 2.5).

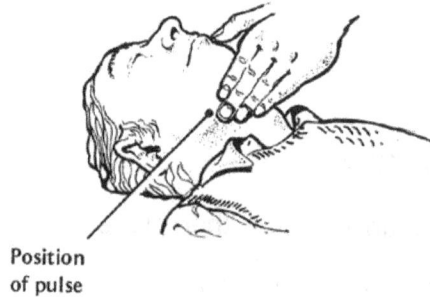

Position of pulse

FIGURE 2.5

Heart beating:

Continue expired air resuscitation for 10 to 14 minutes, and check carotid pulse frequently (Figure 2.5)

Heart not beating:

Supplement expired air resuscitation with external cardiac compression (Figure 2.6).

With one-man operator:

15 compressions, 2 breaths.

With two-man operators:

5 compressions, 1 breath.

Continue above regime until relieved by expert medical assistance.

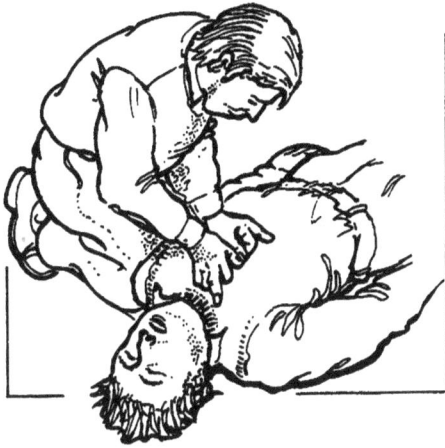

(1)

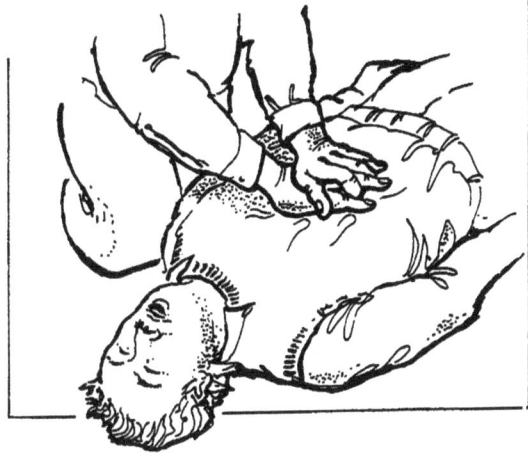

(2)

(4)

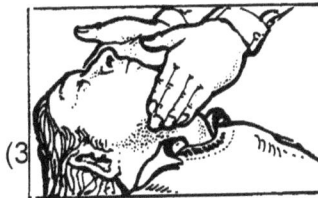

(3

(5

FIGURE 2.6
EXTERNAL CARDIAC COMPRESSION

CHAPTER 3 HAND TOOLS AND PORTABLE APPLIANCES

3.1 GENERAL

Hand tools and portable appliances pose an ever-present hazard of electric shock if there should be a failure of insulation within them which causes the hand-held casing to become live, or if the connecting cord should become damaged. The situation is made worse on an offshore installation where the steel structure offers good body-to-ground contact and a low-resistance earth, and even more so when that structure is wet - a condition which is very normal.

Originally some protection was provided by ensuring that the case of the tool was electrically earthed, so preventing any dangerous voltage building up on it. However, loose or neglected earth connections or damaged connecting cords very easily remove this protection, and this form of tool, known as 'Class I', is no longer permitted on most installations.

Forms of construction of portable tools have been internationally agreed and are classified as follows:

Class I denotes a tool having one or more parts with functional insulation only and required to be earthed.

Class II denotes a mains-operated or transformer-operated tool which is either 'all-insulated' or has double or reinforced insulation. It is not intended to be earthed. Class II tools are subdivided as follows:

Class IIA denotes a tool which is 'all-insulated'. This means that all metallic parts are enclosed within a durable and continuous insulated enclosure. It need not be double-insulated.

Class IIB denotes a tool which is 'double-insulated'. It has not only the normal basic insulation necessary for its proper functioning but also a supplementary insulation for safeguarding against shock if the basic insulation should break down.

In cases where there is not enough space for two separate insulations, a single 'reinforced' insulation electrically equivalent to double insulation may sometimes be accepted.

All Class II tools of accepted design bear the symbol ▢ on the nameplate.

Class III   denotes a tool designed for connection only to extra low voltage. This means a voltage not more than 50V rms (but in practice limited to 42V), obtained from the mains supply by means of a safety isolating transformer. No earth connection is provided.

Class 0   denotes a mains-operated tool without an earth connection. It may only be used in an 'earth-free' environment and is totally excluded from offshore installations.

Unclassified   are tools with their own generation or batteries or which are air-driven.

Under the general heading 'hand-held tools' are included not only portable motor-operated tools but also all portable equipment such as hand lamps, soldering irons etc. which consume mains power. Class I, II and III tools may be operated in a normal or 'polluted', but not hazardous, atmosphere. They are covered by BS 2769.

In many cases the mains outlet sockets to the transformers are provided with sensitive earth-leakage protection, especially where the leads to the transformers are long.

3.2   CLASS I TOOLS

Class I tools operate at mains voltage. Their metal enclosures are earthed through one core of the flexible 3-core connection and a 3-pin plug and socket, as shown in Figure 3.1.

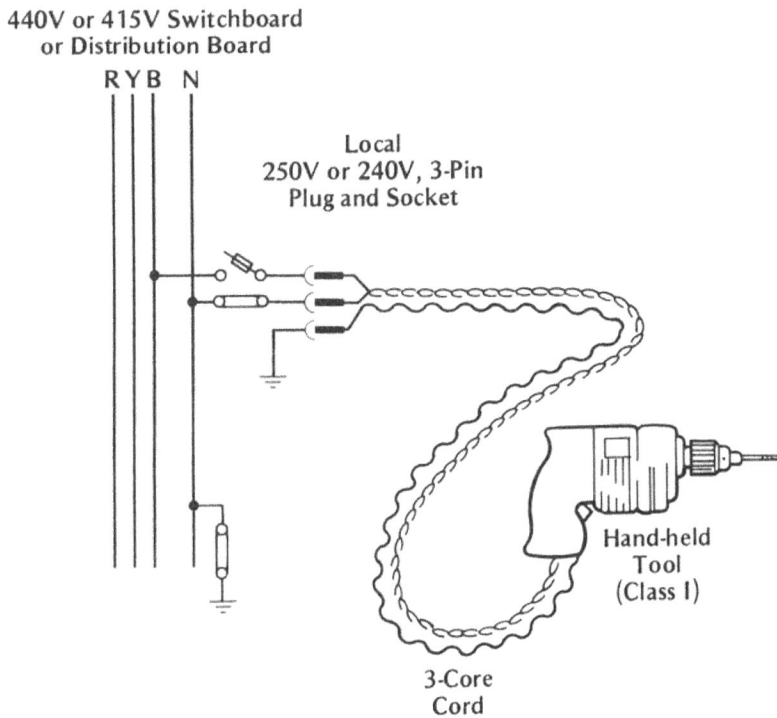

FIGURE 3.1
CIRCUIT FOR CLASS I PORTABLE TOOL (OBSOLESCENT)

There is an inherent danger with this class of tool that, if the flexible connection becomes damaged, the earth connection may be broken. Also the earth connection may become disconnected in the plug or the socket. In either case this condition may not be known to the user, and he would be handling an unearthed metal-enclosed tool. If a line conductor should touch the inside of the case, the operator would receive a full mains shock which could, under adverse conditions such as dampness, prove fatal.

Class I tools have been completely phased out. If any are still in use, special care and regular inspection must be given to ensure that all earth connections are sound.

3.3   CLASS II TOOLS

It is present practice in most companies to operate Class II tools at 110V, and they are supplied through 250/110V (or 240/110V) transformers with their secondaries centre-tapped and earthed at the mains outlet socket through a 3-core cord. This ensures that if, in spite of double insulation, the operator comes in contact with either conductor, the voltage that he suffers will be limited to 55V rms to earth. The electrical arrangement is shown in Figure 3.2. Class II tools are also manufactured to operate at full mains voltage.

Where Class II tools are used, they must comply with certain specified test requirements and the tests are to be performed only by a recognised testing authority.

Special attention must be paid to the connecting cords both between the mains outlet socket and the transformer, and between the transformer and the tool, particularly at the

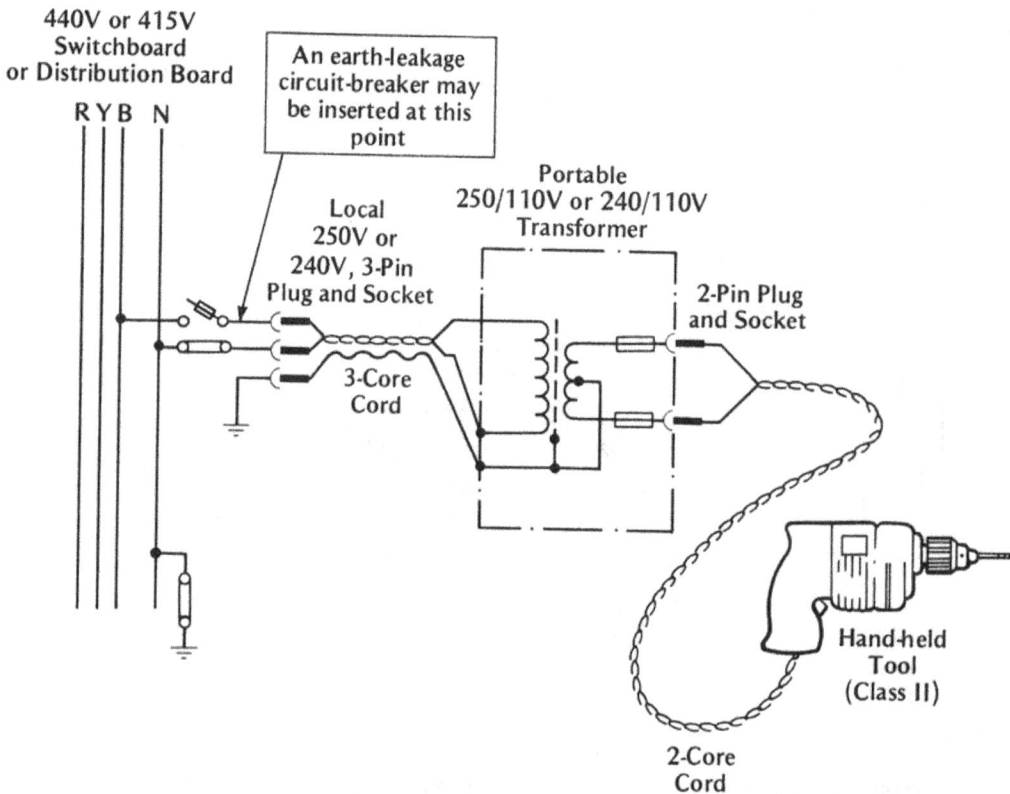

FIGURE 3.2
CIRCUIT FOR CLASS II PORTABLE TOOL

tool end. These must always be in sound condition and not kinked and they must be replaced immediately if they are in any way faulty or worn.

In many installations an earth-leakage circuit-breaker may be installed at the power outlet to the transformer. These special circuit-breakers are described in part 10 'Electrical Protection', Chapter 5.4.

3.4   CLASS III TOOLS

Class III tools are similar to Class II except that they are operated at 42V rms. The electrical arrangement is shown in Figure 3.3.

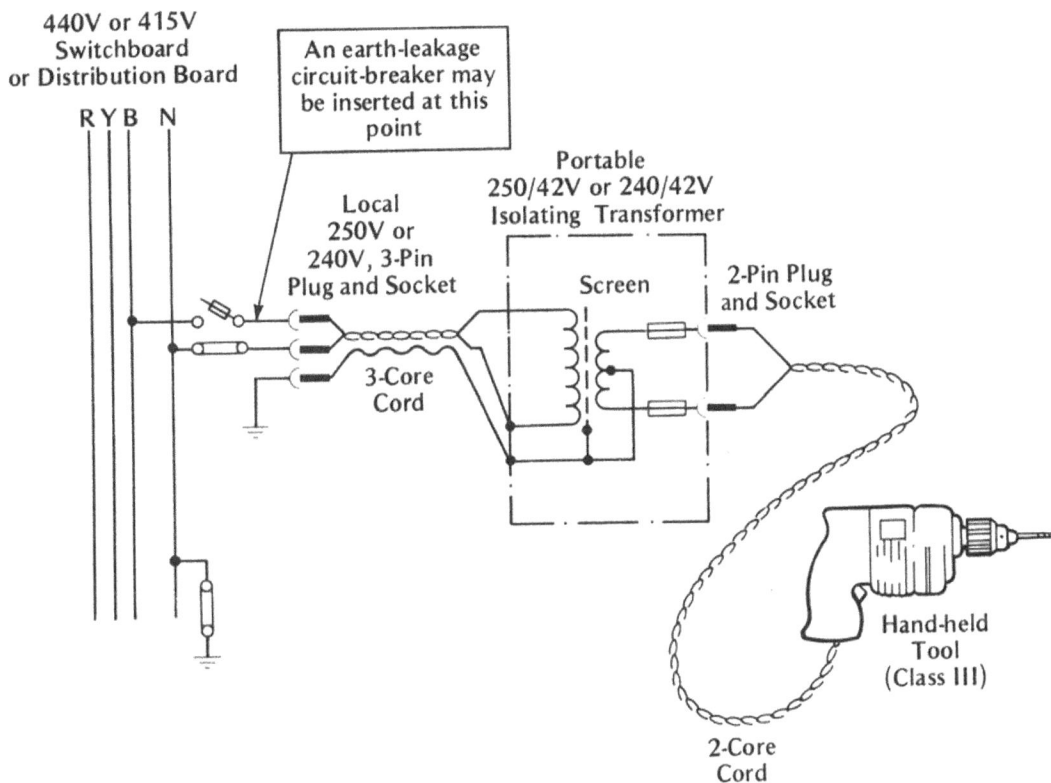

FIGURE 3.3
CIRCUIT FOR CLASS III PORTABLE TOOL

A local 250/42V (or 240/42V) safety isolating transformer is provided and must be either of the inherently short-circuit-proof type or of a non-short-circuit-proof type with a thermal cutout. The former normally has an internal impedance of about 100% so that, even when short-circuited, it only delivers full-load current and can continue to do so indefinitely without risk. The transformer is usually portable and placed near the tool, but it may be permanently mounted in welding-boards or switchgear. Class III tools must never be tested with a megger of voltage greater than 500V.

Class III tools may not always be available, especially in the larger sizes, in which case Class II tools only are to be used.

3.5   TRANSFORMERS

The 250/110V (or 250/42V) portable transformers used with Class II and Class III tools are single-phase, air-cooled, double-wound and are normal in all respects except for the centre-tap on the secondary. They provide electrical isolation between the 110V and 250V (240V) systems and have an earthed barrier screen between the HV and LV windings. The electrical arrangement is shown in Figures 3.2 and 3.3. These devices are referred to as 'Safety Isolating Transformers' and are covered by BS 3535: 1990.

The effect of earthing the centre-point of the transformer secondary is that, should the case of the tool become live in spite of the extra insulation, its voltage to earth is halved and the operator would receive a reduced shock - 55V with a Class II tool or 21V with a Class III.

For use with portable tools safety isolating transformers are classified according to their primary insulation as follows:

Class I   denotes a transformer having some parts intended to be earthed.

Class II   denotes a transformer which is either all-insulated or double-insulated as defined under 'Class II' above for portable tools. Class II transformers are subdivided as follows:

Class IIA   all-insulated

Class IIB   double-insulated or with reinforced insulation

They carry the symbol □ on the nameplate.

Most of those used on offshore installations are of Class IIB.

CHAPTER 4 EARTHING OF FIXED EQUIPMENT

4.1   GENERAL

Every piece of electrical equipment and every principal item of process plant, whether electrical or not, is securely earthed to the frame of the offshore installation:

- to avoid danger to personnel by electric shock resulting from the outer casing of electrical equipment becoming live through an internal fault,

- to avoid risk of explosion by sparking resulting from the vessels and pipework of process plant becoming electrostatically charged by movement of the fluids within them.

Both these risks are avoided by earthing the apparatus or vessels, as explained below.

4.2   ELECTRICAL EQUIPMENT EARTHING

Figure 4.1(a) shows a typical item (a motor) fed from a high-voltage circuit, 6.6kV in this case. The generator star-point is earthed, as is usual, through a neutral earthing resistor (NER), taken in this example as 17 ohms.

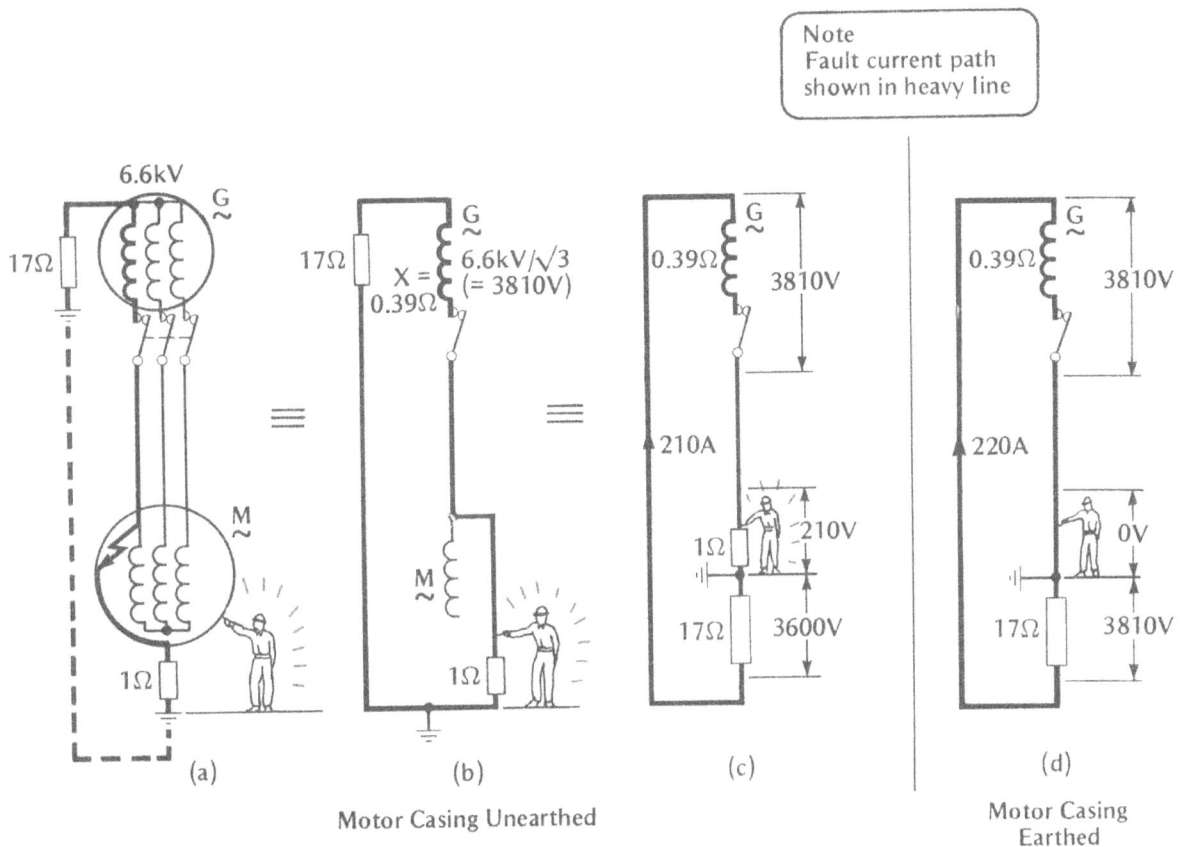

FIGURE 4.1
EARTHING OF HIGH VOLTAGE EQUIPMENT

It is assumed that an internal fault has developed in one phase of the motor, resulting in that line being short-circuited to the motor case. It is further assumed that the motor case is **not** specially earthed and that poor contact (perhaps due to paint) exists between the motor feet and the bedplate, resulting in a contact resistance of 1 ohm. The situation is then as in Figure 4.1(a), and a man standing on the ground and touching the motor case would get a severe shock resulting from the voltage due to the earth-fault current, whose route is shown by heavy line.

Figure 4.1(b) is a simplification of the electrical circuit of Figure 4.1(a) showing only the faulty phase and the resulting current loop. Figure 4.1(c) is a further rearrangement in which the 17-ohm NER and the 1-ohm contact resistance appear as a potentiometer. The earth-fault current is limited to 210A by the combined 17-ohm and 1-ohm resistances. This current, flowing through the 1-ohm contact resistance, causes a 210V difference between the man's hand and feet - enough to cause a severe shock, especially if the deck were wet. The voltage felt by the man is determined by the 1:17 potentiometer ratio and is, in this case, 1/18th of the total phase voltage 3.81kV ($= \frac{6.6}{\sqrt{3}}$). If the contact resistance were higher than 1-ohm, the shock voltage would be greater.

Conversely, if the contact resistance were less, the voltage would be lower. In the ultimate, if the motor case were solidly bonded to the platform frame by an earthing connection or strap of negligible resistance, the contact resistance would be shunted and would become zero. The situation is then as in Figure 4.1(d), where the 1-ohm element has disappeared, and the full phase voltage of 3.81kV appears across the 17-ohm NER only. The earth-fault current has increased slightly due to the reduction of the total circuit resistance from 18 to 17 ohms, but no voltage now appears between the man's hand and feet, and he is protected from shock. (Note that the generator reactance, in this case 0.39 ohms, hardly enters the calculation, since it is negligibly small compared with the 17-ohm resistance of the NER.)

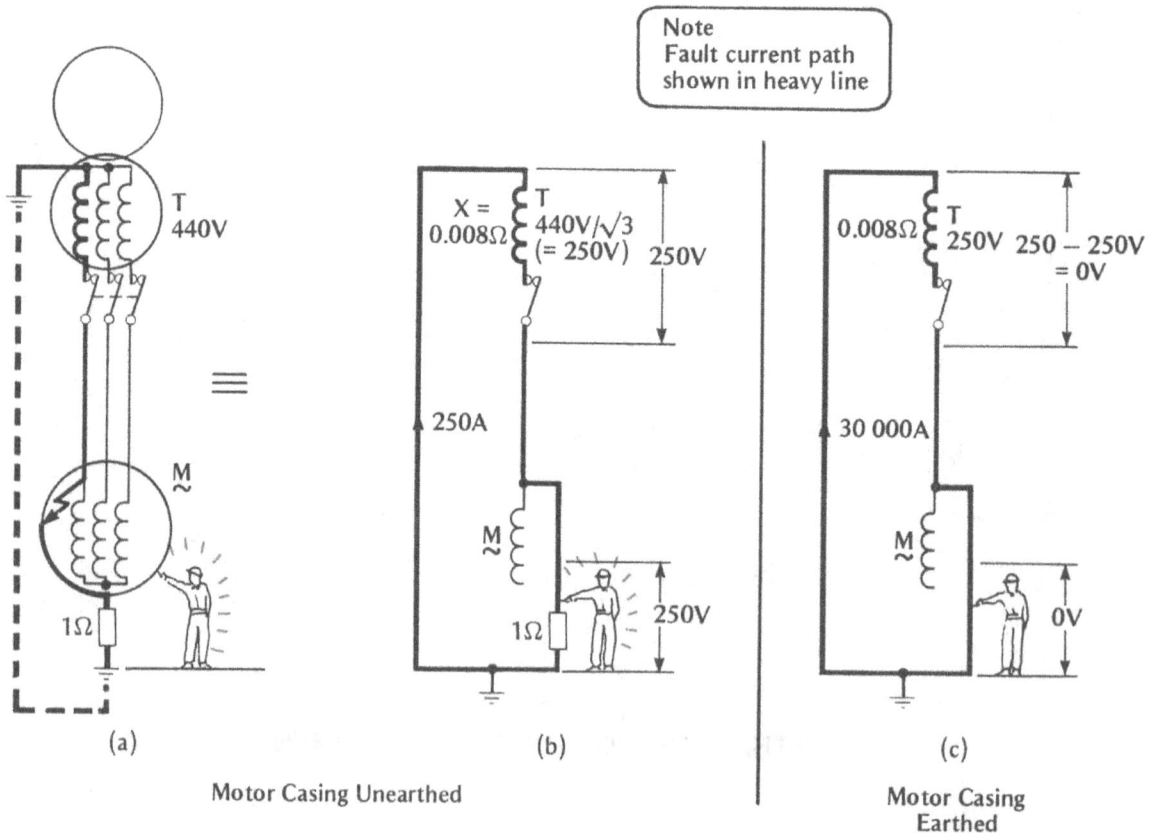

Note
Fault current path
shown in heavy line

(a)

Motor Casing Unearthed

(b)

(c)

Motor Casing
Earthed

FIGURE 4.2
EARTHING OF LOW VOLTAGE EQUIPMENT

A similar situation occurs with a low-voltage motor, as shown in Figure 4.2, except that all LV system neutrals are **solidly** earthed, without any NER. Figure 4.2(a) shows a situation otherwise similar to the high-voltage case, with a phase-to-case fault in the motor, together with a poor contact resistance of 1 ohm between motor and platform. Figure 4.2(b) is the equivalent circuit for the faulty phase. With a typical 2 000kVA, 440V transformer the 1-ohm contact resistance will limit the fault current to about 250A (again ignoring the very small transformer reactance, typically 0.008 ohm), and this will appear as a 250V difference across the 1-ohm between the man's hand and feet - again enough to cause a severe shock.

In Figure 4.2(c) the motor has been solidly earthed and the 1-ohm contact resistance completely shunted, so that it disappears from the figure, leaving the fault current to be limited solely by the reactance of the transformer winding. It will rise to a very high figure of some 30 000A, but no voltage now appears between the man's hand and feet, and again he is protected from shock.

4.3   METHOD OF EARTHING

From the above explanation it is clear that any earthing connection must be capable of carrying the full earth-fault current for the short time before the protection operates to open the circuit-breaker. In the LV system with solidly earthed neutral this current can be very high indeed, although the highest currents will persist only for a very short time where there are HRC fuses in the circuit. Consequently the earthing wires or straps need to be of adequate cross-section and securely fixed.

The explanation also shows that a contact resistance of even 1 ohm can cause risk-voltages of over 200V. Therefore it is essential to keep the resistance of the earthing connections well below this level. A figure of 0.1 ohm is typical of the specified maximum value of the continuity resistance of an offshore earthing connection; it could, and should, be generally less.

The effectiveness of earthing may be checked by a 'continuity tester'. This is a Megger-type instrument which reads low resistance values. Prods are applied: one to a part of the equipment near the earthing point, cleaned of paint or dirt, and the other to a good clean earth.

Earthing may be by welding or, where this is not possible, by thermo-welded copper earthing straps. Where the earthed item is removable, the strap may be bolted to the item. Galvanised clamps, screws, nuts and washers are not permitted. Earthing straps must be of at least 25mm^2 (0.04in^2) cross-section. Exceptionally lighting fittings are not individually earthed but are earthed in groups through an additional conductor in the supply cable, this conductor being earthed to the platform mass at the source. Cable glands and similar devices are bonded to the main frame by heavy wires.

Earth connections should be examined regularly for possible physical damage and measured for continuity resistance. Earthing straps between bedplates and the deck are particularly vulnerable to physical damage.

The subject of earthing is dealt with in detail in British Standard BS 7430:1991.

4.4 ELECTROSTATIC EARTHING OF PROCESS PLANT

All process and similar structures throughout the offshore installation, and their associated pipework and tanks, are solidly bonded together and earthed by bonding to the structural steel mass of the platform. The method of earthing, and the earth continuity resistance, are as described above for electrical equipment.

The flow of gases or oil through pipework and containment vessels can give rise to static charges on the inside, which immediately transfer to the outside of those pipes or vessels, appearing as very high potentials. They will discharge to earth or to each other on contact, or even without contact, by a high energy spark. In a hazardous area and in the presence of gas this could cause a major explosion.

Earthing of the elements of such process plant and similar structures is necessary:

- to prevent the build-up of static charges on those elements with a consequent risk of spark discharge,

- to prevent corrosion due to those static charges.

The bonding and earthing of all pipework is as important as that of the vessels to which it is fitted. Differences of static potential in two connected pipes could lead to sparking and corrosion across flanges and joints. The bonding of such unions is an important part of the structure earthing system.

Even without a discharge spark these static potentials can cause corrosion, especially across flanges and joints.

Earthing and bonding of all vessels and pipework ensures that such static charges do not build up and are carried away as fast as they form.

4.5 INSPECTION OF EARTHING ARRANGEMENTS

One of the duties of the Electrical Authorised Person is to satisfy himself that the earthing arrangements of both electrical and process plant are at all times in a satisfactory state, especially bearing in mind that external earth straps are vulnerable to damage.

This requires a planned programme of inspection over a period, and the results of such inspection should be logged in a register. Apart from these visual inspections there should be a programme of continuity resistance checks to cover the whole installation over a specific period. This is an important safety matter.

CHAPTER 5 OTHER ASPECTS OF SAFETY

5.1 GENERAL

Apart from the specific safety aspects covered elsewhere in this book, there are many less obvious dangers in the use of electrical equipment. Some of these are enumerated here.

5.2 ELECTRIC SHOCK

The Work Permit system described in Chapter 1 is designed specifically to reduce to negligible proportions the risk of electric shock when it is correctly followed. However, familiarity tends to breed contempt. Whereas a man respects high voltage because he knows it is dangerous, he may tend to regard low voltage as comparatively safe and be tempted to bypass normal precautions - for example, inserting a new fuse in a live circuit. Personnel must forever be on guard against such lapses.

5.3 CURRENT TRANSFORMERS

If the secondary circuit of a current transformer whose primary is live and carrying current is broken at any point, a lethal voltage can develop at the point of break. This is fully discussed in the Part 3 'Electrical Distribution', Chapter 9.

5.4 VOLTAGE TRANSFORMERS

When a main circuit is opened, it can sometimes be forgotten that a voltage transformer may remain connected to the live side, causing some parts of the control circuits to be live, albeit to only 110V. VT secondary fuses must always be drawn in these circumstances when maintenance work is to be carried out.

A further danger with voltage transformers is when there may be two working side by side. If the main circuit of one has been opened and isolated and its VT secondary becomes energised from the other, then the first will act as a step-up transformer and will energise that part of the high-voltage system to which it is connected and which was thought to be isolated. Here again, correct practice is always to draw the VT secondary fuses.

5.5 BATTERIES

It must be remembered that batteries are a considerable store of energy. As they have a very low impedance, any accidental short-circuit will produce a fierce current capable of inflicting severe burns. When working on battery-supported d.c. systems the battery fuses should always be withdrawn.

The danger of electrolyte spillage, especially with regard to the eyes and skin, must also be remembered.

5.6   CAPACITORS

Capacitors are a store of electrostatic energy, often at high voltage. Unless steps are taken to discharge them after disconnection, they may retain their charge for a long time. If accidentally touched they can administer a severe shock. Particularly dangerous are those used with VDUs and similar television-type displays; they can be charged up to 25 000 volts.

Before starting work on any piece of apparatus containing a capacitor, the capacitor terminals should be short-circuited (taking care not to touch them while doing so). A permanent short-circuit should then be applied and left in place until work is finished; it must be removed afterwards.

5.7   STORED ENERGY MECHANISMS

Most modern switchgear is actuated by a motor/spring operating mechanism. It is arranged that, after each closing operation, the heavy closing spring is automatically recharged. If the circuit-breaker is withdrawn for maintenance in this state, the closing spring will remain charged and latched-in. Should it be accidentally released, it could cause serious injury to the man working on it.

Before such a breaker is withdrawn after tripping, the control supply to the spring-charging motor must be broken (to prevent its recharging the spring) and the spring released by the mechanical closing button on the mechanism.

5.8   SWITCHBOARD SECTION BREAKERS

There is a particular danger when a part of a switchboard is isolated for maintenance. Such isolation will include the opening and withdrawal of the bus section circuit-breaker.

Figure 5.1 shows a typical switchboard, viewed from the back. Maintenance work is to be carried out on the right-hand side, while the left-hand side continues operational. All incoming circuits on the right-hand side have been opened and their circuit-breakers withdrawn, including the section breaker.

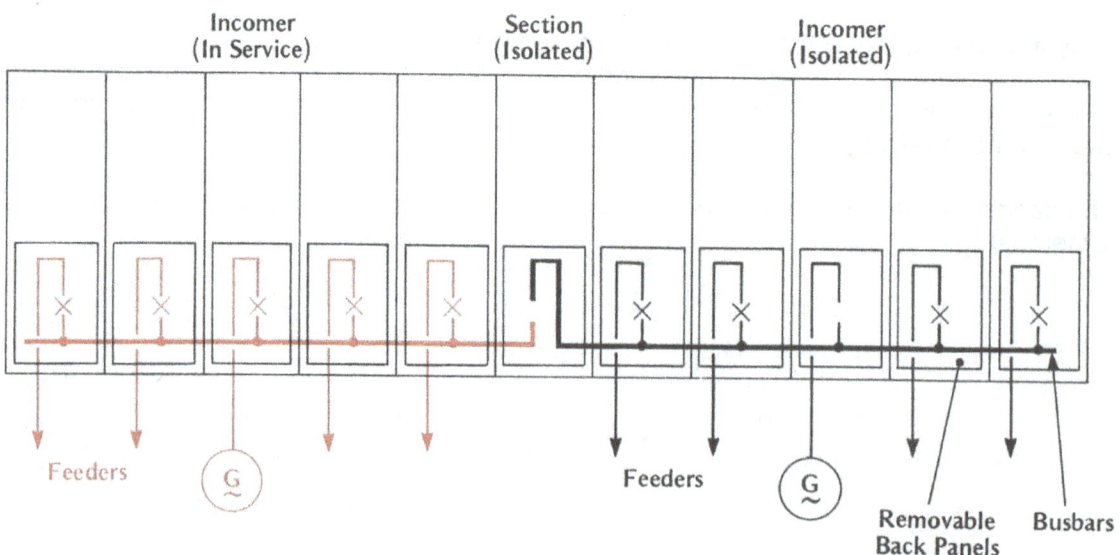

FIGURE 5.1
SWITCHBOARD ISOLATION

The maintenance work will involve removing the back panels of each right-hand side cubicle to expose the busbars. In Figure 5.1 those busbars which are dead are shown black, whereas those still live are shown red. It can be clearly seen that, if the back panel of the section breaker is removed, there is immediate access to live conductors. It is imperative that this danger be brought out in the Work Permit and that removal of the bus section panel is placed strictly outside the limits of the work.

5.9 RADIOACTIVE MATERIAL

Certain electronic equipment can contain radioactive elements (a smoke detector is an example). These are always identified by a radioactive warning symbol and are normally safe unless the container is broken. Care should be exercised when handling such units, and, if breakage occurs, appropriate disposal and cleaning procedures must be followed.

Broken fluorescent lighting tubes or cathode-ray tubes can present a health hazard to personnel and should be carefully disposed of. If a cut is received to the skin, medical attention should be sought.

CHAPTER 6 EQUIPMENT IN HAZARDOUS AREAS

6.1 HAZARDOUS AREAS

Hydrocarbon oils and gases are handled in large quantities on all offshore installations, and they are highly combustible. Moreover gas or vapour, mixed with air within the flammable range, forms a dangerously explosive mixture which is readily ignited by a spark or even excessive heat. The situation with gas is aggravated when it is compressed, where even the smallest escape can release considerable quantities of gas which will form an explosive mixture in the area of the leak and in all affected confined spaces.

Areas where gas may collect or be present at any time are called 'hazardous areas'. For the purpose of assessing the degree of risk they are classified in BS 5345:1989 into three zones:

Zone 0 an area in which an explosive gas/air mixture is continuously present, or present for long periods.

Zone 1 an area in which an explosive gas/air mixture is likely to occur in normal operation.

Zone 2 an area in which an explosive gas/air mixture is not likely to occur in normal operation and, if it occurs, it will exist only for a short time.

Other areas are termed 'non-hazardous' and include exterior areas which are at an adequate distance from any possible gas or vapour escapes so that the gas or vapour will be dispersed before reaching this area. Areas which, are pressurised or force-ventilated with air from a non-hazardous area are also non-hazardous, but they are classified as Zone 2 if the ventilation or pressurisation is shut down. Similarly normal Zone 2 areas become Zone 1 from an electrical point of view on loss of ventilation, and such areas would be equipped with Zone 1 apparatus.

No electrical equipment would be fitted in Zone 0 areas. Such areas are very rare and are not found in most installations outside process vessels.

Gas detection sensors are installed in all areas and give an alarm in the platform control room if gas is detected above a certain minimum concentration. If the concentration is above a specified higher level in a non-hazardous area, a complete electrical shutdown follows automatically.

Special attention is given in hazardous areas to the installation of certain equipments - for example cables and cable glands which must be bonded and insulated to prevent circulating currents (see Part 3 'Electrical Distribution Equipment', Chapter 7).

6.2 ELECTRICAL IGNITION OF GAS

When concentrated above the 'Lower Explosive Limit' (LEL), gas will be ignited by sufficient heat. Such heat arising from an electrical source can occur due to:

- electrostatic spark discharge between charged bodies or between a charged body and earth,

- arcing between the enclosure of a faulty piece of electrical equipment and earth,

- arcing between switching contacts of electrical equipment (switch, contactor or relay),

- an internal arcing fault in a piece of electrical equipment,

- 'Hot spots' due to local overheating.

The first situation can occur with process (i.e. non-electrical) plant due to the flow of fluids in pipes and vessels, which can charge the containers up to a considerable potential. If such a charge were allowed to build up, a spark discharge could occur spontaneously to earth or between vessels - for example across a pair of pipe flanges; this would manifest itself as a miniature form of artificial lightning. It could also occur if such a vessel were accidentally touched by personnel. It is prevented by earthing and bonding all such structures as described in Chapter 4.

The second situation can occur if an internal fault in a piece of electrical equipment caused a live conductor to touch, or arc-over to, the inside of the case. This would raise the potential of the case and could cause a discharge to earth similar to that described above, except that it would take the form of a continuing arc rather than an electrostatic spark discharge. It is prevented by earthing all electrical equipment with wires or straps capable of carrying the full earth-fault current, as described in Chapter 4.

The remaining situations cannot be prevented and must therefore be contained. Arcing between switching contacts is indeed a normal operating requirement in all non-electronic control equipment, whereas an internal arcing fault, though not normal, can always occur without warning. A local hot-spot could develop from bad contact pressure or high contact resistance across a switch, plug or other connection.

Since any of these conditions will release energy which could ignite gas if it is in contact with the arc or hot spot, such equipment in hazardous areas must be contained within an enclosure that will not transmit that energy to the outside gas mixture. Such an enclosure is called 'Ex-protected'. (It is also sometimes referred to as 'Explosion Proof'.)

6.3   EX-PROTECTION

Ex-protection may take several forms, depending on the type of equipment enclosed and especially on the zone classification of the area in which it is to be used. The various forms are defined in detail in BS 5345:1989, and the ones most likely to be encountered offshore and onshore are:

- **Ex-d Flameproof.** This is an enclosure which will withstand an internal explosion of an explosive mixture without suffering damage or propagating the internal flammation through any joints or structural openings to an external explosive mixture against which it is designed. A flameproof enclosure is usually certified with the protected electrical apparatus in place. It is permitted in all zones except Zone 0. Flame-proof enclosures are further discussed in para. 6.4.

- **Ex-e Increased Safety.** This is a type of protection giving an increased security against the possibility of excessive temperatures and the occurrence of arcs or sparks. It is permitted in all zones except Zone 0.

- **Ex-h Hermetic Sealing.** An hermetically sealed apparatus designed to confine potential means of ignition within a gastight shell protected against mechanical damage. It is permitted in all zones except Zone 0.

- **Ex-i Intrinsically Safe.** An intrinsically safe apparatus contains only intrinsically safe circuits. These are discussed more fully in para. 6.9.

- **Ex-n Type 'n' Protection.** A type 'n' apparatus, when operating normally within its rated duty, will not ignite a surrounding explosive mixture. (Note that it does not cater for abnormal electrical conditions such as overload or fault.) This may include non-sparking apparatus such as squirrel-cage motors or purely solid-state relays.

- **Ex-o Oil-filled.** Although used onshore, this type of equipment is little used offshore.

- **Ex-p Pressurised.** An enclosure into which the entry of an explosive mixture or flammable substance is prevented by maintaining the internal pressure (of air or other protective gas) above that of the surrounding atmosphere. The minimum overpressure is normally 0.5 mbar before and during the operation of the enclosed electrical apparatus. Such apparatus is normally purged before the enclosed electrical circuits are energised. To guard against a hazardous situation arising if the overpressure is not maintained, an alarm is usually fitted to signal if the pressurisation falls to less than 0.5 mbar above ambient. Pressurised protection (with alarm) is permitted in Zone 2, and also in Zone 1 if the enclosed electrical apparatus itself is suitable for use in Zone 2. If operation of the 'overpressure failed' alarm also automatically isolates the apparatus from the electrical supplies, Ex-p protection may be used in all zones except Zone 0.

- **Ex-q Sand-filled.** This includes all solid-particle-filled apparatus, of which the HRC fuse is the most common.

- **Ex-s Special.** This covers apparatus not included in the above definitions but which, as a result of test, has been proved to be equally safe.

6.4   FLAMEPROOFING (EX-D ENCLOSURE)

It should be particularly noted that 'flameproof' is only one of many forms of explosion-proofing, and the word should only be used for the particular type as defined for 'Ex-d'. In this case no particular attempt is made to seal joints of the enclosure against entry of outside gas into the enclosure, and it is accepted that it may, with time, enter through the flanges and shaft or spindle openings. Once inside the gas may be ignited by any arcing, and an internal explosion will take place. However, it is on a small scale, and all Ex-d enclosures are designed to withstand it without damage. The hot explosion products escaping through the long, narrow flanges and shaft and spindle opening gaps are so choked and cooled that they cannot ignite any gas mixture **outside** the enclosure. Every design of flameproof equipment is type-tested and receives a 'BASEEFA* Certificate' (formerly called 'Buxton Certificate') of flameproof worthiness.

In order to achieve the necessary choking and cooling effect, all gaps in the enclosure, whether due to cover flanges, window inserts, motor shaft or control spindle openings, must be such that the flamepath is long and very narrow. The minimum length is normally 25mm (1in), and with this length there is an associated maximum permitted gap between the flange surfaces. This maximum depends on the type of gas against which the enclosure is designed. In certain circumstances a 12.5mm (0.5in) flange length is allowed, and in this case the permitted gap is smaller to provide the same choking effect.

*BASEEFA stands for 'British Approvals Service for Electrical Equipment in Flammable Areas'.

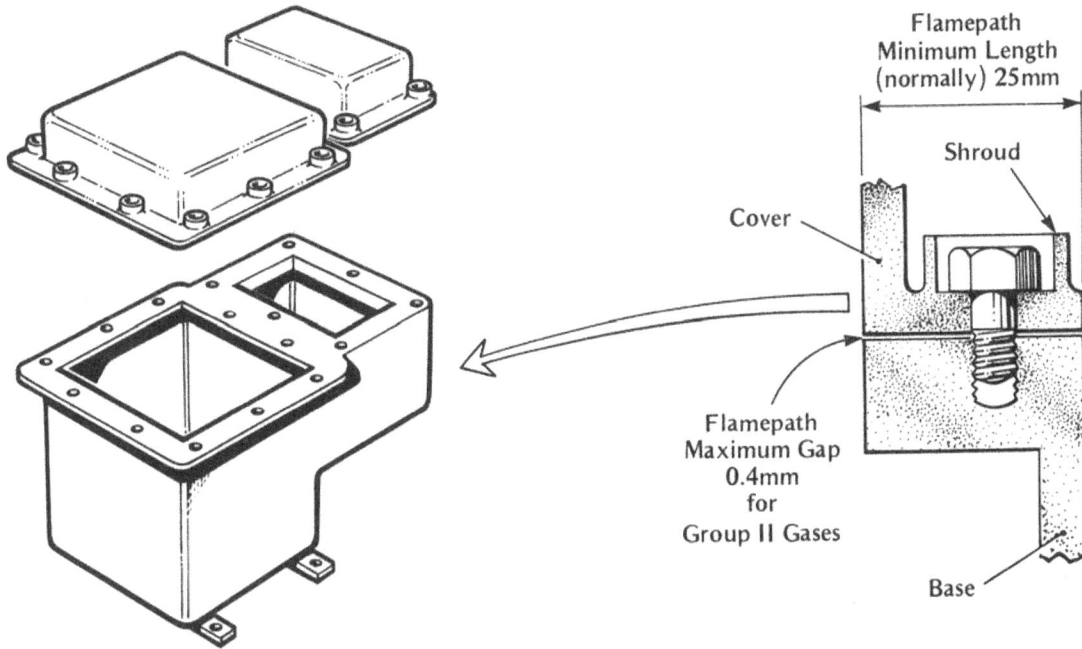

FIGURE 6.1
TYPICAL FLAMEPROOF ENCLOSURE

A typical flameproof enclosure which illustrates these points is shown in Figure 6.1.

The relation between the type of joint, the minimum length of flamepath (25mm or 12.5mm) and the maximum permitted gap (in millimetres) is given in the table below. There are two sets of figures: one for enclosures with volumes between 100cm^3 and 2 000cm^3, and one for enclosures with volumes greater than 2 000cm^3. All the figures are for flameproofing against those hydrocarbon gases (Group II) which are found in offshore installations and refineries. Different figures apply to other gases, such as free hydrogen.

MAXIMUM PERMISSIBLE GAPS WITH HYDROCARBON GASES (GROUP II)

Type of Opening	Min Length of Flamepath	Maximum Gap	
		Enclosure Volume 100 - 2 000cm3	Enclosure Volume Above 2 000cm^3
Flanged Joints	25mm (1in)	0.4mm	0.4mm
	12.5mm (0.5in)	0.3mm	0.2mm
Operating Rods and Spindles	25mm (1in)	0.4mm	0.4mm
	12.5mm (0.5in)	0.3mm	0.2mm
Shafts with Sleeve Bearings	40mm (1.6in)	0.5mm	0.5mm
	25mm (1in)	0.4mm	0.4mm
	12.5mm (0.5in)	0.3mm	0.2mm
Shafts with Ball or Roller Bearings	40mm (1.6in)	0.75mm	0.75mm
	25mm (1in)	0.6mm	0.6mm
	12.5mm (0.5in)	0.45mm	0.3mm

In the case of flanged joints the gap is the maximum linear distance between the metal flanges. In the case of operating rods, spindles and shafts the gap refers to the maximum diametral (twice the radial) clearance. Where these rods or shafts are large - that is, if the diameters are greater than the minimum length of the flamepath - special figures apply. Further details may be obtained from BS 5345.

6.5 TYPES OF COVER JOINTS

Joints may be broadly classed as:

(a) Butt joints.

(b) Spigoted joints.

(c) Screwed joints.

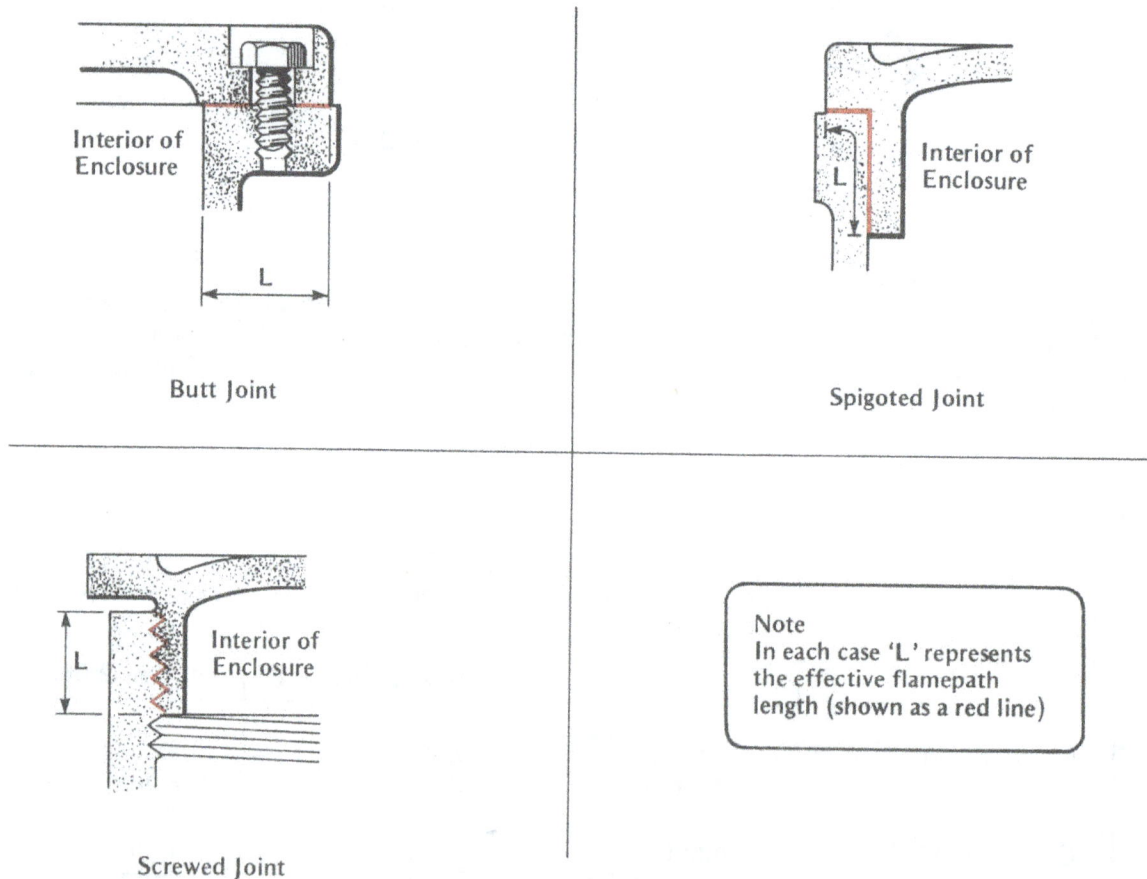

FIGURE 6.2
FLAMEPROOF JOINTS

These three types are shown in Figure 6.2. The length of the flamepath, indicated by red line, is the dimension 'L' in each case. For the purpose of measuring the minimum L, bolt-holes are ignored provided that they are not excessively large. In the case of screwed joints the overall length of the mating thread is taken, disregarding the thread faces.

6.6 WEATHERPROOFING OF FLAMEPROOF ENCLOSURES

Because of the flange gaps flameproof equipment is not of itself weatherproof or even dustproof. If this feature is required special steps must be taken, and they are usually confined to the spigoted type of joint.

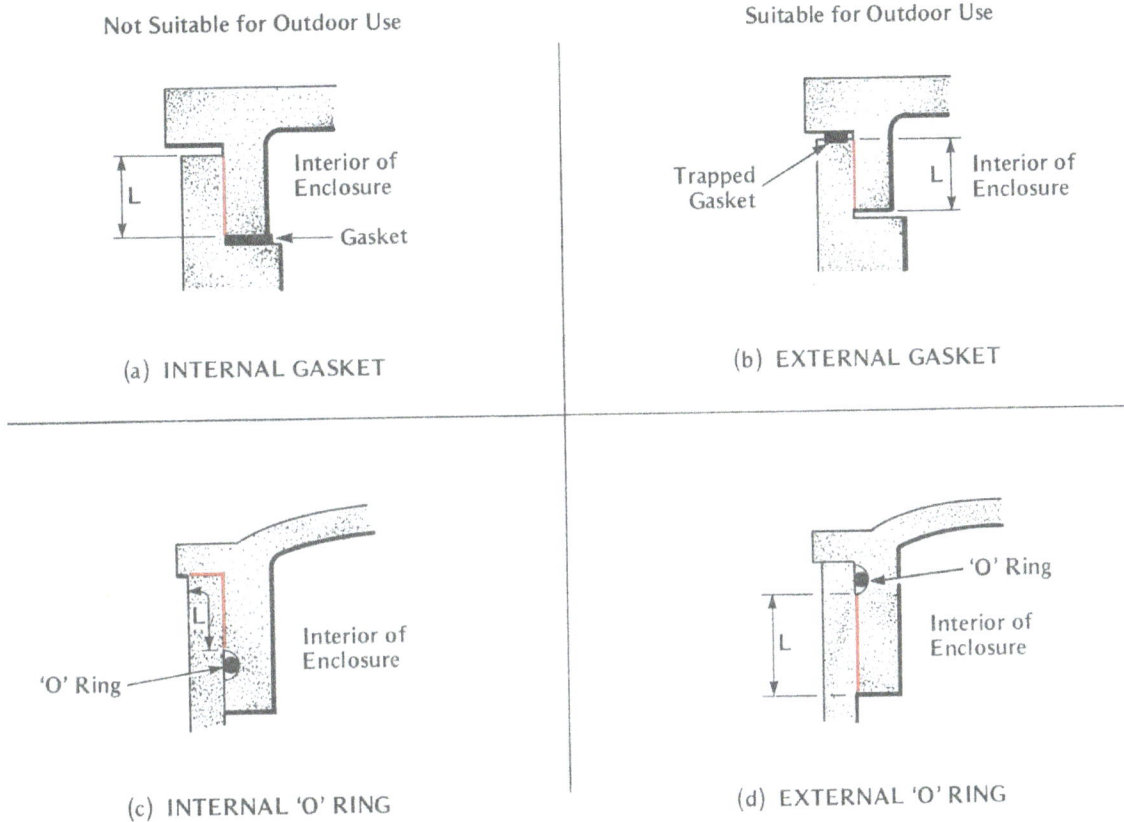

FIGURE 6.3
WEATHERPROOFING OF SPIGOTED JOINTS

Either a normal gasket or an O-ring may be used, as shown in Figure 6.3. Such a seal however must be independent of, and additional to, the special flameproof gap. It may be either 'external' or 'internal' - that is, on the outside or inside of the flameproof gap. Figures 6.3(a) and (b) show respectively an internal and an external gasket joint, and Figures 6.3(c) and (d) an internal and external O-ring joint. All four types will exclude dust, but both the 'internal' types may allow moisture to seep through the flameproof gap and eventually corrode and possibly widen it. These types are not therefore suitable for outdoor use or where heavy moisture is present.

6.7 TYPES OF SEALING FOR MOTOR SHAFTS

The last two groups in the table show the diametral clearances between motor shaft and casing when used with a flameproof (Ex-d) motor. Figure 6.4(a) shows how this can be achieved with a roller bearing. Note that normal practice is to place the flameproof gap behind the bearing so that the latter can be serviced without disturbing the assembly.

If there is not sufficient axial space for a flameproof gap, an alternative method is to use a labyrinth gland in conjunction with a sleeve bearing as shown in Figure 6.4(b). In this case

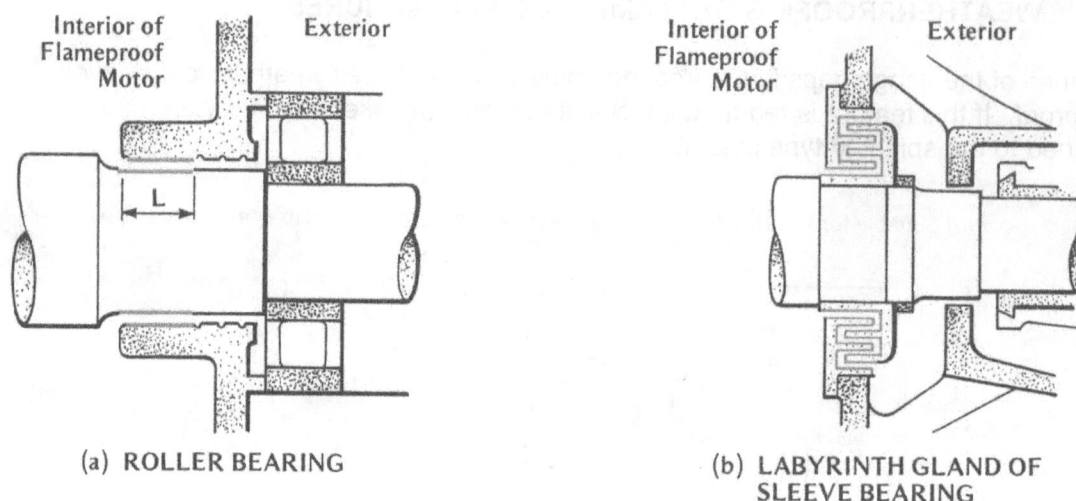

(a) ROLLER BEARING   (b) LABYRINTH GLAND OF SLEEVE BEARING

FIGURE 6.4
FLAMEPROOFING OF MOTOR SHAFTS

the flamepath is the tortuous route through the labyrinth. As the gap may have to be wider because of some rotor radial movement, the table on page 30 gives additional figures for a 40mm long flamepath (easily achievable with a labyrinth) and a correspondingly wider maximum gap.

In some larger motors, particularly those with sleeve bearings which are subject to wear, the radial movement of the rotor may be too large to be accommodated within the allowable maximum gap. In such cases (not shown here) the gap assembly is mounted on the shaft and moves radially with the shaft to take up bearing wear.

6.8 OPENING UP, INSPECTING AND REASSEMBLING FLAMEPROOF EQUIPMENT

Unless it is part of an intrinsically safe circuit, equipment containing energised components located in hazardous areas must on no account be opened without prior isolation of all incoming connections, including the neutral conductor. If essential work has to be carried out on equipment with exposed live components, it may not begin until the area is first designated 'gas-free' and the work is authorised by the Gas Test section of the Preparation! Reinstatement Permit.

The boltheads which secure covers on flameproof equipment are shrouded (see Figure 6.1) or countersunk (see Figure 6.2(a)) against damage and may require special tools to release them.

Once the equipment is open, loose parts such as covers must be very carefully handled to avoid damage or distortion. Opportunity should be taken from time to time to check the flatness of all mating surfaces by using a straightedge. Even a slight distortion could prevent correct closing up on reassembly, thereby causing in places a larger gap than the maximum permitted. This could destroy the flameproof-effectiveness of the whole assembly. Shaft and spindle gaps should also be checked, especially those in constant use.

Before reassembly, mating surfaces should be checked to make sure they are clean. A single speck of foreign matter could effectively open a flange gap and destroy the flameproof property of the equipment. Flamepath surfaces should be preserved using a **light** application of Chemodex or Molypol PCB grease. Also, the equipment must be reassembled **exactly** as it was, down to the last washer. Incorrect reassembly could invalidate the equipment's certificate.

It is absolutely essential that, after covers have been replaced, all bolts be in place and tightened down (it is good practice to do this in sequence to prevent distortion). If a bolt is left out, not only could the gap between the flanges under it be widened but the bolt-hole itself would 'shunt' the flamepath and effectively shorten it. It is all too easy, when servicing a piece of flameproof apparatus, to lose a bolt and intend to replace it later - only to be forgotten. All flameproof equipment must be regularly checked to see that all covers are on and all bolts are in place and tightened down.

All inspections and maintenance procedures must be carried out in accordance with British Standard 5345 - 'Code of Practice for the Selection and Maintenance of Electrical Apparatus in Potentially Explosive Atmospheres'.

6.9   INTRINSIC SAFETY ('EX-i')

Some circuits may be such that even switching will not create an arc with sufficient energy to cause ignition, nor would any internal fault create one. Such circuits are termed 'intrinsically safe' and are invariably of low voltage. Examples include some electronic and most telephone speech circuits (but not the telephone power signalling circuits). The British Standard definition is:

> 'An intrinsically safe electrical **circuit** is one in which any spark or thermal effect, produced either when it conforms electrically and mechanically with its design specification or in specified fault conditions, is incapable of causing ignition of a given explosive mixture.'

It should be noted that a circuit which itself is intrinsically safe may be rendered unsafe by connection to an external element - for example inductive apparatus such as a choke, which greatly increases the stored energy. By the same token a circuit which is intrinsically safe must on no account be electrically connected to a circuit which is not intrinsically safe. Any coupling must be by non-electrical means.

For an **apparatus** to be classed as intrinsically safe it must contain **only** intrinsically safe circuits. Apparatus is defined in two levels as follows:

Ex-ia   apparatus in this category is incapable of causing ignition in normal operation, or with a single fault, **or with any combination of two faults applied**. This offers the highest degree of protection, and such apparatus may be used in all zones.

Ex-ib   apparatus in this category is incapable of causing ignition in normal operation or with any single fault applied. It may be used in all zones except Zone 0.

All electrical plant, equipment and apparatus installed in hazardous (Zones 0, 1 or 2) areas on a platform must either be intrinsically safe or be ex-protected and used only within the limits of the definitions above.

6.10   SUMMARY

The difference between zones in practice seldom makes any difference to electrical equipment, since the risk of transmitting a spark is the same whether the presence of gas is likely under normal or abnormal conditions. Most ex-protected electrical equipment is consequently rated as 'suitable for Zone 1 or Zone 2 areas'.

Not all types of ex-protection will be found on any one installation offshore, but the most common are:

Ex-d   for motors, junction boxes, switches, loudspeakers and some lighting fittings,

Ex-e   for lighting fittings generally,

Ex-i   for instrumentation and speech circuits,

Ex-p   for very large motors and equipment in drilling areas.

Main switchboards (HV and LV) and generating sets are normally installed in non-hazardous areas. Therefore there is generally no need for them to be ex-protected, and they are of normal industrial or marine construction.

Particular care must be exercised when making temporary connections or disconnections for maintenance on any apparatus in a hazardous area. On no account may the above rules be breached, nor may any ex-protected equipment be opened up unless the surrounding atmosphere has been confirmed gas-free.

6.11   MODIFICATIONS TO EX-PROTECTED EQUIPMENT

All ex-protected equipment used on offshore installations is certificated by BASEEFA as conforming with the current rules. The issue of the certificate is based on a type-test of the equipment in question.

Any alteration or modification, however minor, not only invalidates the certificate but could unwittingly lead to a dangerous situation.

On no account may any modification be made, or any but the correct part be fitted, to any ex-protected equipment by platform or station personnel. If a modification is made by the manufacturer, a new type-test and a new certificate are required.

CHAPTER 7 QUESTIONS AND ANSWERS

7.1   QUESTIONS

1.   In which document would you expect to find detailed electrical safety rules and instructions?

2.   Which Work Permits will be required to carry out work on a high-voltage motor cable box?

3.   Who may issue an Electrical Permit to Work?

4.   What safety aspects are dealt with by the Company's Standing Instructions Electrical when work is to be carried out on high-voltage apparatus?

5.   Where are padlocks used for the security of electrical spaces and apparatus, and what is the purpose of each?

6.   Which type of padlock has unique numbered keys?

7.   What is the least electric shock current that is regarded as possibly fatal to a human?

8.   What conditions make a man particularly vulnerable to electric shock?

9.   What action would you take on seeing a man unconscious and touching a live conductor?

10.   Name, and briefly describe, the three classes of portable electric tool.

11.   If a Class I tool is still in use, what steps would you take to check its safety?

12.   How would you recognise a Class II tool or safety isolating transformer?

13.   Why is the centre point of a safety isolating transformer secondary winding tapped and connected to earth?

14.   What is a typical company policy on the use of electric portable tools?

15.   Why are the following types of equipment earthed:

    (a)   all electrical equipment
    (b)   some process equipment?

16.   Why are the flanges of some pipework bonded?

17.   How is equipment earthed? What is the maximum acceptable earth resistance, and how is it checked?

18.   How would you organise the inspection of earthing and bonding?

19.   What special dangers do current transformers present?

20.   When disconnecting a piece of apparatus containing a high-voltage capacitor, what precaution would you take?

21.   Hazardous areas are divided into three Zones. Describe them briefly.

22.   Describe the manner in which gas may become ignited from an electrical source.

23.   Name, and briefly describe, how the methods 'Ex-d', 'Ex-i' and 'Ex-p' render electrical equipments safe in a hazardous area.

24.   What precautions would you take when opening up, and reclosing, flameproof equipment for inspection or maintenance?

25.   Why are local modifications to ex-protected equipment not allowed?

7.2 ANSWERS

(Figures in brackets after each answer refer to the relevant chapter and paragraph in the text).

1. Company Standing Instructions Electrical. (1.1)

2. Hot Work permit; possibly a Preparation/Reinstatement permit; and an Electrical Permit to Work. (1.4)

3. Only a Senior Authorised Person. (1.4)

4. - Ensuring that the apparatus is dead,
    - Isolation,
    - Earthing-down,
    - Screening, CAUTION and DANGER notices,
    - Arrangements for issue of Permit to Work,
    - Accompanied Working,
    - Switching. (1.5)

5. In switchrooms — for general use for preventing access.

    On HV equipment — to prevent unauthorised interference with HV equipment.

    On LV equipment — to prevent unauthorised interference with LV equipment.

    For isolation — to prevent unauthorised interference with electrical equipment which is not available for operational use and which is under cover of an appropriate permit. (1.6)

6. Safety padlock. (1.6)

7. 15mA. (2.1)

8. (a) Standing on metal deck or near metal structure.
    (b) Wet deck.
    (c) Perspiration. (2.1)

9. (a) Switch off current (if possible without delaying).
    (b) Pull victim away using dry material and standing on dry floor.
    (c) Send someone to fetch medical assistance and notify the OIM or Superintendent.
    (d) If victim is not breathing, start respiratory resuscitation at once.
    (e) If victim's heart has stopped, start cardiac massage. (2.2.1)

10. (a) Class I: operates direct from mains and has an earthed metal casing.

    (b) Class II: operates either direct from mains or through a safety isolating transformer and is not earthed. Class IIA is 'all-insulated'; all metallic parts are enclosed in an insulated covering. Class IIB is 'double-insulated' or has reinforced insulation.

    (c) Class III: similar to Class II but must use a safety isolating transformer and operates at 42V rms or less. (3.1)

11. Check the state of the flex and replace if necessary. Check the continuity of the earth connection between the tool casing and the platform or station earth. Maximum value 0.1 ohm. (3.2)

12. By the symbol ▢ on the nameplate. (3.1)

13. To ensure that, if any part of the tool should become live from either secondary terminal, the voltage to earth - and therefore the risk to the operator - is only half the actual secondary voltage. Thus a transformer with 110V secondary has only 55V to earth, and a Class III tool operating at 42V has only 21V to earth. (3.5)

14. Most companies policy is to phase out Class III tools completely in favour of Class II, and eventually to adopt Class when they become available. (3.6)

15. (a) The casing of any electrical equipment can become live, and therefore a danger to personnel, due to a fault developing inside the casing. Earthing the case prevents it from becoming live and discharges the resulting fault current direct to earth. (4.2)

    (b) Vessels and pipework of process plant can become electrostatically charged by movement of the fluids within them, resulting in the risk of a spark discharge and ignition of any gas present outside. Earthing the vessel prevents the build-up of such charges. (4.4)

16. The mechanism described in Answer 15(b) can also build up charges on adjoining, but otherwise insulated, pipes, and there could be a spark discharge between adjacent flanges. (4.4)

17. By using an 'earth strap' of strip or wire of sufficient section to carry the earth-fault current for a limited time and which has a resistance of not more than 0.1 ohm between the equipment and the platform or station earth. The ohmic value of the earth strap is checked by use of a continuity tester. (4.3)

18. All earth straps of electrical and process equipments throughout an offshore or onshore installation should be listed in a register. Systematically they should each be visually inspected for physical damage and measured for continuity resistance. The results should be logged. (4.5)

19. If the secondary circuit of a current transformer is opened while the primary is carrying current, a possibly lethal voltage could appear at the point of break. (5.5)

20. A short-circuit should be applied to the capacitor terminals and left in place until work is finished. (5.6)

21. Zone 0: An area in which an explosive gas/air mixture is continuously present, or present for long periods.

    Zone 1: An area in which an explosive gas/air mixture is likely to occur in normal operation.

    Zone 2: An area in which an explosive gas/air mixture is not likely to occur in normal operation, and, if it does occur, it will exist only for a short time. (6.1)

22. Gas can become ignited by:

    (a) Electrostatic spark discharge between charged bodies or between a charged body and earth.

    (b) Arcing between the enclosure of a faulty piece of electrical equipment and earth.

    (c) Arcing between switching contacts of electrical equipment (switch, contactor or relay).

    (d) An internal arcing fault in a piece of electrical equipment.

    (e) 'Hot spots' due to local overheating.   (6.2)

23. Ex-d is otherwise known as 'flameproof'. It is an enclosure designed to withstand a limited internal explosion and not to propagate the explosion products through joints or seals so as to ignite an explosive gas/air mixture outside the enclosure.

    Ex-i is known as 'intrinsically safe'. A **circuit** is intrinsically safe when the energy stored in it is insufficient to produce a spark or thermal effect which can cause ignition of a given gas/air explosive mixture. An intrinsically safe **apparatus** may contain only intrinsically safe circuits.

    Ex-p is 'pressurised'. Such equipments maintain a small internal overpressure (of the order of 0.5 mbar minimum) which prevents the entry of a gas/air explosive mixture or flammable substance.   (6.3)

24. (a) Before any flameproof equipment is opened up the area must be certified gas-free.

    (b) Once open, loose parts such as covers must be carefully handled to avoid damage or distortion.

    (c) Make sure that equipment is reassembled **exactly** as it was to the smallest detail.

    (d) Before reclosing, ensure that all mating surfaces are absolutely clean.

    (e) Ensure that all cover bolts are in place, and that all are tightened down. (6.8)

25. Any change made to a BASEEFA-certificated equipment invalidates the certificate. It may also leave the equipment in an unsafe condition. All modifications must be official and in most cases require a new certificate. (6.11)

PART 10 ELECTRICAL PROTECTION

CHAPTER 1 INTRODUCTION

1.1   THE REASONS FOR PROTECTION.

Any consideration of an Electrical Distribution System is not complete without some thought being given to Protection.

Electrical Equipment that is correctly installed and maintained is normally very reliable, but the consequence when this equipment does become faulty can be out of all proportion in terms of danger, extensive damage and loss of production unless it is adequately "protected".

Electrical plant, machines and distribution systems must be protected against damage that may occur through abnormal conditions arising.

Abnormal conditions may be grouped into two types:
- Operation outside the designed ratings due to overloading or incorrect functioning of the system.
- Fault conditions due usually to breakdown of some part of the system.

The first condition is usually 'chronic' - that is, it may persist for some time and is often acceptable for a limited period. It may give rise to temperatures outside the design limit of the machines and equipment, but, unless these are very excessive or very prolonged, it seldom causes sudden or catastrophic failure. It can usually be corrected before it leads to breakdown or a fault condition.

The second condition on the other hand is 'acute' and arises from electrical or mechanical failure that, once established, produces a condition beyond control. It usually gives rise to very severe excess currents that will quickly cause catastrophic failure of other electrical and mechanical plant in the system unless the fault is rapidly isolated. It may be caused by a breakdown of insulation due to a material failure or overheating or to external conditions such as weather, or it may be due to physical damage to an item of plant or cable.

Automatic protection against these conditions is possible in electrical installations because it is easy to measure various parameters, to detect abnormalities, and to set in motion the protective action the instant an abnormality arises.

An electrical network normally operates within its designed rating. Generators, transformers, cables, switchboards, busbars and connected apparatus are each designed to carry a certain maximum current. Most can carry a moderate overload for a short time without undue overheating.

However, if a fault should develop somewhere in the system, that is to say a phase to phase short circuit or a phase to earth breakdown, then all connected generators will feed extremely high currents into that fault, which will be limited only by the impedance of the complete circuit from generator to fault. Fault currents can be ten to twenty times the normal full load current.

Such currents will quickly cause intense overheating of conductors and windings, leading to almost certain breakdown unless they are quickly disconnected; they will also give rise to severe mechanical forces between the current carrying conductors or windings. All such apparatus must be manufactured to withstand such forces.

The purpose of automatic protection is to remove the fault from the system and so break the fault current as quickly as possible. Before this can be achieved, the fault current will have flowed for a finite, if small, time, and much heat energy will have been released. Also the severe mechanical forces referred to above will already have occurred and subjected all conductors to intense mechanical stress.

1.2 ROLE OF PROTECTION.

Protection is needed to remove, from the system, as speedily as possible any part of the equipment in which a fault has developed. So long as it is connected the whole system is in jeopardy from three main effects of the fault, namely:
- a risk of extended damage to the affected plant.
- a risk of damage to healthy plant.
- a risk of extending the outage to other plant on the system, with resultant loss of protection and interruption of vital processes.

It is the function of protective equipment, in association with the automatic switch fuse, contactor or circuit breaker to avert those effects.

1.3 PROTECTION PRINCIPLES

Protection of an electrical system is provided for one or more of the following principles:
- To maintain electrical supplies to as much of the system as possible after a fault has been isolated.
- To protect the generators and other plant against damage due to abnormal conditions and faults.
- To protect the consumer equipment against damage due to abnormal conditions (e.g. overload).
- To isolate faulty equipment to limit the risk of fire locally.
- To limit damage to the cable system resulting from a fault.

These principles will determine the type of protective equipment fitted in any installation. It will be noted that the first principle conflicts with the other requirements to some extent. For example, the best way to protect a generator against damage by fault currents is to disconnect it, but it would not then be available to supply other consumers.

Where continuity of supply is considered essential alternative feeds are necessary. But, if full advantage is to be gained from this additional capital outlay, the protection must be highly 'selective' in its function.

For this it must possess the quality known as 'discrimination' whereby it is able to select and disconnect only the faulty element leaving all others in normal operation so far as it is possible.

1.4 DISCRIMINATION

If we consider a simple typical electrical layout the need for some form of discrimination will become clear.

Figure 1.1 shows an 11kV Oil fuse switch (OFS) controlling a transformer beyond which there are a bank of Low Voltage (LV) fuses. Clearly a fault as indicated must be interrupted by fuse B so that supply may continue to the other circuits. The 11kV OFS, fuse A must not trip.

11kV

415/240V

A

B

➤ **Fault Current**

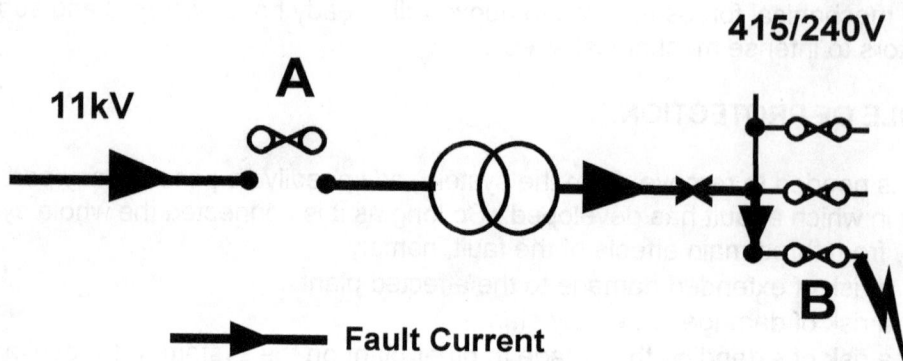

FIGURE 1.1
11kV OIL FUSE SWITCH

1.5   FAULTS AND FAULT LEVELS

Before selecting a protection system we must consider the kind of fault that may occur.

The principal types are:
- 3 Phase (with or without earth)
- Phase - to - phase
- Phase - to -Earth
- Double Phase - to - Earth

Sometimes there are open-circuits involved. Transformers and motors are also subject to short circuits between turns of the same winding.

Only the 3-phase short circuit is a balanced condition. The others are unbalanced and require a knowledge of the method known as 'symmetrical components' before they can be fully analysed.

This analysis is necessary if the amount of fault current that will flow is to be correctly predicted but is beyond the scope of this summary.

However, if a fault should develop somewhere in the system, that is to say a phase-to-phase short-circuit or a phase-to-earth breakdown, then all connected generators will at first feed extremely high currents into that fault, which will be limited only by the impedance of the complete circuit from generator to fault. Fault currents can be ten or more times the normal full-load current.

Such currents will quickly cause intense overheating of conductors and windings, leading to almost certain breakdown unless they are quickly disconnected. They will also give rise to severe mechanical forces between the current-carrying conductors or windings. All such apparatus must be manufactured to withstand these forces.

> **A fault current of 50000A (rms) flowing in two busbars 3 inches apart will produce between them a peak mechanical force of nearly half a ton-force per foot run of bars.(or approx. 500Kg per metre)**

The purpose of automatic protection is to remove the fault from the system and so break the fault current as quickly as possible. Before this can be achieved, however, the fault current will have flowed for a finite, if small, time, and much heat energy will have been released. Also the severe mechanical forces referred to above will already have occurred and will have subjected all conductors to intense mechanical stress.

CHAPTER 2 OVERVIEW OF PROTECTION COMPONENTS

2.1   FUSES

Modern H.R.C. fuse-links are manufactured to the highest standard to ensure reliability. The ceramic cartridge usually holds a fuse element of pure silver surrounded by powdered quartz. The quartz filler is required to condense the metal vapour, produced by the element under short circuit conditions, as quickly as possible. In addition, due to the high temperatures produced by the fault condition, the quartz forms a glass insulating barrier between the contact points. The greater the surface area in contact with the filler the faster is the heat dissipation. This is why the element is made up of fine wires or strips. By modifying the shape of these strips the speed at which a fuse will blow with any particular current flowing can be controlled. This relationship between speed or time and current is known as the time- current characteristic. In selecting a fuse you must ensure that:
- The fault level of the circuit does not exceed the fuse 'Breaking Capacity'
- The maximum load will not exceed the Fuse Rating.
- The fuse will rupture with an earth fault at the remotest point of the apparatus controlled.

Discrimination will be obtained with protection both preceding and following the fuse.

2.2   THERMAL TRIP UNITS

Lower levels of fault current from a little over the rated load current of the circuit-breaker up to about 10x load current setting the protection is by a thermal tripping device. This consists of a bimetallic strip that is deflected by the heat generated in the strip by the fault current and eventually trips the circuit- breaker or contactor. The arrangement provides a time delay that is inversely proportional to current

These are normally used on Low Voltage motors up to approx. 50 kW. They are usually an integral part of the contactor and cause tripping by breaking the contractor holding coil circuit. The heaters are connected one in each phase in the main circuit and carry the actual motor current and only for large motors would current transformers be used. The indirectly heated device usually has a heater winding wrapped around the bimetal whereas the bimetal itself is shaped so that it becomes the path for the current in the case of a directly heated device.

2.3   ELECTROMECHANICAL RELAYS

When two protection devices are required to discriminate the chosen settings will depend on how closely the devices can be guaranteed to conform to their characteristic curves. Most devices have fairly generous tolerances in both operating levels and time and therefore if close discrimination is required then protection relays would have to be used.

A relay is a device, which makes a measurement or receives a signal, which causes it to operate and to effect the operation of other equipment.

A protection relay is a device, which responds to abnormal conditions in an electrical power system to operate a circuit-breaker to disconnect the faulty section of the system with the minimum interruption of supply.

Many designs of relay elements have been produced but these are based on a few basic operating principles. The great majority of relays are in one of the following groups.
- Induction Relays
- Attracted-armature relays.
- Moving-coil relays.
- Thermal relays.
- Timing Relays.

2.4   STATIC OR ELECTRONIC RELAYS

Relays based on electronic techniques offer many advantages over the older electromechanical type. Apart from the obvious advantage of no moving parts the power requirements are low and therefore smaller current and voltage transformers can be used to provide the input. Additional benefits are improved accuracy and a wider range of characteristics.

The invention of the transistor and the microprocessor has allowed the development of static relays but difficulties were experienced because the high voltage substation proved to be a very hostile environment to the device. The close proximity of high voltage heavy current circuits produces conditions which could damage the transistor because of its low thermal mass or cause mal-operation of the relay because of the electromagnetic or electrostatic interference.

A lot of research and development has taken place and commercial relays, which meet very exacting standards, have been produced. Electromechanical relays will still represent a large proportion of relays remaining in service. However as new equipment and systems are designed it is likely that there will be a change-over to static relays and most of the future development in protection will be in static relays.

The large application potential of the digital integrated circuit has led to enormous expenditure on research and development which has resulted in microprocessors with spectacular computing capabilities at a low cost. It is fairly certain that microprocessors will ultimately dominate protection and control systems.

The utilisation of microprocessors in the field of protection means that a programme held in the microprocessor memory can replace the logic part of the relay. This enables a relay function to be specified by software, which widens the scope of the relay and allows a single relay to be provided with a number of characteristics.

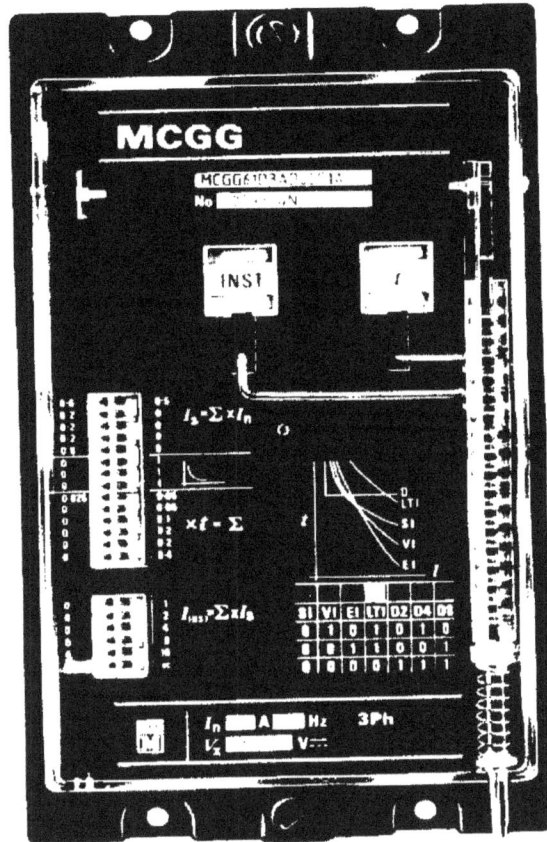

FIGURE 2.1
MICROPROCESSOR-BASED OVERCURRENT RELAY

Experience has been gained with microprocessors in high voltage substations over a number of years by using them for voltage control, automatic switching and reclosing and other control functions. Therefore the difficulties, which arise in this environment, have been overcome.

An example of the versatility of the microprocessor is demonstrated in one of the first protection applications. This is an overcurrent relay which has a setting range of 10% to 200%, an extremely wide range made possible by the low power requirements of the relay, and a choice of five different characteristics. Figure 2.2 shows the block diagram of the relay.

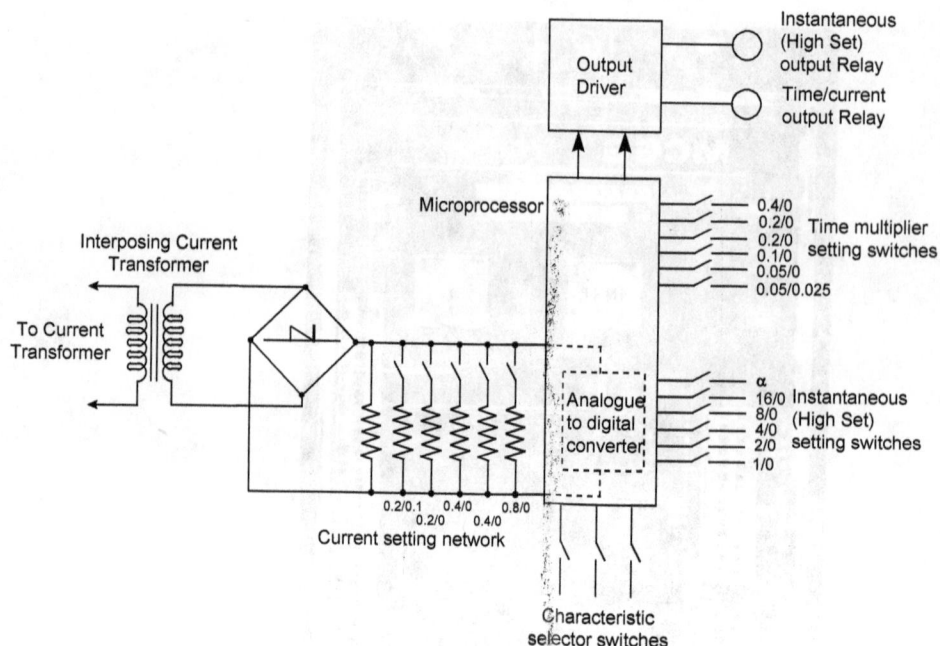

FIGURE 2.2
SIMPLIFIED BLOCK DIAGRAM

The CT current is reduced to a more suitable level by an interposing current transformer in the relay. The current is rectified and passed through a resistance network which produces a voltage output which is proportional to current. The network provides the current setting control by switches mounted on the front of the relay and its output is fed into the analogue-digital converter which is part of the micro-processor.

Three other banks of switches are mounted on the front of the relay. The switches are connected to separate input ports on the microprocessor and control the time multiplier setting, the high-set relay setting and selection of the type of characteristic required, i.e. normal, very or extremely inverse, long time inverse or definite time.

2.5 ELECTRONIC INVERSE TIME OVERCURRENT RELAY

To a considerable extent protection relays of the electromagnetic type, in which a moving armature or disc is actuated by some kind of electromagnet, are being superseded by electronic types. In these the functions of signal detection and processing are carried out by entirely static circuits, and only the final operation of contacts is done by electromechanical relays, which can be of any suitable but simple control type. The advantages of this technique include a greater flexibility in providing virtually any desired function, however complex, better accuracy, ease of adjustment, and the usual benefits of static circuits with regard to reliability and freedom from regular servicing requirements.

The diversity of functions and principles to be found in static protection relays is such that no comprehensive discussion is possible here. The various characteristics and adjustments established in electromagnetic relay practice are readily reproduced electronically. Figure 2.3, without exactly representing any actual apparatus, illustrates as an example the application of analogue principles to inverse-time overcurrent protection. (An analogue system is one in which continuously variable internal signals are used to represent external quantities such as current and time.)

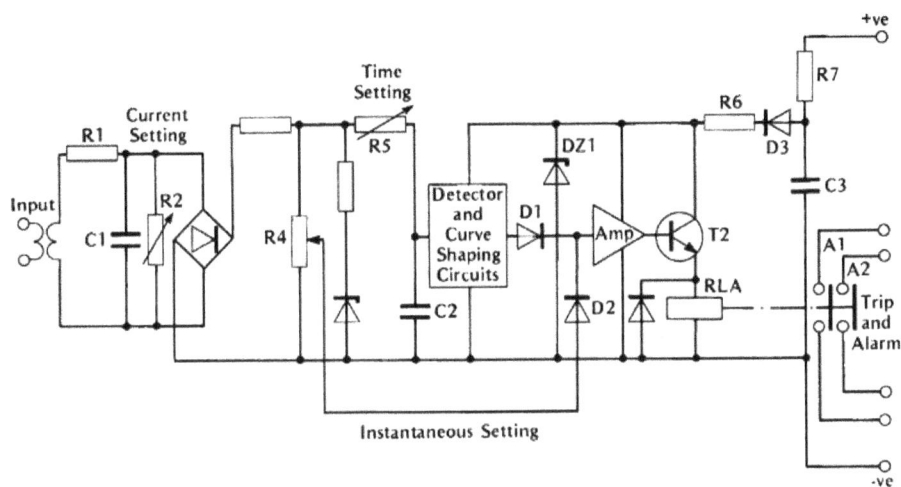

(a) Circuit Diagram

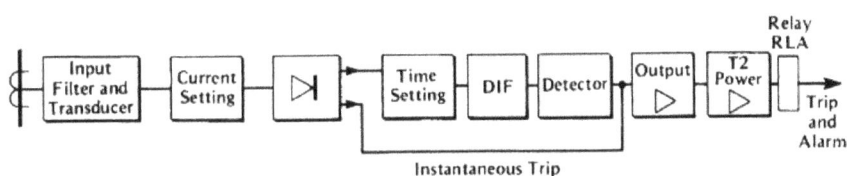

(b) Corresponding Block Diagram

FIGURE 2.3
ELECTRONIC INVERSE TIME OVERCURRENT RELAY

In Figure 2.3(a) the input from the line current transformer is fed through a small matching transformer to a low-pass filter R1/C1, which suppresses transient voltage, surges. A voltage proportional to the input current is developed across the current-setting potentiometer R2. This voltage is applied to the bridge rectifier.

The d.c. output voltage, which is proportional to the line current, is used to charge the capacitor C2 through the potentiometer R5. The setting of this potentiometer determines the rate at which the voltage across C2 increases and hence the timing of the inverse-time operating characteristic of the relay. When the voltage across C2 reaches a predetermined value, the detector circuit operates to switch the electromechanical relay RLA through the output amplifier and power transistor T2.

'Instantaneous' operation is obtained by applying the output voltage of the bridge rectifier directly to the input of the amplifier through R4. Thus, for higher values of fault current, the inverse-time delay circuit is bypassed.

The power supply for the solid-state circuits is applied through D3 and R6. It is stabilised by Zener diode DZ1, and R7 and C3 afford spike protection. The diode B3 guards against reversed polarity of the d.c. power supply.

Figure 2.3(b) shows the corresponding circuit in block form.

The flexibility and scope of present-day electronics enables a very wide variety of characteristics to be created with relative ease. While a simple analogue overcurrent circuit has been described above for the purpose of illustration, digital techniques have latterly been adopted very widely as a result of the availability of microprocessors and other digital integrated circuits.

2.6 CURRENT AND VOLTAGE TRANSFORMERS

Current transformers

The current transformer is well established but it is generally regarded as merely a device which reproduces a primary current at a reduced level. A current transformer designed for measuring purposes operates over a range of current up to a specific rated value, which usually corresponds to the circuit normal rating, and has specified errors at that value. On the other hand, a protection current transformer is required to operate over a range of current many times the circuit rating and is frequently subjected to conditions greatly exceeding those which it would be subjected to as a measuring current transformer. Under such conditions the flux density corresponds to advanced saturation and the response during this and the initial transient period of short-circuit current is important.

It will be appreciated, therefore that the method of specification of current transformers for measurement purposes is not necessarily satisfactory for those for protection. In addition an intimate knowledge of the operation current transformers is required in order to predict the performance of the protection.

Current transformers have two important qualities:

- They produce the primary current conditions at a much lower level so that the current can be carried by the small cross-sectional area cables associated with panel wiring and relays.
- They provide an insulating barrier so that relays which are being used to protect high voltage equipment need only be insulated for a nominal 600V.

Current transformers are usually designed so that the primary winding is the line conductor which is passed through an iron ring which carries the secondary winding. They are mostly of this type and are known as bar-primary or ring-wound current transformer.

Voltage transformers

The voltage transformer in use with protection has to fulfil only one requirement, which is that the secondary voltage must be an accurate representation of the primary voltage in both magnitude and phase.

To meet this requirement, they are designed to operate at fairly low flux densities so that the magnetising current, and therefore the ratio and phase angle errors, is small. This means that the core area for a given output is larger than that of a power transformer, which increases the overall size of the unit. In addition, the normal three-limbed construction of the power transformer is unsuitable as there would be magnetic interference between phases. To avoid this interference a five-limbed construction is used, which also increases the size. The nominal secondary voltage is sometimes 110V but more usually 63.5V per phase to produce a line voltage of 110V.

CHAPTER 3 PROTECTION OF GENERATOR/TRANSFORMER UNITS

Where a generator is connected to the power system by means of a generator transformer it is usual to protect the generator and transformer as a single unit using biased differential protection.

The current transformer balance is produced in terms of both phase and magnitude, i.e. in the arrangement shown in Figure 7.5 there is an overall phase change of 30° that is corrected by connecting a set of auxiliary current transformers in delta. Because of the difference in current transformer ratios the settings of the generator transformer protection has to be somewhat higher than the settings of the generator protection. Because of this the generator is sometimes protected separately but is also included within the zone of the generator-transformer protection as an extra insurance. The transformer is connected directly to the generator and so no harmonic restraint circuit is required in the protection as no switching can occur. There is a low level of magnetising inrush current following a fault when the voltage is restored from being depressed but this is usually insufficient to unbalance the protection. Figure 5 shows a complete protection system for a generator.

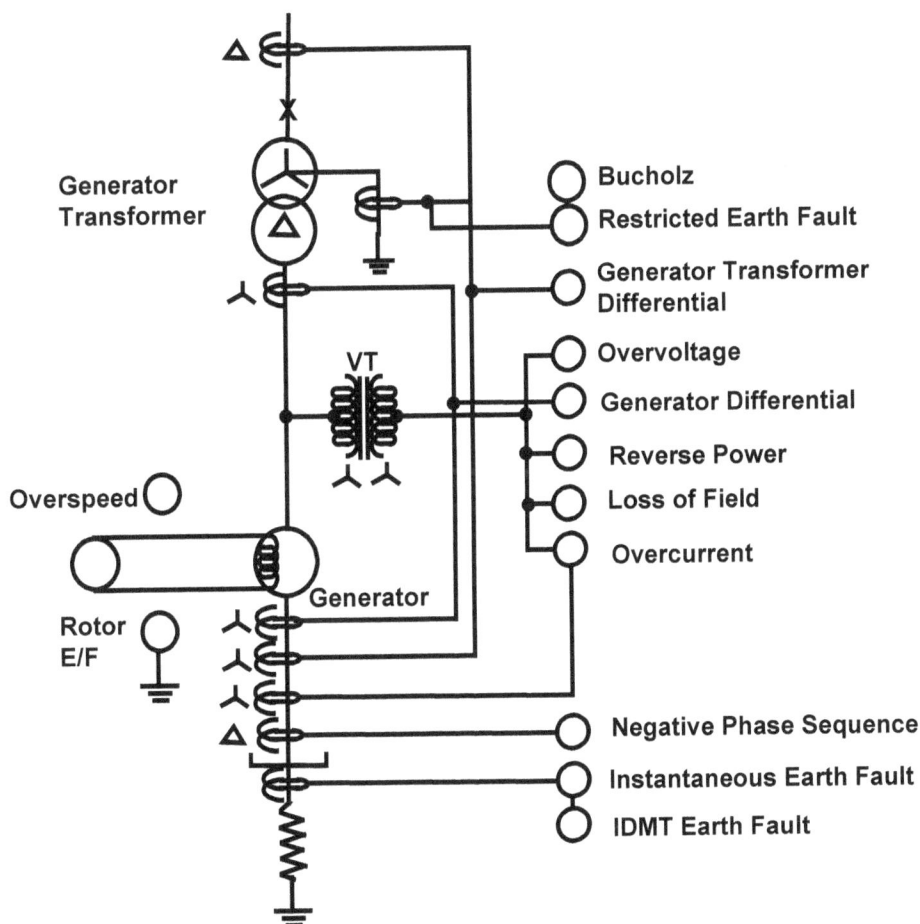

FIGURE 3.1
GENERATOR PROTECTION

CHAPTER 4 TRANSFORMER PROTECTION: GENERAL

All main transformers that transmit bulk power between the generators and the low-voltage distribution system of an offshore installation, and between the Supply Authority's system and the low-voltage equipment in onshore installations, have their own individual protective systems. This is to protect the transformer against damage due to electrical faults arising both outside and inside it.

A typical transformer protection scheme is shown in Figure 4.1, which also shows associated instrumentation. Many of the general protection measures described earlier are applied also to transformers, but in addition there are some more specific ones.

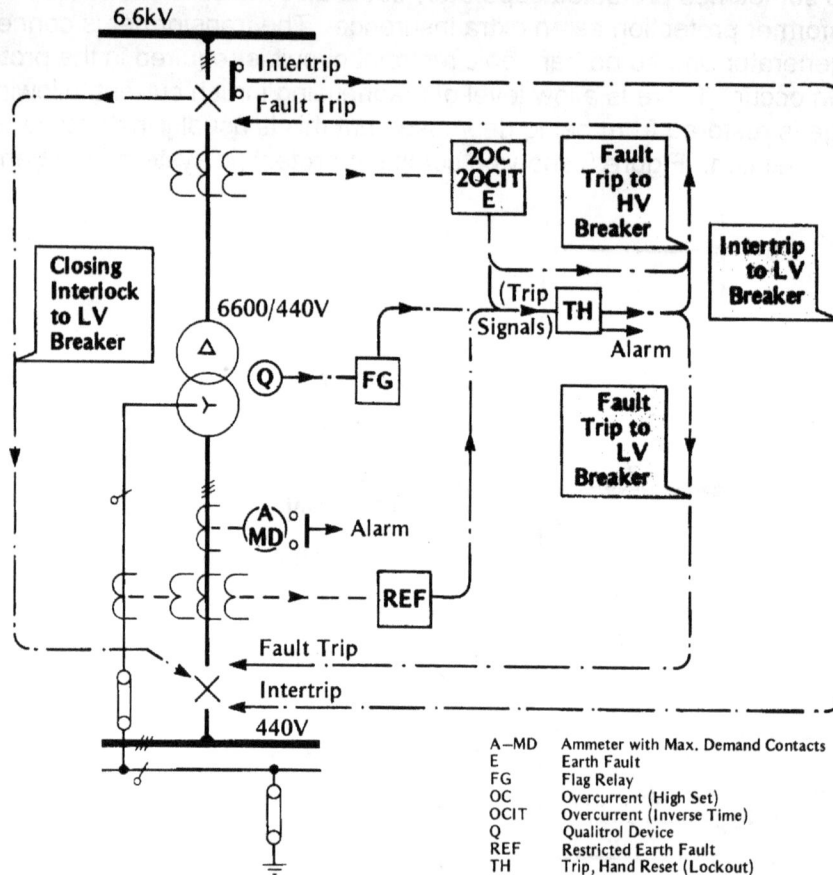

A–MD	Ammeter with Max. Demand Contacts
E	Earth Fault
FG	Flag Relay
OC	Overcurrent (High Set)
OCIT	Overcurrent (Inverse Time)
Q	Qualitrol Device
REF	Restricted Earth Fault
TH	Trip, Hand Reset (Lockout)

FIGURE 4.1
TYPICAL TRANSFORMER PROTECTION

Points worthy of note in Figure 4.1 include the following:

- Overcurrent protection is on the HV side only. It is provided by two inverse- time elements combined with an earth-fault element (2OCIT/E) together with two instantaneous high-set overcurrent elements (2OC), all in the same case. The relay operates to trip the HV circuit-breaker directly and both the HV and the LV breakers through the lock-out relay (TH). The time and current settings will be determined by the overall discrimination plan. Overcurrent on the LV side causes corresponding overcurrent on the HV side, which therefore takes care of both overloading and LV short-circuits.
- Restricted earth-fault protection is used on the secondary side (it is the only secondary-side protection), with four protective type CTs. The relay operates instantaneously to trip both the HV and the LV breakers through the lock-out relay.
- Lock-out hand-reset relay (TH).
- There is interlocking and intertripping from the HV to the LV circuit-breakers (but not in reverse).
- Instrumentation includes a maximum-demand ammeter with an alarm contact.

CHAPTER 5 MOTOR PROTECTION

The motor is protected by an inverse-time overcurrent device which will cause the contactor to trip if the overcurrent is sufficiently high and persists. The device usually takes the form of a thermal element in each phase, either directly or CT-operated. It has an inverse-time characteristic, which is more nearly matched to the thermal behaviour of the motor itself than that of the inverse-time electromagnetic overcurrent relay described earlier. It must allow the large starting current (up to five times full-load current) to flow during the run-up period without operating, but it must trip the motor if even a small overcurrent persists for a longer period. A typical setting of such a device would be 110% full-load continuous current with the appropriate time setting. For short starting times the inverse-time characteristic must be such that the starting current and run-up time are taken into account. In this respect it must be remembered that high-inertia loads such as a motor-generator set or a compressor take much longer to run up than, say, a centrifugal fan.

For the majority of LV motors and a few HV motors the inverse-time device is thermal. For the smaller LV motors it is in series with the motor itself, but for the others it is a separate relay operated through CTS. For most HV motors on the later platforms however, the device is wholly electronic but with a similar characteristic; it too is a separate relay, CT-operated (see Figure 5.1). Where these relays are separate the overcurrent device is combined with certain other features into a single `Motor Protection Relay'.

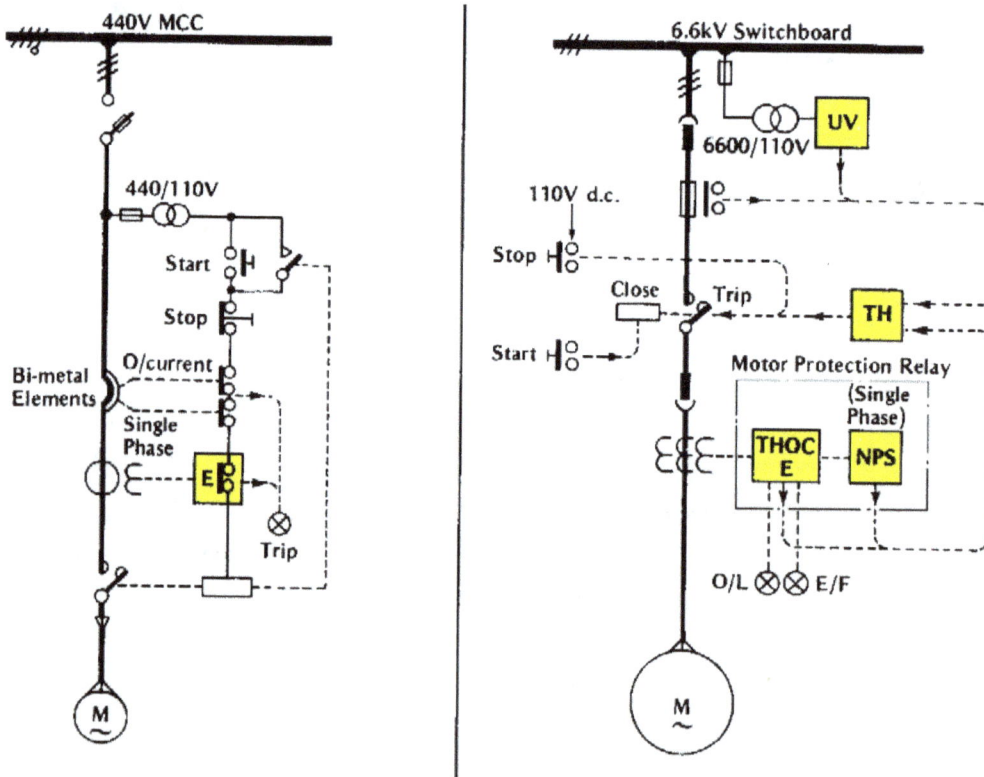

FIGURE 5.1
LV & HV MOTOR WITH PROTECTION RELAYS

A characteristic of inverse-time relays which is particularly noticeable in thermal relays, and which has to be taken into account in allowing for starting current, is `overshoot' (or `overrun'). This means that if the relay is energised with something more than its minimum operating current it may close its contacts even after the current has subsequently fallen below the operating level. For example, a motor could be tripped after it had safely started and reached full speed, even though the relay had not operated during the starting period. This can have a considerable effect on the discrimination that can be achieved between starting and overload currents, unless complications are added to the protection scheme. Whereas the contactor with its inverse-time overcurrent device (thermal or electronic) provides overload protection for the motor, such contactors cannot in general clear a fault of short-circuit proportions. For this they must be backed up by series HRC fuses. When used with motors such fuses must have special characteristics. They must have a continuous rating which will allow them to pass the full-load current of the motor continuously, and they must also allow the considerably greater starting current to pass for the period of the run-up time without melting the fuses.

There are a number of electronic relays that protect the motor in the same way as a thermal relay, which are capable of matching the motor characteristic more accurately. In addition to adjustments for current level the operating time can be adjusted as well as the settings for unbalanced current and earth fault.

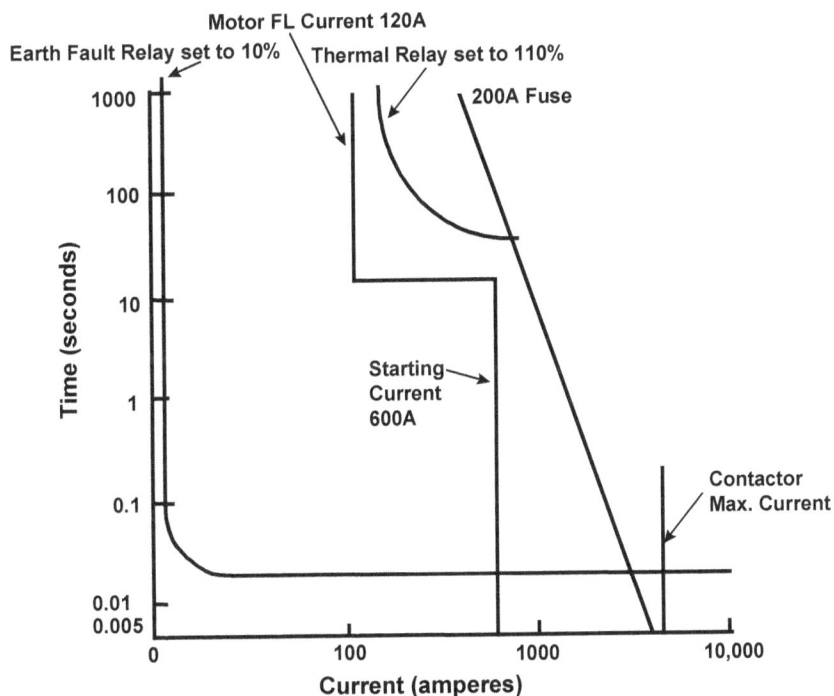

FIGURE 5.2
COMPOSITE CHARACTERISTIC THERMAL RELAY AND 200A FUSE

The type of protection described would only be applied where the motor is supplied via a circuit-breaker. In many cases the switching of the motor is by a contactor which, although capable of making fault current, cannot interrupt fault current. Therefore fuses are used to clear the fault instead of the instantaneous overcurrent relays. The earth-fault relay is used as it will detect low-level faults and trip the contactor but for high-fault levels the fuse 'would operate faster than the earth-fault relay. It may be that a time delay is to be introduced into the earth-fault relay circuit to ensure that the fuse operates faster. Figure 8 shows a typical composite characteristic where 160 A fuse is used in conjunction with a relay thermal element and earth-fault relay to protect a 550kW motor with a full-load current of 120 A. From 132A, 110% full load current to 700A the motor is protected by the thermal relay which would trip the contactor. Above 700A the 200A fuse would clear the fault. Similarly earth fault from 12A to 3000A would be cleared via the contactor and above 3000A by the fuse. The contactor will never be called upon to break a fault current beyond its capability.

CHAPTER 6 OVERVIEW OF PROTECTION TESTING AND MAINTENANCE

6.1   WHY PROTECTION RELAYS NEED MAINTENANCE.

Protective relays are normally inactive, although energised and possibly under some degree of stress, or can possibly be subject to continuous low amplitude vibration. Over a long period the effects of environmental conditions such as temperature, humidity and atmospheric pollution can deteriorate parts of the relays.

A relay may be only be called upon to `work' at very infrequent intervals, as normally failures of modern electrical equipment are comparatively rare, but a long period of inactivity may allow `sticking effects' to develop due to the forming of sticky substances evaporating from the insulating materials used in the relay's construction.

Many other conditions can affect the response of a relay when it detects a system fault. A short list of the more common conditions follows:-

- Continuous vibration at power frequency can cause some contact adhesion.
- Corrosion can also cause considerable problems in some types of relays.
- Silicon oils can be the cause of problems in protection relays and may contaminate the contacts.
- Dust can be a hazard, as it acts as an insulating material and can cause contact failure if allowed to enter a relay case.
- Relay coils are subject to corrosion from electrolytic action. (due to leakage current with respect to earth)

All parts of a relay may be affected by these causes, but the two most vulnerable parts are the contacts and coils. In polluted atmospheres, silver contacts can become coated with black tarnish or silver sulphide which may in time cause protection failure due to high contact resistance.

Particularly inexperienced personnel should avoid over-maintenance of protection relays. Relay adjustment requires personnel with the experience of this type of work.

6.2   IMPORTANCE OF PROTECTION RELAY SETTING DATA.

Complete schedules of all the original commissioning data supplied by the manufacturer/contractor shall, where available, be given to the personnel who are to carry out the preventative maintenance of the protection equipment. This shall include the protective relay settings, the function and details of all such settings and any basic fault calculations used.

6.3   FREQUENCY OF MAINTENANCE

The recommended frequency for maintaining different types of protection equipment is to be found in the manufacturers information. However, individual companies will also have to take into account their own operational experience and the importance of the electrical equipment `protected' by the overall protection scheme, when deciding on how often to maintain relays, etc.

The following is a suggested list of inspections and maintenance activities that should be carried out at differing pre-determined intervals :-

- Inspection of relays and checking of relay settings.
- Trip tests including intertripping tests.
- Insulation resistance checks on all small wiring, etc.
- Secondary injection tests on all protection relays such as overcurrent and earth fault induction type, but possibly less frequently on thermal or electronic units.
- Inspection and testing of transformer's Buchholz Relays.
- Inspection and testing of transformer's Winding or Oil temperature instruments.

CHAPTER 7 PRACTICAL TESTING

7.1   INTRODUCTION

The importance of maintenance and testing of protective gear will be appreciated if its role in the power system is considered. It is different from all other equipment in that it is operative for a very small proportion of the time. It is therefore most important that when operation is required it will function correctly.

To ensure this, regular maintenance and testing of the relay and its associated equipment is required. It is not possible to specify the frequency of testing this depends on the location and the importance of the equipment. An important piece of equipment with protection mounted in a location where conditions are poor would require some attention every 12 months, or maybe more often in the light of experience, whereas a less important unit with protection in a good location would require testing every 4 years. There is no hard-and-fast rule - a case for judgement and common sense and experience.

7.2   WORKS TESTS

To appreciate the aims of site testing it is necessary to consider the tests to which a relay is subjected in the manufacturers works.

During the development stage of a relay, or protection scheme, many tests are performed to achieve the desired characteristics and performance. When development is completed the equipment is subjected to a type test. This is in two parts. The tests which are common to every relay, impact, vibration, insulation tests, etc. and the tests which are particular to the relay to prove its characteristics, speed of operation, stability level, CT requirements, etc.

It is information derived from these tests that will ultimately be used in leaflets describing the relay and specifying its performance. These type tests would be performed on only a few of the relays and none of these would be supplied to customers. Relays, which are supplied against customer's orders, are subjected to a series of tests to prove that their characteristics conform, within limits, to the specification and that it will perform in the manner described in the technical literature describing the relay. The tests required are enumerated in a test specification that usually culminates in a pressure test to check the insulation.

7.3   TESTS ON SITE

The relays are delivered to the site either mounted on a switchgear panel or as "loose" relays for mounting on the panel on site. In the latter case a check should be made before mounting to confirm that the relay has not been damaged in transit.

Tests conducted on the relays when they are installed in their final location are to prove that the connections to the relay are correct and that there is no damage or foreign matter introduced into the relay during installation and, in the case of relays delivered already mounted on the switchgear panel, in transit.

It is desirable to have a wiring diagram of the equipment which is to be tested as this will reduce considerably the time to perform the tests. All the wiring should have ferrule numbers and it is a simple matter to relate these to the wiring diagram.

Another aid is the use of standard wiring numbers which are used by many switchgear manufacturers. These follow the recommendations of BS 158:1961 even though this particular British Standard Specification is now withdrawn. From a knowledge of the nomenclature the function of much of the panel wiring can be deduced without reference to a wiring diagram.

The ferruling consists of a letter that refers to a function and a number that in the case of CT and VT circuits refers to a phase. A, B, C and D are CT circuits, A, B and C are for differential, busbar and overcurrent circuits respectively whilst D is for metering circuits. E is associated with VT circuits. The numbers 10 to 29, 30 to 49 and 50 to 69 refer to R-, Y- and B-phases respectively whilst 70 to 89 are for neutral and residual circuits whilst 90 is for connections made directly to earth. Figure 12 is an example of the use of these numbers. In addition the letter K is used for tripping circuits and L for indication and alarm circuits. The usual practice is to use odd numbers for connections on the positive side of the supply and even numbers for connections on the negative side.

When tests are conducted it is most important to record the results in a clear and legible manner. This not only allows the results to be examined later, but provides a permanent record of the state of the equipment at that time and provides a basis for comparison for future tests.

7.4   COMMISSIONING TESTS

During commissioning a comprehensive series of tests are required to check the whole installation from the current transformers to the tripping circuit.
The tests can be divided into five parts:
- CT Polarity Check
- CT Magnetising Curves
- Relay Characteristic Check
- Insulation Tests
- Tripping Circuit Check.

7.5   CT POLARITY CHECK

In many protection schemes the relative polarity between current transformers is important and therefore tests must be carried out to ensure that they are correctly connected. Figure 7.1 shows the diagram of a current transformer with the current flow convention which is when primary current flows from P1 to P2, secondary current flows from S1 to S2 in the external circuit connected to the current transformers. A simple way of checking the relative polarities is by the flick test which uses a battery to send a pulse of current through the current transformer as shown in Figure 7.2. If a d.c. is passed through the CT from P1 to P2 then there will be a momentary deflection of a voltmeter connected across the secondary winding terminal S1 being momentarily positive. When the current is removed, terminal S2 will become momentarily positive. The usual method is, however, by primary injection.

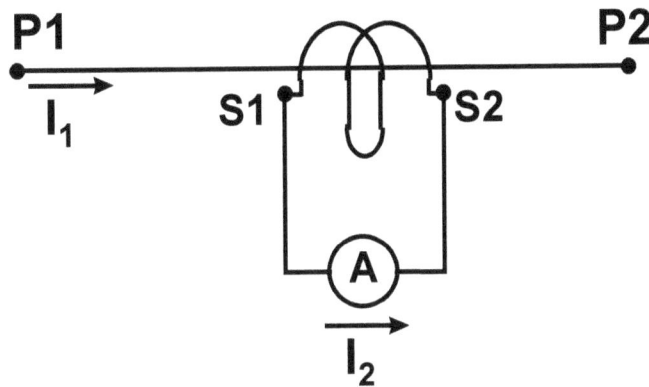

FIGURE 7.1
CURRENT TRANSFORMER SHOWING CONVENTIONAL CURRENT FLOW

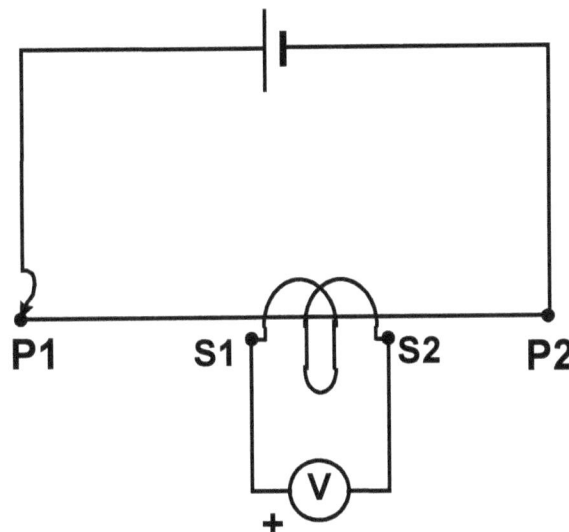

FIGURE 7.2
CHECKING CT POLARITY BY FLICK TEST

7.6   PRIMARY INJECTION

Primary-injection testing involves the passing of heavy currents through the current transformers to establish firstly the ratio and then the relative polarity. A short-circuit is placed as near as possible to the current transformers and current injected. The usual method of injection is into the switchgear feeder orifices by means of expandable rods and placing the short-circuit in the cable connecting box. If the latter is compound filled or if the connections are sleeved then the short-circuit would have to be placed in the CT chamber itself. The arrangement is shown diagrammatically in Figure 7.3.

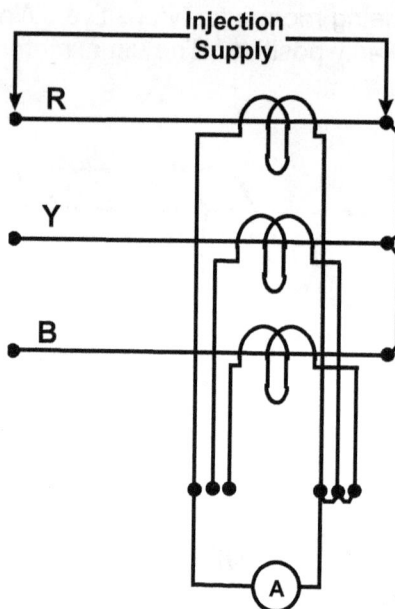

FIGURE 7.3
RATIO CHECK ON RED PHASE CURRENT TRANSFORMER

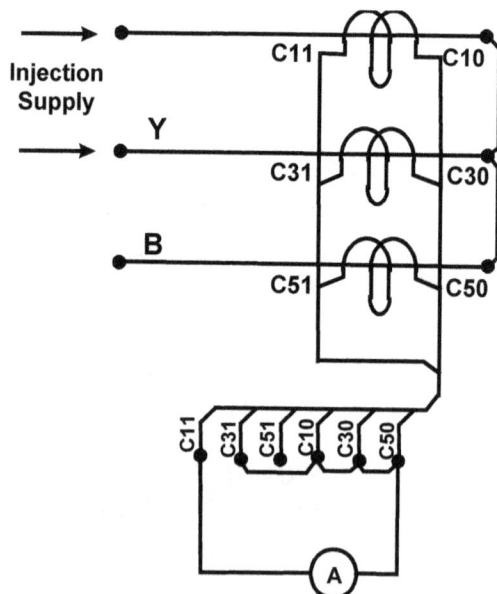

FIGURE 7.4
RATIO CHECK WITH INJECTION AT FEEDER ORIFICES

After making the heavy current connections to inject current in, say, the red phase an ammeter is connected across the CT secondary and a short-circuit to any other CT which will be subjected to primary circuit during the test. The secondary connections are usually made at the main connecting block.

As an alternative the current could be injected via two feeder orifices, say the red and the yellow phases. In this case ammeters could be connected to both red and yellow current transformers or one CT short-circuited whilst the other CT ratio was checked. This arrangement using wiring numbers is shown in Figure 7.4.

Current is injected via the two feeder orifices to check that the polarity of the current transformer is correct. The current which will pass through two current transformers in opposite directions and will produce a secondary current in each secondary winding. If the polarity of the current transformers is correct this current will circulate around the two secondary windings and very little current will flow in the ammeter which is connected as shown in Figure 7.5. If the polarity is incorrect then the sum of the currents in the secondary windings will pass through the ammeter.

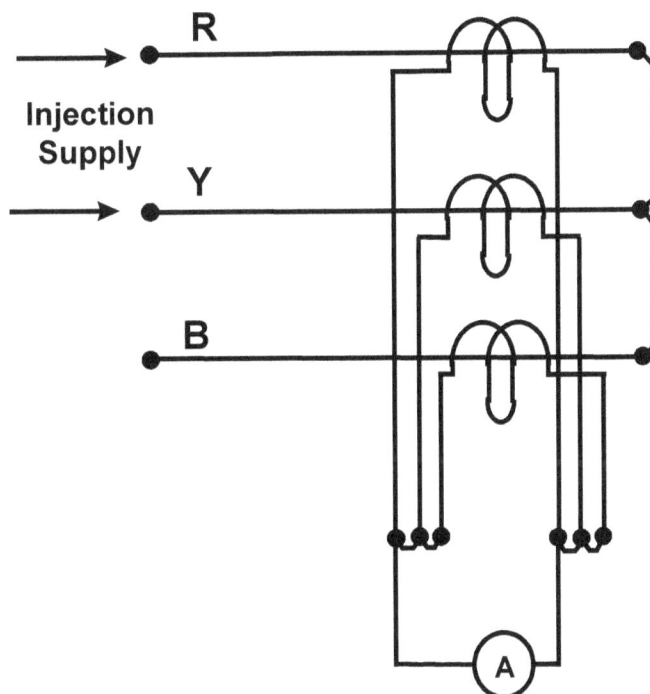

FIGURE 7.5
POLARITY CHECK, RED AND YELLOW CURRENT TRANSFORMERS

Should an incorrect polarity be indicated then the injection through a different phase will reveal the incorrect current transformer, e.g. if an incorrect polarity is shown when the injection is via R and Y and also when the injection is via Y and B then the Y current transformer is incorrect. If the R-Y injection shows an incorrect polarity and a Y-B injection does not the R current transformer is faulty.

In all cases, whether a correct or incorrect polarity is shown in the two tests, the third test should be performed as a double check.

Having checked the current transformer ratios and that the relative polarity is correct the main function of primary-injection testing has been fulfilled and further testing can be performed by simulating the current transformer secondary current by injecting a current at the CT secondary terminals. This current will, of course, be much lower than the primary-injection current and the equipment smaller. It is not possible to test the relays completely by the primary-injection method owing to the difficulty in producing high multiples of setting current. Therefore further primary-injection testing is confined to injecting sufficient current to produce movement of the relay.

Primary injection tests are performed only when the equipment is being commissioned or if for any reason one or more current transformers are changed.

7.7   CT MAGNETISING CHARACTERISTIC CURVES

This test is conducted on all current transformers and is intended to prove that they are suitable for the protection with which they are associated by determining the knee-point voltage, i.e. the voltage at which saturation starts. It is also intended to show that all current transformers in a group are similar and to test for open-circuited secondary windings or short-circuited turns. A variable voltage supply is connected across the secondary terminals of the current transformer and the current measured at different voltages. The circuit is as shown in Figure 7.6. Note that the voltmeter is connected so that the ammeter does not read voltmeter current which in some cases could be of the same order in the CT magnetising current.

It is useful to establish roughly the voltage at which saturation starts by increasing the voltage until there is a large increase in current for a small change in voltage. It can then be decided at what values to take readings to give sufficient points to plot a clear graph - too many points would be tedious.

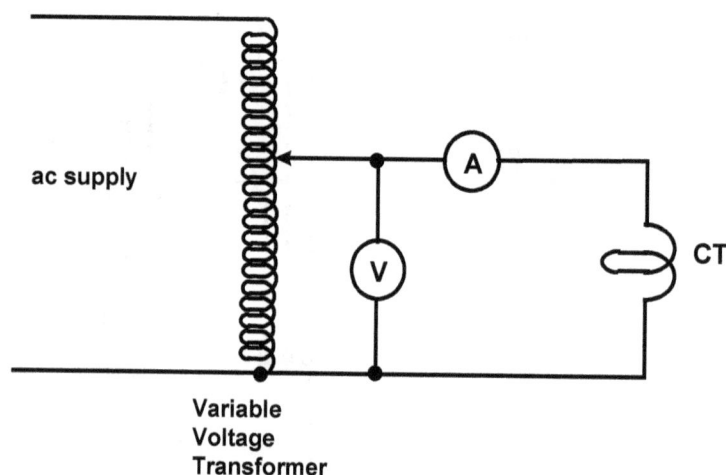

FIGURE 7.6
CIRCUIT TO DETERMINE CT MAGNETISING CHARACTERISTIC

As an example take the following tests on a set of three current transformers. The test equipment is connected to the R-phase current transformer. The voltage is increased steadily until there is a rapid increase in current. In this case the increase was from about 0.05A to 0.1A when the voltage increased from about 100 to 120V. Ten readings would seem to be a reasonable number between 0 and 120V say 20V, steps initially and smaller steps when saturation starts. The Y and B-phases are checked in turn.

Voltage	Current		
	R	Y	B
0	0	0	0
20	0.01	0.01	0.011
40	0.016	0.016	0.017
60	0.023	0.024	0.025
80	0.031	0.031	0.032
90	0.037	0.037	0.039
100	0.046	0.047	0.049
110	0.065	0.066	0.069
120	0.105	0.107	0.111

The knee-point voltage, i.e. the voltage at which an increase of 10% will result in a 50% increase in magnetising current, is about 100 V, the point in our preliminary test where the current started to increase rapidly.
As a rough check on this

@ 100V $\qquad$ $I_e = 0.046$

@ 100V +10% = 110V $\qquad$ $I_e = 0.065$

increase $\dfrac{0.065 - 0.046}{0.046}$ $\qquad = 0.41$

a 41% increase.

@ 110V $\qquad$ $I_e = 0.065$

@ 110V +10% = 121V $\qquad$ $I_e = 0.105$

increase $\dfrac{0.105 - 0.065}{0.065}$ $\qquad = 0.62$

a 62% increase.

Hence the knee-point voltage is between 100 and 110 V. If a more accurate result is required curves may be plotted as shown in Figure 7.6. The knee point is 104 V for the R current transformer. Following these tests the resistance of the CT secondary winding is checked.

This is best done by a bridge or by d.c. voltage and current measurement but a multimeter reading is better than nothing. All results must, of course, be recorded.

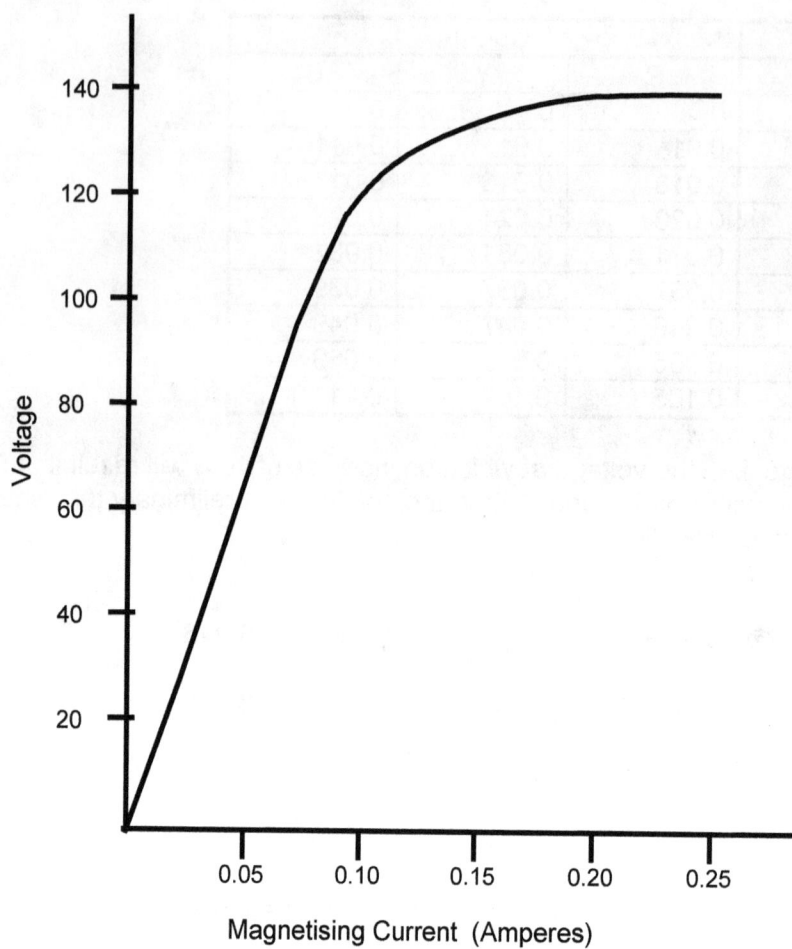

FIGURE 7.6
CT MAGNETISING CHARACTERISTIC

CHAPTER 8 GEC RELAY IDENTIFICATION CODE

All GEC Measurements relays bear an identification or model number and a serial number that must be quoted with all relative after-sales correspondence.

The model number of standard relays is contrived as follows:

Example **CMM42PF3A5**

```
C M M 4 2 P F 3A 5
a b c d e f g h  j
```

a.	OPERATING QUANTITY	b.	BASIC MOVEMENT
A	Phase angle comparison	A	Attracted armature
B	Balanced current	B	Buchholz
C	Current	C	Induction cup
D	Differential	D	Induction disc
E	Direction	G	Galvanometer
F	Frequency	I	Transactor
G	Gauss	J	Mixed types
I	Directional current	M	Sensitive balanced armature
K	Rate of rise of current	P	Plug
M	Manual	R	Rectifier
O	Oil pressure	S	Synchronous motor
P	Polyphase volt-amperes	T	Static circuit
R	Reactive volt-amperes	W	Weight or gravity
S	Slip frequency		
T	Temperature		
V	Voltage		
W	Power		
X	Reactance		
Y	Admittance		
Z	Impedance		

c.	INDICATION OF APPLICATION				
A	Auxiliary		L	Load limiting	
B	Testing		M	Motor or semaphore	
C	Carrier or counting		N	Negative phase sequence	
GB	Capacitor bank		O	Out-of-step	
D	Directional		P	Potential or fuse failure	
E	Earth		Q	Alarm	
EF	Earth fault		R	Reclosing	
V	Flag or indicator		S	Synchronising	
G	General or generator		T	Transformer or timer	
GF	Generator field		U	Definite time	
H	Harmonic restraint		V	Voltage control	
I	Interlocked or industrial		W	Pilot wire	
IG	General or generator (instantaneous)		WA	Interposing auxiliary	
J	Tripping		WJ	Intertripping	
JE or JX	Tripping (Elec-reset)		X	Supervisory	
JH or JY	Tripping (Hand-reset)		Y	Flashback or backfire	
JA, JS or JZ	Tripping (Self-reset)		Z	Special application	
JB	Control (Tripping)		ZS	Zero phase sequence	
K	Check alarm				

d. NUMBER OF UNITS

The first number indicates the number of relay units, excluding seal in and reinforcing auxiliary units.

For example, the CMM42 relay consists of a thermal overcurrent unit, an instantaneous overcurrent unit, an earth fault unit and an instantaneous unbalance unit.
A triple pole overcurrent relay has the number 3.

e. CHARACTERISTIC

The second number indicates the particular characteristic of one of a group of relays. For example CMM41 and CMM42 relays are of the same type and composition but differ in their time delay characteristics.

f. **CASE SIZE**

A	Size 1 drawout single ended	10 terminals
B	Size 1 drawout double ended	20 terminals
C	Size 2 drawout single ended	10 terminals
D	Size 2 drawout double ended	20 terminals
E	Size 3 drawout single ended	10 terminals
F	Size 3 drawout double ended	20 terminals
L	Size ½ non-drawout (type VAK relay)	12 terminals
N	Size ½ drawout single ended	10 terminals
P	Size 1½ drawout single ended	10 terminals
R	Size 1½ drawout double ended	20 terminals
S	Size 1½ drawout double ended	40 terminals
T	Size 4½ drawout single ended	30 terminals
U	Handle-lock plug-in	
Y	Size ¼ moulded non-drawout	10 terminals
Z	Size ½ moulded non-drawout	12 terminals

g. **CASE MOUNTING**

F	Flush vertical
G	Flush horizontal
P	Projecting vertical
B	Projecting horizontal

h. **IDENTIFICATION**

This may be one or more numbers or letters and enables our engineers to identify rating, contact arrangement etc.

i. **SUFFIX**

Some relay types, for example auto reclose relays (VAR82A, VAR82B etc.), vary only in detail and are identified by a suffix letter.
Where applicable this identification is given on the data sheet. Otherwise the last letter must be regarded as part of the identification h above.
Where the last digit is a number 5 or 6, this indicates a relay specifically for use on a 50 or 60 Hz supply, respectively.

DISTANCE PROTECTION SCHEMES

Exceptions to the rule are distance schemes, which have a group of prefix letters to identify type, measurement etc.

The initial letter 'M' or 'B' indicates mho or reactance measurement respectively and, where two initial letters are used, the first refers to phase faults and the second to earth faults. For switched schemes '5 'or' SS' precedes the initial letter. The digit denotes the number of time distance steps and the final letter 'V' or 'T' indicates that the scheme is electromagnetic or static in operation, respectively. For example, MR3V indicates a three-step electromagnetic distance scheme with three mho type measuring units for phase faults and three reactance type measuring units for earth fault.

SPECIAL

Instead of a type number (CMM42), special relays are prefixed by the letters SPEC, followed by one or more numbers or a combination of numbers and letters.

CHAPTER 9 CEE RELAYS IDENTIFICATION CODE

The relays are identified by a code made up from 3 or possibly 4 letters and 4 numbers.
Example ITMF 7721

FIRST LETTER: OPERATING QUANTITY

D	Differential value	V	Vibration
F	Flux	W	Power
G	Slip	S	Temperature sensor or synchronism
H	Frequency		
I	Current	T	Voltage
P	Power	Y	Admittance (mho)

SECOND LETTER: MEASURING PRINCIPLE

A	Attracted armature unit	M	Microprocessor
D	Digital	T	Transistorised

THIRD LETTER: APPLICATION

A	Auxiliary	J	Tripping auxiliary
B	D.C. networks	L	Line or cable protection
D	Directional	M	Rotating machine protection
	Discordance	R	Reclose
	Unloading	S	Sensitive
	Trip	T	Transformer protection
E	Insulation		Time measurement
G	General use	V	Protection with voltage control
H	Earth/ground fault	X	Supervision
I	Negative sequence		

FOURTH LETTER: USED WHEN NECESSARY TO DISTINGUISH A PARTICULAR APPLICATION FOR RELAYS IN THE SAME FAMILY

A	Incoming feeder or Active power	F	Protection for motors having fuses
D	Outgoing feeder	H	Hand reset
d	Positive sequence	R	Reactive power
E	Electrical reset	S	Self reset

FIRST NUMBER: IDENTIFICATION OF THE SERIES

7	Equipment in type R modular case

SECOND NUMBER: OPERATING TIME CHARACTERISTIC

0	Instantaneous	5	⎫
1	Independent or definite time	6	⎬ Thermal curve
2	Inverse time	7	⎭
3	Very inverse time	8	
4	Extremely inverse time	9	Programmable curve

THIRD NUMBER: NUMBER OF DIFFERENTMEASUREMENTS OF THE OPERATING QUANTITY

1   One measurement
2   Two measurements etc ...

FOURTH NUMBER: IDENTIFICATION OF A PARTICULAR TYPE OF RELAY IN THE SERIES

1   2   etc ......

CHAPTER 10 CODES USED ON NAMEPLATES OF CEE RELAYS

A		Phase 1
B		Phase 2
C		Phase 3
—		Single Phase
=		Two Phase
≡		Three Phase
	{ N	Nominal
	{ D	Differential
	{ P	Primary (ring CT)
I (current)	{ R	Restraint
	{ th	Thermal
	{ ↺	Negative sequence
	{ ↻	Positive sequence
	{ 0	Zero sequence
P		Active power
-P		Reverse power (active)
P_n		Nominal active power
Q		Reactive power
R		Resistance
U		Voltage
U_n		Nominal voltage
U_d		Positive sequence voltage
U_0		Zero sequence voltage
>		Low set unit, overvoltage or overcurrent
>>		High set unit, overvoltage or overcurrent
<		Undervoltage or Undercurrent
N		Number
T or t		Time delay
T_m or T_s		Temperature
θ_c		Characteristic angle
K		Coefficient
F		Operating frequency
F_n		Nominal frequency
f_o		0 or 2 Hz
f		Setting in multiples of 0.1 Hz within a range of 0 to 2 Hz
Z		Impedance

CHAPTER 11 ELECTRICAL POWER SYSTEMS; DEVICE NUMBERS AND FUNCTIONS

The devices in switching equipment are referred to by numbers, with appropriate suffix letters when necessary, according to the function they perform.

These numbers are adopted as standard for automatic switchgear by the IEEE, and are incorporated in American Standard C37.2-1970. This system is used in connection diagrams, instruction books and specifications.

Device No.	Device	Definition and Function
1	Master Element	This is the initiating device, such as a control switch, voltage relay, float switch etc., which serves either directly, or through such permissive devices as protective and time-delay relays to place an equipment in or out of operation.
2	Time-Delay Starting or Closing Relay	This is a device which functions to give a desired amount of time delay before or after any point of operation in a switching sequence or protective relay system, except as specifically provided by device functions 48, 62, and 79 described later.
3	Checking or Interlocking Relay	This operates in response to the position of a number of other devices, (or to a number of predetermined conditions), in an equipment, to allow an operating sequence to proceed, to stop, or to provide a check of the position of these devices or of these conditions for any purpose.
4	Master Contactor	This device, generally controlled by Device No.1 or equivalent, and the required permissive and protective devices, that serves to make and break the necessary control circuits to place an equipment into operation under the desired conditions and to take it out of operation under other or abnormal conditions.
5	Stopping Device	This control device is used primarily to shut down an equipment and hold it out of operation. [This device may be manually or electrically actuated, but excludes the function of electrical lockout (see device function 86) on abnormal conditions]
6	Starting Circuit Breaker	The principle function of this device is to connect a machine to its source of starting voltage.
7	Anode Circuit Breaker	Is used in the anode circuits of a power rectifier for the primary purpose of interrupting the rectifier circuit if an arc back should occur.
8	Control Power Disconnecting Device	This is a disconnecting device - such as a knife switch, circuit breaker or pullout fuse block, used for the purpose of connecting and disconnecting the source of control power to and from the control bus or equipment. Note! *Control power is considered to include auxiliary power which supplies such apparatus as small motors and heaters*
9	Reversing Device	This is used for the purpose of reversing a machine field or for performing any other reversing functions.
10	Unit Sequence Switch	Used to change the sequence in which units may be placed in and out of service in multiple-unit equipments.

11	Reserved for future application	
12	Over-Speed Device	This is usually a direct connected speed switch which functions on machine over-speed.
13	Synchronous-Speed Device	For example centrifugal-speed switch, a slip-frequency relay, a voltage relay, an undercurrent relay or any type of device, operates at approximately synchronous speed of a machine.
14	Under-Speed Device	Functions when the speed of a machine falls below a predetermined value.
15	Speed or Frequency, Matching Device	Functions to match and hold the speed or the frequency of a machine or of a system equal to, or approximately equal to, that of another machine, source or system.
16	Reserved for future application	
17	Shutting or Discharge Switch	This switch serves to open or to close a shutting circuit around any piece of apparatus (except a resistor), such as a machine field, a machine armature, a capacitor or a reactor. NOTE! *This excludes devices which perform such shutting operations as may be necessary in the process of starting a machine by devices 6 or 42, or their equivalent, and also excludes device 73 function which serves for the switching of resistors.*
18	Accelerating or Decelerating Device	This is used to close or to cause the closing of circuits which are used to increase or to decrease the speed of the machine.
19	Starting-to-Running Transition Contactor	This device initiates or causes the automatic transfer of a machine from the starting to the running power connection.
20	Electrically Operated Valve	This is an electrically operated, controlled or monitored valve in a fluid line. NOTE! *The function of the valve may be indicated by the use of the suffixes*
21	Distance Relay	This device functions when the circuit admittance, impedance or reactance increases or decreases beyond predetermined limits.
22	Equaliser Circuit Breaker	This breaker serves to control or to make and break the equaliser or the current-balancing connections for a machine field, or for regulating equipment, in a multiple-unit installation.
23	Temperature Control Device	Its function is to raise or lower the temperature of a machine or other apparatus, or of any medium, when its temperature falls below, or rises above, a predetermined value. NOTE! *An example is a thermostat which switches on a space heater in a switchgear assembly when the temperature falls to a desired value as distinguished from a device which is used to provide automatic temperature regulation between close limits and would be designated as 90T.*
24	Reserved for future application	

25	Synchronising or Synchronism-Check Device	This device operates when two ac circuits are within the desired limits of frequency, phase angle or voltage, to permit or to cause the paralleling of these two circuits.
26	Apparatus Thermal Device	This device functions when the temperature of the shunt field or the armature winding of a machine. or that of a load limiting or load shifting resistor or of a liquid or other medium exceeds a predetermined value; or if the temperature of the protected apparatus, such as a power rectifier, or of any medium decreases below a predetermined value.
27	Undervoltage Relay	This device functions on a given value of undervoltage
28	Flame Detector	Monitors the presence of the pilot or main flame in such apparatus as a gas turbine or a steam boiler.
29	Isolating Contactor	Expressly used for disconnecting one circuit from another for the purposes of emergency operation, maintenance, or test.
30	Annunciator Relay	This non-automatic reset device gives a number of separate visual indications upon the functioning of protective devices, and which may also be arranged to perform a lockout function.
31	Separate Excitation Device	Connects a circuit such as the shunt field of a synchronous converter, to a source of separate excitation during the starting sequence; or one which energises the excitation and ignition circuits of a power rectifier.
32	Directional Power Relay	This functions on a desired value of power flow in a given direction or upon reverse power resulting from arc back in the anode or cathode circuits of a power rectifier.
33	Position Switch	Makes or breaks contact when the main device or piece of apparatus, which has no device function number, reaches a given position.
34	Master Sequence Device	This device such as a motor operated multi-contact switch, or the equivalent, or a programming device, such as a computer, that establishes or determines the operating sequences of the major devices in an equipment during stopping or during other sequential switching operations.
35	Brush-Operating, or Slipring Short Circuiting Device	This is used for raising, lowering, or shifting the brushes of a machine, or for short circuiting its sliprings, or for engaging or disengaging the contracts of a mechanical rectifier.
36	Polarity or Polarising Voltage Device	Operates or permits the operation of another device on a predetermined polarity only or verifies the presence of a polarising voltage in an equipment.
37	Undercurrent or Underpower Relay	Functions when the current or power flow decreases below a predetermined value.
38	Bearing Protective Device	Functions on excessive bearing temperature, or on other abnormal mechanical conditions, such as undue wear, which may eventually result in excessive bearing temperature.
39	Mechanical Condition Monitor	This device functions upon the occurrence of an abnormal mechanical condition (except that associated with bearings as covered under device function 38), such as excessive vibration, eccentricity, expansion, shock, tilting, or seal failure.

40	Field Relay	Functions on a given or abnormally low value or failure of machine field current, or on an excessive value of the reactive component of armature current in an ac machine indicating abnormally low field excitation.
41	Field Circuit Breaker	Is a device which functions to apply, or to remove, the field excitation of a machine.
42	Running Circuit Breaker	The principle function of this device is to connect a machine to its source of running or operating voltage. This function may also be used for a device, such as a contactor, that is used in series with a circuit breaker or other fault protecting means, primarily for frequent opening and closing of the circuit.
43	Manual Transfer or Selector Device	This transfers the control circuits so as to modify the plan of operation of the switching equipment or of some of the devices.
44	Unit Sequence Starting Relay	Is a device which functions to start the next available unit in a multiple-unit equipment on the failure or on the non-availability of the normally preceding unit.
45	Atmospheric Condition Monitor	This functions upon the occurrence of an abnormal atmosphere condition, such as damaging fumes, explosive mixtures, smoke, or fire.
46	Reverse-Phase, or Phase Balance, Current Relay	This relay functions when the polyphase currents are of reverse-phase sequence, or when the polyphase currents are unbalanced or contain negative phase sequence components above a given amount.
47	Phase Sequence Voltage Relay	Functions on a predetermined value of polyphase voltage in the desired phase sequence.
48	Incomplete Sequence Relay	This relay generally returns the equipment to the normal, or off, position and locks it out if the normal starting , operating or stopping sequence is not properly completed within a predetermined time. If the device is used for alarm purposes only, it should preferably be designated as 48A (alarm).
49	Machine or Transformer, Thermal Relay	This relay functions when the temperature of a machine armature, or other load carrying winding or element of a machine, or the temperature of a power rectifier or power transformer (including a power rectifier transformer) exceeds a predetermined value.
50	Instantaneous Overcurrent, or Rate-of-Rise Relay	This functions instantaneously on an excessive value of current, or on a excessive rate of current rise, thus indicating a fault in the apparatus or circuit being protected.
51	AC time overcurrent Relay	Is a relay with either a definite or inverse time characteristic that functions when the current in an ac circuit exceeds a predetermined value.
52	AC Circuit Breaker	This is used to close and interrupt an ac power circuit under normal conditions or to interrupt this circuit under fault or emergency conditions.
53	Exciter or dc generator relay	This forces the dc machine field excitation to build up during starting or which functions when the machine voltage has built up to a given value.
54	Reserved for future application	
55	Power Factor Relay	This operates when the power factor in an ac circuit rises above or below a predetermined value.

56	Field Application Relay	Is a relay that automatically controls the application of the field excitation to an ac motor some predetermined point in the slip cycle
57	Short Circuiting or Grounding Device	This primary circuit switching device functions to short circuit or to ground a circuit in response to automatic or manual means.
58	Rectification Failure Relay	Functions if one or more anodes of a power rectifier fail to fire, or to detect an arc-back or on failure of a diode to conduct or block properly.
59	Overvoltage Relay	Functions on a given value of overvoltage.
60	Voltage or Current Balance Relay	Operates on a given difference in voltage, or current input or output of two circuits.
61	Reserved for future application	
62	Time Delay Stopping or Opening Relay	Serves in conjunction with the device that initiates the shutdown, stopping, or opening operation in an automatic sequence.
63	Pressure Switch	Operates on given values or on a given rate of change of pressure.
64	Ground Protective Relay	Functions on failure of the insulation of a machine, transformer or of other apparatus to ground, or on flashover of a dc machine to ground. **NOTE!** *This function is assigned only to a relay which detects the flow of current from the frame of a machine or enclosing case or structure of a piece of apparatus to ground, or detects a ground on a normally ungrounded winding or circuit. It is not applied to a device connected in the secondary circuit or secondary neutral of current transformer, connected in the power circuit of a normally grounded system.*
65	Governor	Is the assembly of fluid, electrical or mechanical control equipment used for regulating the flow of water, steam, or other medium to the prime mover for such purposes as starting, holding speed or load, or stopping.
66	Notching or Jogging Device	Functions to allow only a specified number of operations of a given device, or equipment, or a specified number of successive operations within a given time of each other. It also functions to energise a circuit periodically or for fractions of specified time intervals, or that is used to permit intermittent acceleration or jogging of a machine at low speeds for mechanical positioning.
67	AC Directional Overcurrent Relay	Functions on a desired value of ac overcurrent flowing in a predetermined direction.
68	Blocking Relay	Initiates a pilot signal for blocking of tripping on external faults in a transmission line or in other apparatus under predetermined conditions, or co-operates with other devices to block tripping or to block reclosing on an out-of-step condition or on power swings.

69	Permissive Control Device	Generally a two position, manually operated switch that in one position permits the closing of a circuit breaker, or the placing of an equipment into operation, and in the other position prevents the circuit breaker or the equipment from being operated.
70	Rheostat	This variable resistance device used in an electric circuit, which is electrically operated or has other electrical accessories, such as auxiliary, position, or limit switches.
71	Level Switch	Operates on given values, or a given rate of change, of level.
72	DC Circuit Breaker	Used to close and interrupt a dc power circuit under normal conditions or to interrupt this circuit under fault or emergency conditions.
73	Load Resistor Contactor	Used to shunt or insert a step of load limiting, shifting, or indicating resistance in a power circuit, or to switch a space heater in circuit, or to switch a light, or regenerative load resistor of a power rectifier or other machine in and out circuit.
74	Alarm Relay	Is a device other than an annunciator as covered under Device No.30, which is used to operate in connection with, a visual or audible alarm.
75	Position Changing Mechanism	Used for moving a main device from one position to another in an equipment; as for example, shifting a removable circuit breaker unit to and from the connected, disconnected, and test positions.
76	DC Overcurrent Relay	Functions when the current in a dc circuit exceeds a given value.
77	Pulse Transmitter	Used to generate and transmit pulses over a telemetering or pilot-wire circuit to the remote indicating or receiving device.
78	Phase Angle Measuring, or Out-of-Step Protective Relay	Functions at a predetermined phase angle between two voltages or between two currents or between voltage and current.
79	AC Reclosing Relay	Controls the automatic reclosing and locking out of an ac circuit interrupter.
80	Flow Switch	Operates on given values, or on a given rate of change, of flow.
81	Frequency Relay	Functions on a predetermined value of frequency - either under or over or on normal system frequency - or rate of change of frequency.
82	DC Reclosing Relay	Controls the automatic closing and reclosing of a dc circuit interrupter, generally in response to load circuit conditions.
83	Automatic Selective Control or Transfer Relay	Operates to select automatically between certain sources or conditions in an equipment, or performs a transfer operation automatically.
84	Operating Mechanism	This is the complete electrical mechanism or servo-mechanism, including the operating motor, solenoids, position switches, etc., for a tap changer, induction regulator or any similar piece of apparatus which has no device function number.
85	Carrier or Pilot Wire Receiver Relay	Operated or restrained by a signal used in conjunction with carrier current or dc pilot-wire fault directional relaying.

86	Locking Out Relay	Operated hand or electrically reset, relay that functions to shut down and hold an equipment out of service on the occurrence of abnormal conditions.
87	Differential Protective Relay	Functions on a percentage or phase angle or other quantitative difference of two currents or of some other electrical quantities.
88	Auxiliary Motor or Motor Generator	Used for operating auxiliary equipment such as pumps, blowers, exciters, rotating magnetic amplifiers etc.
89	Line Switch	Used as a disconnecting load interrupter, or isolating switch in an ac or dc power circuit, when this device is electrically operated or has electrical accessories, such as an auxiliary switch, magnetic lock etc.
90	Regulating device	Regulates a quantity, or quantities, such as voltage, current, power, speed, frequency, temperature, and load, at a certain value or between certain (generally close) limits for machine, tie lines or other apparatus.
91	Voltage Directional Relay	Operates when the voltage across an open circuit breaker or contactor exceeds a given value in a given direction.
92	Voltage and Power Directional Relay	Permits or causes the connection of two circuits when the voltage difference between them exceeds a given value in a predetermined direction and causes these two circuits to be disconnected from each other when the power flowing between them exceeds a given value in the opposite direction.
93	Field Changing Contactor	Functions to increase or decrease in one step value of field excitation on a machine.
94	Tripping or Trip-free relay	Functions to trip a circuit breaker, contactor, or equipment, or to permit immediate tripping by other devices; or to prevent immediate reclosure of a circuit interrupter, in case it should open automatically even though its closing circuit is maintained closed.
95		Used only for specific applications on individual installations where none of the assigned numbered functions from 1 to 94 is suitable.
96		Used for 'trip circuit supervision' monitoring tripping supplies and (sometimes) circuit continuity
97		Used only for specific applications on individual installations where none of the assigned numbered functions from 1 to 94 is suitable.

PART 11
COMMISSIONING
ELECTRICAL EQUIPMENT

CHAPTER 1 INTRODUCTION

The objective of this part of the book is to give general guidance on the types of tests commissioning and maintenance procedures that need to be carried out prior to the commissioning of electrical installations and equipment. Many of these tests and procedures also are necessary for the re-commissioning of electrical equipment following major maintenance overhauls.

All inspections, maintenance work and testing mentioned should be carried out strictly in accordance with the relevant company's Electrical Safety Rules, relevant Standing Instructions and Permit to Work System and be under the direction of an Authorised Electrical Person. Procedures, methods, examples and data provided in this part are for guidance only, Company documentation will provide specific data.

CHAPTER 2 GENERAL

One of the main objectives of any set of commissioning or maintenance documents is to give a commissioning engineer clear guidance relating to the series of tests and checks he should carry out on items of electrical equipment. This should cover items of equipment that are to be commissioned on site for the first time, or are to be returned to service after extensive maintenance overhauls have been carried out on them.

In brief, this guidance should be sufficient so a commissioning engineer can be satisfied that the tests and checks he carries out are the correct ones for a particular type of electrical equipment. Additionally, that the results obtained are considered acceptable so he can be reasonably assured that such items of electrical equipment will give satisfactory performance, be safe to operate and be reliable in service.

Another objective of the documentation is to specify the different maintenance procedures necessary for specific types of electrical equipment and the recommended frequency that these should be carried out.

The importance of recording all test and maintenance results is vital for giving a ready comparison for future commissioning. Possibly more importantly, maintenance results indicate the known state of the plant item and the probable frequency it may need to be inspected or maintained in the future.

CHAPTER 3 TESTING AND COMMISSIONING METHODS

Only the more significant tests will be referred to in this here, but some practical guidance and general information has been added, which may be helpful to engineers who have not yet had a great deal of experience of field testing and commissioning.

3.1 INSULATION TESTING

The two types of insulation tests that can be carried out before electrical equipment is commissioned are:-
1. Insulation Resistance (IR) Tests - principally for motors, generators, transformers and LV switchgear.
2. High Voltage (HV) a.c. or d.c. testing - for HV cables and HV switchgear.

3.2 INSULATION RESISTANCE (IR) TESTS

The quality of the electrical insulation is determined by measuring the value of the insulation resistance. The testing is to be carried out using a d.c. insulation tester instrument, such as a 'Megger', at a voltage recommended as being suitable for testing a particular type of electrical equipment.

An insulation tester instrument is a sensitive instrument that compares d.c. voltage and current to provide a reading in ohms, or more usually in megohms. During testing, current leakage across the insulation surface, due to moisture or dirt, will give a false low IR reading. All insulation surfaces should therefore be dry and clean before testing. For the same reason the leads of the instrument used should have a high IR and must be maintained in good condition.

Carrying out the test with an insulation tester of lower than the recommended voltage will give an increased value of resistance. The insulation tester's operating voltage should always be recorded on any test sheet so the results can be assessed in an objective way.

When measuring insulation resistance, a record must be made of the temperature of the insulation at the same time. A figure of 100 megohm (100MΩ) measured with the insulation cold (20°C) falls to 70MΩ when the figure is corrected for 25°C. The change in value of insulation resistance with temperature is a logarithmic function. IR values can be corrected for variations in temperature using a nomogram.

It is very important to ensure that immediately after insulation testing, the equipment should be discharged to earth before it is touched by hand. The charge stored in the insulation during testing, particularly when using higher voltage test instruments on cables, can be dangerous when touched.

3.3   HIGH VOLTAGE TESTS

Insulation resistance (IR) tests should be carried out on site before applying HV tests to either cables or HV switchgear units. HV testing is normally restricted to HV switchgear and cables only and for practical purposes the HV tests can be carried out using d.c. test equipment.

HV tests should be applied on site prior to commissioning HV switchgear units that have been assembled into a complete switchboard. Although the individual units will have been subject to HV tests in the manufacturer's factory, the assembly on site of the whole switchboard's busbars and the busbar 'insulated joints', between adjacent units, needs to be fully tested in their installed position.

The same principle applies to HV cables for although individual section lengths should have been fully tested in the factory, they are then jointed on site, possibly under damp conditions, to the required route length and jointed at the each end to different types of electrical equipment. HV testing is necessary to 'prove' the total cable circuit insulation is acceptable.

HV testing at a voltage higher than the normal system voltage, particularly when using d.c. test equipment, could 'overstress' the electrical equipment's insulation so repeated 'over voltage' testing is not considered good engineering practice.

HV tests are recommended to be limited to three minutes duration for cables and one minute for switchgear. Company documentation will provide the recommended values of the HV test voltages that may be used for HV switchgear and those for new cables. Reduced values of HV test voltages are also given that should be used on older cables that have been repaired following faults, or have been re-jointed in a new position.

3.4   POLARISATION INDEX (PI) TESTING

This type of insulation testing is normally used to determine the insulation resistance value of the windings of motors and generators. The PI is the ratio of insulation resistance of the winding measured after one minute and after ten minutes of continuous testing at the recommended voltage.

The current that flows during insulation resistance testing is made up of three components and so does not immediately assume its final value. These components are: -
- Capacitance current which may be large but only takes a few seconds to decay.
- Leakage current over the surface and through the volume of the insulation - this is constant.
- Dielectric absorption current which decays exponentially.

So the true value of the insulation resistance cannot be obtained until these three components have stabilised, but in practice it is sufficient to apply the insulation test over a ten minute period to achieve a reasonably acceptable assessment of PI value.
The polarisation index is the ratio of:

<u>10 Minute Resistance</u>
1 Minute Resistance

and for a clean dry winding, the index value should exceed 2. The polarisation factor is independent of the winding temperature.

If the insulation is damp or dirty, the leakage current will be large in relation to charging currents and the insulation resistance will assume a final value very quickly, whereas if the insulation is dry the dielectric absorption current will be high and take some time to decay.

This will result in the insulation resistance value increasing gradually over the duration of the test.

3.5   INSULATION TESTING OF ELECTRONIC EQUIPMENT

Equipment involving electronic semi-conducting components require special care to see that no damage results from any insulation resistance testing and the manufacturer's instructions and recommendations should be closely followed.

3.6   INTERPRETATION OF THE RESULTS OF INSULATION TESTING

Acceptable values of IR and PI test results for particular types of electrical equipment will be provided in appropriate company documentation. If the test result on an item of electrical equipment does not meet these acceptable values it should not normally be put back into service.

Should a 'flash-over' occur while HV testing is being carried out then the insulation of the item of electrical equipment has failed the test and the equipment should not be put into service.

3.7   CONDUCTIVITY AND EARTH RESISTANCE TESTS

3.7.1 Conductivity tests

A continuity check using a low resistance test set [e.g. Ductor] should be carried out across switchgear busbar joints. These joints are normally assembled on site between the individual units that make up a complete switchboard.

The measured voltage drops should be interpreted on a comparative basis. For identical connections the figures obtained should not differ more than 20% from each other.

Operational experience of poorly made HV joint connections has been that these can lead to 'hot spots', generation of gases and eventual failure of insulation.

Conductivity tests should also be carried out across the 'closed contacts' of circuit breakers to ensure that all main contacts are firmly 'made' and any internal connections are tight. The recorded figures should be approximately the same for all three phases.

3.7.2 Earth resistance tests

To safeguard electrical equipment and to assist protection relays to operate correctly the earthing connections to both the 'earth electrodes' and the earthing connections to the electrical equipment must be of an acceptable low resistance.

The Megger Earth Tester is the only really satisfactory method of testing the resistance of an earth electrode to earth. It is basically a d.c. ohm meter with an inbuilt source of supply, it rapidly reverses both current and potential, so that the reading is independent of d.c. stray currents in the ground and of a.c. currents differing in frequency from the speed of winding the handle.

The resistance measurement increases initially according to the distance away from the earth electrodes and increases again the closer one gets to the temporary remote current electrode. In between there is an area over which the instrument sees the resistance of the earth electrode only to earth.

It is usual to use one current electrode driven into the ground at some distance from the earth electrode under test and to drive in the potential electrode half way between.

The resistance areas of the earth electrode under test and of the auxiliary current electrode should not overlap. Shifting the intermediate potential will show whether or not the distance between the two is sufficient electrode 2m either way and the results obtained should be virtually identical.

Instructions should be found also with the Megger Earth Tester itself.

CHAPTER 4 TESTING AND COMMISSIONING PROCEDURES.

4.1   GENERAL

These guidelines are not dealt with in this document in any detail, except when a cross-reference to certain technical guidance is considered to be helpful explaining why certain specified tests are carried out on particular types of electrical equipment.

4.2   HAZARDOUS AREA NOTES

There are three prime considerations associated with the severity and extent of a hazardous area, they are:
1.   the ease of ignition of the hazard (defined by the Gas Group)
2.   the area over which the hazard will or may extend (defined by the Zone)
3.   the temperature at which the hazard will auto-ignite (identified by T-Class)

4.2.1 Ease of ignition

The IEC recommendations, which are published in IEC 79-0, and which are followed by CENELEC and by UK national practice, splits all gases and vapours into two groups:
Group I   -   mining - typified by the gas Methane (Firedamp)
Group II -   surface industry

Group II is sub-divided into three sub-groups:
IIA   typical gas Propane
IIB   typical gas Ethylene
IIC   typical gas Hydrogen

These IEC definitions are used throughout Europe and in most other countries.

Group IIC (Hydrogen) covers the most hazardous gases: that is to say the most easily ignitable gases. (Under worst case conditions of flammable mixture of air and gas, Hydrogen will ignite with as little as 20μJ of energy.)

It should be understood at this stage that there is no connection between ignition energy and ignition temperature. They are two distinct mechanisms for causing ignition.

Although North America recognises the IEC definitions, and indeed aims to follow the IEC guidance and Standards, they have traditionally used their own classification and definition system in which there are three Classes defining the hazard:
Class I   Gases and vapours
Class II   Dusts
Class III   Fibres
Within these Classes there are several sub-groups:

Class I	Group A	typically Acetylene
	Group B	typically Hydrogen
	Group C	typically Ethylene
	Group D	typically Methane
Class II	Group E	combustible metal dusts
	Group F	typically carbon black, charcoal
	Group G	typically flour or grain dust
Class III		No sub-groups

4.2.2 Extent of hazard and likelihood of existence

Having defined the type of hazard, the next concern is the likelihood or possibility of the hazard being present in concentrations (i.e. between the lower and upper flammable limit of concentration) for ignition to occur if a source of ignition is present. This is the Zone (Europe), or Division (North America).

4.2.3 Zones and divisions

The IEC recognises three zones of risk within a hazardous area. In Europe these zones are used for area classification. (In North America, although part of the IEC organisation, the concept of zones is not used at present and instead the area classification is by reference to divisions. To a great extent zones and divisions are equivalent.)

There are three zones:

Zone 0   The worst case: hazard is continuously present
              (Normally stemming from a continuous source of release)

Zone 1   Hazard likely to be present under normal operation
              (Normally stemming from a primary source of release)

Zone 2   Hazard unlikely to be present. Present only for short periods or
              under fault conditions.
              (Normally stemming from a secondary source of release)

Clearly if none of the definitions for Zone 0 or 1 or 2 apply then it must be a non-hazardous or safe area.

It may be helpful to put some guideline figures to the zone concept. These are not published in any Standard, but are widely used as a rule of thumb.

Zone 0   Hazard present for more than 1000 hours per year
Zone 1   Hazard present for less than 1000 and more than 10 hours per year
Zone 2   Hazard present for less than 10 hours per year

Whilst the **zone number** may be readily defined, the **extent of zone is** a more complex matter. In general, the extent of a zone will be a distance (possibly different in each dimension) from the source of release to the point where, with the release occurring, the concentration will have fallen to a level below the Lower Explosive Limit (LEL).

Clearly, to define this boundary, properties of the hazard e.g. Vapour Density, LEL% by volume etc. need to be considered.

In North America the categorisation of risk is the Division. There are two divisions:

Division 1
A location in which ignitable concentrations of flammable gases or vapours can exist under normal operating conditions or in which ignitable concentrations of such gases or vapours may exist frequently because of repair or maintenance operations or because of leakage: or in which breakdown or faulty operation of equipment or processes might release ignitable concentrations of flammable gases or vapours, and might also cause simultaneous failure of electric equipment.

Division 2
A location in which volatile flammable liquids or flammable gases are handled, processed or used, but in which the liquids, vapours or gases will normally be confined within closed containers or closed systems from which they can

escape only in case of accidental rupture or breakdown of such containers or systems, or in case of abnormal operation of equipment: or in which ignitable concentrations of gases or vapours are normally prevented by positive mechanical ventilation, and which might become hazardous through failure or abnormal operation of the ventilating equipment; or that is adjacent to a Division 1 location, and to which ignitable concentrations of gases or vapours might occasionally be communicated unless such communication is prevented by adequate positive pressure from a source of clean air, and effective safeguards against ventilation failure are provided.
(from Article 500-4 of the National Electrical Code USA)

It is important to realise that unless the hazardous area classification is correct, then the intended safety, levels from the different methods of protection may not be achieved.

4.3   PROTECTION AND RELAY TESTING

During initial pre-commissioning, primary injection testing should be used, where practicable, as this test includes the whole circuit. Since, however, the amount of primary current available is normally limited by the size of the available 'power' source, relays cannot always be checked by providing high currents through the primaries of the current transformers. Secondary injection testing then is normally carried out.

The results of secondary injection testing will, in addition, provide useful reference data for future routine protection testing, for which secondary injection testing only is carried out.

An insulation resistance (IR) test on the secondary wiring using, a 500V Megger, can be carried out on one circuit at a time, with all others earthed. The relay or auxiliary contacts being closed as necessary, to ensure that all the wiring is included in the test. These tests are to prove the integrity of the insulation of the circuit to earth and to all other circuits. [See 3.5 of this document regarding IR testing of electronic relays]

The results obtained may vary slightly with ambient conditions, but it is the trend in insulation values that are more important than the actual value. If the records taken in future, show the insulation resistance is progressively falling investigation of the wiring circuits will be necessary. An insulation resistance value in excess of 1 megohm will normally be considered acceptable.

A complete sequence of tripping tests should be carried out from the protection relay to the tripping of the circuit breaker. The relays can be operated manually, but with some 'delicate' relays it may be preferable to operate them electrically.

Modern static relays with reed outputs cannot be tripped manually and these relays can only be operated by secondary injection. When trip testing is taking place the relay contacts should be checked to see if the contact wipe is correct. The dropping of the relay flags and remote/local alarms should be properly checked, for correct indications, at the commissioning stage.

Intertripping tests, where applicable, should be carried out to ensure the overall protection scheme is working correctly.

4.4   FAULT FINDING

4.4.1 Tips to help with fault diagnosis

1. Believe that relays have operated correctly even if at first there seems to be some inconsistency.
   Relays are very reliable - any problems are normally due to incorrect inputs.
2. Do not reset any flags until their operations have all been noted down.
3. Once all operations have been noted down, make sure all flags are reset before re-energising.
4. Do not alter any relevant relay settings before the originals have been noted down.
5. With second or third stage protection, it is nor uncommon to find more than one relay has operated due to the very quick clearance times. However at times of system high voltage faults, severe stresses are placed on otherwise healthy circuits and it is feasible for two "more or less" simultaneous faults to occur, one brought on by the other, so two relay operations may truly represent two faults.
6. Make full use of high voltage and protection schematic diagrams. They can be a great help in piecing together the jigsaw particularly when under pressure to restore supplies. Keep them handy.
7. Do not wait for a fault to prove your protection. Regular trip testing of circuit breakers and regular battery maintenance can prove their worth many times over.

4.4.2 When things don't add up

Although very reliable, when properly commissioned and maintained, protective systems do sometimes fail. The problem is rarely the fault of the relay. Here are a few pointers that may help diagnose the problem.

1. Is the battery OK? Healthy DC volts are clearly vital to correct tripping and closing.
2. If the battery is OK. is there a full DC voltage on the switchgear buswire?
3. Is the tripping fuse OK?
4. If the relay has an auxiliary flag contactor, no flag will show if the DC is dead, even if the disc has turned. A mechanical flag will of course operate even though no tripping will occur.
5. Is the DC supply fed through a high resistance contact (say a corroded battery terminal). To a voltmeter, full volts will appear but an applied load will drop the volts to a much lower value.
6. Are the secondary wiring isolating contacts making soundly?
7. The trip coil is fed via the auxiliary switch. Is this OK?
8. Is the trip coil or mechanism stuck? Some trip coils have a sensitivity adjusting screw that alerts the start position of the moving core. Occasionally the screw is loose and the core falls to the bottom of its travel, preventing operation.
9. Is there a sensible setting on the relay?
10. Have metal filings or swarf fallen into the relay? The brake magnet will trap steel shavings and prevent the disc from turning. It is a good practice to always wipe over the case before opening the relay. A bird's feather is an ideal tool to use to clear away debris and dust off the disc.
11. If unit protection is employed, it is vital that the pilots are working well. With Solkor B. for instance, crossed pilots will cause operation for out - of-zone faults, shorted pilots will cause operation for both in and out - of zone faults, whilst open circuits will prevent any operation.
12. Some rotary wiping contacts on relays have a tendency to stick closed. This could lead to the auxiliary tripping contactor being left in circuit and burning out. Gentle

cleaning of the wiping contacts and lightly rubbing with a soft lead pencil can help.

13.　Most electronic relays such as the GEC "Midos" types require a separate DC supply to operate their circuitry. This is often wired via a separate fuse. If this has blown, the relay cannot work even though ct currents will still flow correctly.

14.　Shunt trip coils and spring release coils are highly inductive. To break the DC current presents an onerous task to contacts. Usually heavy-duty auxiliary switch contacts perform this function but if these fail to operate then light-duty relay contacts try to break the inductive current and are usually destroyed in the process. This tends to apply more readily to electronic relays since their output contacts are generally smaller than in the electromechanical types.

15.　Some circuits are protected with time - limit - fuse ct - release protection. An open-circuit or failed fuse may not manifest itself until load current has become high enough to cause the trip coil to operate, misleading the operator into thinking there is a fault. Earth faults operate the trip coil with no shorting fuse, often leaving no real evidence of what has caused the circuit breaker to open.

CHAPTER 5 MAINTENANCE OF PARTICULAR TYPES OF ELECTRICAL EQUIPMENT

5.1 GENERAL INSPECTION PROCEDURES

Company electrical procedures deal with the maintenance procedures that are considered necessary for particular types of electrical equipment and recommends, in general terms, the frequency they should be carried out.

However, the scope of these procedures will also recommend that in addition to the main maintenance tasks that are necessary on items of electrical equipment there is also a need for regular inspections of all electrical equipment, where such inspection visits are practicable.

The basis for the sound appreciation of all inspection is the availability of original design and the recorded results of earlier inspection, against which comparison can be made and developing trends determined for closer investigation during subsequent inspection and maintenance works.

An essential requirement for an inspection system is the keeping of up to date detailed records of all activities. This may be by equipment record cards, data bank or computerised record systems. Provision of an automatic reminder system to give notice when inspections are due is recommended.

For effective inspection work, the 'inspector' needs to have an extensive experience and knowledge concerning the acceptable condition of the equipment to enable him to make a satisfactory subjective judgement using his senses of sight, smell, hearing and touch (where this is safe) to recognise equipment condition.

Mechanical breakdown due to wear, tear and atmospheric conditions are the major condition changes that the inspection report will cover: -
- a) Overheating of bearings and casings
- b) Excessive vibration of equipment
- c) Metallic rubbing noises
- d) Cracks in metal, cracked or broken glasses and failure of cement round cemented glasses in flameproof enclosures.
- e) The effects of corrosion
- f) Slackness of joints in conduit runs and fittings
- g) Defects to cables and the condition of flexible cables used with portable equipment
- h) Leakage of oil/compound
- i) Earthing connections in poor condition

Test equipment is not intended to be used on inspection visits.

The inspection procedure record should detail the applicable safety requirements and the inspector should simply be required to identify a 'correct' or 'incorrect' condition by means of a tick on a check list. It is essential that remedial measures be taken quickly when the inspectors' reports indicate that they have identified potential defects or safety hazards.

Most Companies have produced check lists and forms for recording the results of inspections and tests. Such detailed forms will obviously be most helpful to both the inspector and to the Company supervisory staff who have to check the results of inspections and possibly initiate remedial actions.

The provision of a clear equipment inspection data record system is also essential to the development of a sound system that leads directly to the clear interpretation of the collected results.

5.2   MAINTENANCE RECOMMENDATIONS

Inspection and maintenance frequencies can only be recommendations in this document and that each Company will have to consider its own operational experience of each type of electrical equipment. The service performance of specific manufacturers' units must be reviewed, when deciding how often maintenance is to be carried out on the different types of plant on their electrical systems.

Additionally, the importance of particular units of electrical equipment in terms of the Company's safety critical systems and essential continuity of supply to specific core activities also determines how often electrical equipment is inspected, tested or maintained. The economics of unexpected plant failure has to be assessed when deciding the overall costs of a planned maintenance budget.

Company electrical procedures will recommend the frequencies for carrying out tests and maintenance procedures for different types of electrical equipment. The tables containing the recommendations are divided into four columns headed as below:-

TYPE   DESCRIPTION   INTERVAL   EXTENT

In addition to these frequency tables for each type of electrical equipment there are usually notes explaining how they should be applied, in some instances excluding smaller size units and certain specialised equipment from the recommended tests.

5.3   ELECTRIC MOTORS AND GENERATORS.

Obviously some of the maintenance procedures and the testing required for both motors and generators are similar but the main ones are as follows.

5.3.1 Insulation resistance (ir) tests

Before dismantling, all motors should be given an insulation resistance test for the Polarisation Index. This test is described in the section on insulation resistance testing. Before reconnecting the motor, a further insulation resistance test for the Polarisation Index is required.

Anti-condensation heaters, temperature detectors and other auxiliary electrical plant require an insulation resistance test carried out with a 500V Megger.

In the case of an insulation test on an electric motor, which may be required during major shutdown, the practical details of the test procedure can be as follows:
   a) Before commencing the insulation test, ensure that the machine frame, the parts of the winding not being tested and all associated auxiliary circuits are fully earthed. For reasons of safety and accuracy, it is essential that the winding under test is adequately earthed both before and after the application of the test potential.
   b) Using a d.c. insulation tester, of the recommended voltage, measure the resistance between each individual phase and earth with the remaining phases earthed.
   c) The measured resistance value is also dependent upon the temperature of the insulation humidity and the condition of the winding. Test reports should therefore include these details.
   d) Measured insulation resistance values are compared with the recommended minimum winding resistance value.

e) The insulation resistance testing of electric motors also includes the determination of the Polarisation Index.

If the test provides results which comes below the recommended minimum insulation resistance values, the windings should be dried out in accordance with the manufacturer's specified drying out procedure.

5.4   HV AND LV SWITCHGEAR

5.4.1 Testing procedures

The following tests should be carried out, where practicable, on HV and LV switchgear during routine maintenance prior to re-energising.

a) Insulation Resistance Tests
- Busbars, measured phase to earth and phase to phase.
- Circuit Breaker spouts and receptacles, measured phase to earth and phase to phase.
- Secondary wiring.
- Instrument transformer secondaries.
- Voltage transformer primaries.

The insulation resistance of the busbars and circuit breakers should be in excess of 1000MΩ and of the secondary wiring in excess of 1MΩ.

A low IR value indicates that dampness is present and warm air should be used to dry out the moisture.

b) Switchgear Oil Test
The sample of oil to be tested should be drawn from the lowest part of the circuit breaker tank or voltage transformer. The sample should be given tests for dielectric strength and for moisture content.

5.4.2 Special tests for vacuum circuit breakers

A HV test is required to test the condition of the sealed vacuum bottles and reference should be made to the manufacturers' instructions for details of the special precautions to be taken when carrying out this 'over voltage' test.

5.5   TRANSFORMERS (AND REACTORS)

5.5.1 Testing procedures

The following tests should be carried out on power transformers, during a maintenance outage:-

a)   Insulation resistance (IR) tests.
b)   Oil sample tests (breather type transformers only).
c)   Tap-changer oil sample and contact resistance tests, operation checks.
d)   Earthing resistance test for neutral and earthing connections.
e)   Operation checks of cooling fan or pump circuits.
f) Buchholz relay alarm and trip operation check. [if one fitted]
g)   Oil and winding temperature indicators, calibration check. [if fitted]

Some of these tests are described in more detail below:-

5.5.2 Insulation Resistance (IR) tests

The following measurements of IR should be made on the main structure of the transformer using a Megger for 60 seconds.

a) Core to tank with test link open.
b) Primary winding to secondary winding.
c) Primary winding to tank.
d) Secondary winding to tank.

Winding IR measurements should be similar to those made when the transformer was installed. Values of 50 per cent or less than this figure should be investigated and, if no cause can be found, should be referred to the manufacturer for his advice.

Very low values, in the region of 10 megohms are almost certainly indicative of excessive moisture content and, in conjunction with oil tests, should be referred to the manufacturer for a decision of whether to dry-out the transformer.

5.5.3 Transformer oil tests

Oil from both the top and bottom of the transformer's tank should be checked for odour, appearance and colour checks, acidity, dielectric strength and moisture content.

Oil sample testing is not applicable for hermetically sealed units not designed for oil sampling or for dry type transformers.

5.5.4 Tap-changer tests and checks

Tap-changer oil tests should be carried out as for the main transformer.

Off-load tap-changers should be operated several times to ensure freedom of movement.

On-load tap-changers should be checked for correct sequence, auto and manual initiation, alarms, limit switches, indication, local/remote control, interlocks, secondary wiring IR, parallel operation schemes, etc.

The contact resistance of each tap position should be measured with a Ductor and, where practicable, compared with the readings taken at commissioning. Any significant increase in resistance value requires investigation.

5.5.5 Buchholz relays

The Buchholz relay is designed partly to give warning of insulation deterioration within the transformer. When insulation is 'breaking down' electrically, it gives off various hydrocarbon and acetylene gases. These gases bubble up into the top section of the Buchholz relay and operate an 'alarm contact', giving warning that something is going wrong inside the transformer.

If an electrical fault occurs inside the transformer tank, the resulting explosion causes a blast of gas and oil up through the Buchholz relay, operating the 'trip, or surge contact' and a protection relay should make the transformer dead by tripping the HV and LV circuit breakers.

The operation of the alarm and trip contacts should be checked, preferably using a compressed air bottle to inject air via the petcocks. A slow injection is necessary to prove the alarm (gas accumulation) contacts and a rapid injection to prove the trip (oil surge) contacts. At the completion of the tests, care must be taken to ensure all air is removed from the Buchholz relay.

5.5.6 Oil and winding temperature indicators

Any cooling system should be run and the fan motor overloads checked. A winding temperature instrument at a pre-set temperature usually activates the cooling system. The winding temperature relay can also be fitted with a 'trip contact' set at a higher pre-set temperature.

The temperature sensing elements of oil and winding temperature indicators can be removed, where practicable, from their pockets and calibrated by immersion in hot water and comparison with a mercury/glass thermometer.

Readings should be made as the water cools since there will be a different speed of response between the thermometer and the sensing element.

The heater coil of the winding temperature device should be proved by injecting into the test winding a steady current of the value indicated in the manufacturer's instructions and comparing this reading on the instrument with the manufacturer's recommendations.

5.5.7 Breathers

Check the condition of silica gel breathers regularly. A strong blue colour indicates dry and effective crystals. As the moisture is absorbed the blue crystals turns white and a pink colour indicates saturated damp crystals. These should be replaced with dry [blue] silica gel crystals. Oven drying can recycle damp crystals.

5.5.8 Dry air-cooled transformers

HV distribution transformers that are of the dry air-cooled or cast in resin air-cooled type require some maintenance items. The external cleaning and checking of cable boxes, protection testing, etc. need doing. However, internal cleaning and checking the tightness of connections is of prime importance for these installed transformers.

In addition to the normal circuit protection relays these dry air-cooled transformers are often installed with cooling fans that are designed to disperse the heat given off from the transformer's windings and assist in improving its service rating. These fans normally come into service when the 'temperature instrument' fitted to transformer reaches a pre-set level. These fans and their 'initiating device' require regular maintenance.

5.6   CABLES

5.6.1 Cable testing

Not considered necessary as part of routine maintenance, but is required after replacing or repairing a cable.

Cables should normally be tested after installation and jointing. Power cables in 600/1000V and lower voltage circuits need be tested for insulation resistance only, but in 3.3kV and higher voltage, circuits should be tested with high voltage d.c. Before testing with HV d.c. the terminal equipment may need to be disconnected to ensure the cable test voltage would not damage the equipment. Transistorised solid state equipment **must** be disconnected.

HV tests should normally be made between each core and between each core and earth.

Immediately after testing, particularly with HV d.c. but also after Megger testing, the conductors should be earthed to discharge them. It can take up to at least 5 minutes to discharge a cable after HV testing, depending on its total capacitance and the test voltage used.

As a final check all power cables should be tested with a 500V Megger immediately before they are re-energised.

5.7 HIGH VOLTAGE OVERHEAD LINES

5.7.1 Visual inspections

HV overhead lines are installed in comparatively few OpCos (Operating Companies), but company electrical procedures will give a list of the checks that need to be made when carrying out maintenance.

Some OpCos have installed wood or steel pole overhead lines for distribution and transmission circuits, whereas others have single and double circuit steel tower overhead lines for transmission purposes.

A regular inspection of all the conductors, towers or poles, insulators and fittings is required along the whole route of the line. This check should be carried out, preferably every six months, unless unusual environmental conditions make more frequent inspections necessary.

In addition a complete inspection may be required before and after each rainy season (in tropical climates) and each winter season (in Europe).

Defective insulators and fittings must be reported immediately. Insulators must be cleaned when dust layers or salt crusts are visible to the naked eye, or if 'flashovers' resulting in circuit outages are experienced, due to pollution problems.

5.7.2 Earthing arrangements

Steel towers are directly earthed via earthing bar or tape. Wood poles are not normally earthed, but it is normal practice to earth pole mounted transformers and isolators, etc. Concrete poles may or may not be earthed depending on whether an overhead earth wire is installed.

a) For earthed towers or poles visual checks should include looking to see if: -
   - the earth tape connection to the tower or pole - is in good condition, free from corrosion and if bolted, tightened correctly.
   - earth tape has not have been damaged or the ground surrounding it disturbed
   - the overhead earth wire - must be securely connected to the tower or pole
   - all steelwork on wood pole structures should be bonded together, even if the steelwork is not connected to earth.

b) Earthing resistance measurement tests should be made on all overhead installations.

c) An earthing resistance test using an 'Earth Megger' should be used on the earth connection, at the same time ensuring the overhead earth wire is insulated from the tower or pole. The measured value should compare with the previous value and be 5 ohms or less.

d) This value may be exceeded in certain rocky terrains but the previous taken measurement [when installed] will indicate if the figure is reasonably acceptable. The condition of the ground, e.g. wet or dry should be recorded at the time of measurement.

CHAPTER 6 NOMOGRAM CORRECTION

NOMOGRAM FOR TEMPERATURE CORRECTION

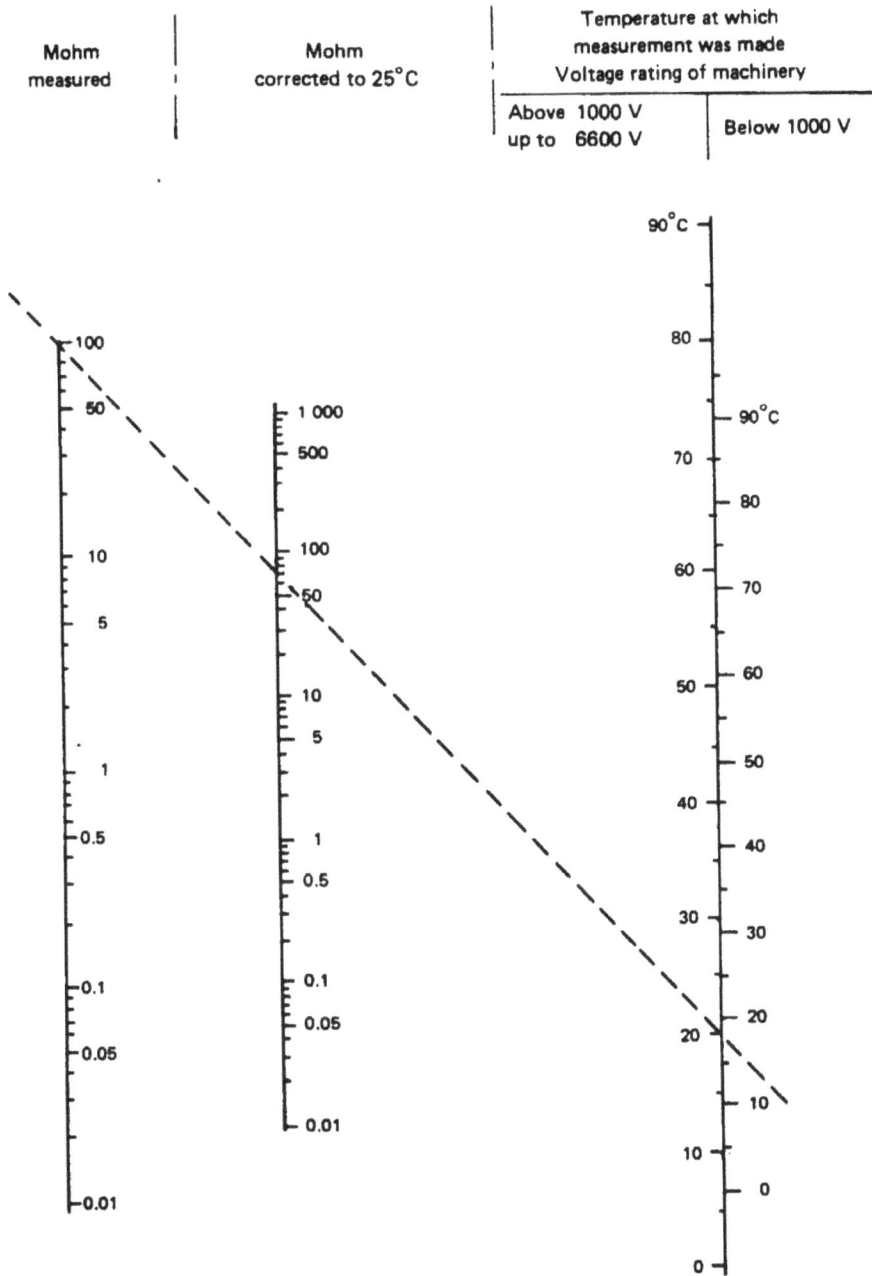

Example: Measured resistance: 100°C
Temperature at measurement: 20°C (1000-6600 V)
Corrected resistance: 70°C

CHAPTER 7 RECOMMENDED TEST VOLTAGES FOR COMMISSIONING AND MAINTENANCE

7.1 CABLES

Insulation resistance tests and high voltage tests should be carried out between each phase and earth with the remaining phases connected to earth and loads disconnected.

High voltage tests:
a) New cables
Duration: 3 minutes

Cable designation voltage kV (a.c.) UO /U (Um)	Test voltage kV (d.c.)
< 1.0	IR test only
1.8/3 (3.6)	10
3.6/6 (7.2)	15
6/10 (12)	25
8.7/15 (17.5)	35
12/20 (24)	50
18/30 (36)	70

NOTES: 1. Cables specifically manufactured for unearthed systems, i.e. $U_o = U$, should be tested at the value given above for U_o. E.g. cable type 3/3 (3.6) should be tested at 11 kV.
2. U_o is the rated power-frequency voltage between conductor and earth or metallic screen, for which the cable is designed.
U is the rated power frequency voltage between conductors, for which the cable is designed.
Um is the maximum value of the 'highest system voltage' for which the equipment may be used.

b) Repaired cables
It is widely accepted that the repeated application of DC test voltages to XLPE cables shortens the life of the insulation. Therefore, routine maintenance tests should not be carried out.

Repaired cables should be tested at reduced voltages, as follows:

Age of cable	Test voltage
0 to 2 y	75% of above
2 y to 10 y	50% of above
over 10 y	Soak test using AC system voltage only

Insulation resistance tests:

System voltage	Test voltage
LV 50 V - 1 kV	500 V (d.c.)
HV up to 4.6 kV	2500 V (d.c.)
HV above 4.6 kV	5000 V (d.c.)

7.2 MOTORS, GENERATORS, TRANSFORMERS (COIL WOUND EQUIPMENT).

Insulation resistance tests:

System voltage	Test voltage
LV below 1 kV	500 V (d.c.)
HV up to 4.6 kV	2500 V (d.c.)
HV above 4.6 kV	5000 V (d.c.)

7.3 SWITCHGEAR

High Voltage tests and insulation tests on busbar systems should be carried out between each phase and earth with the remaining phases connected to earth (voltage transformers and load disconnected).

High Voltage tests and insulation resistance tests on circuit breakers and contactors may be carried out together with or separately from the busbar and should be carried out with the breaker/contactor closed, with loads disconnected.

High Voltage test voltages:
Duration: 1 minute

max system voltage kV	3.6	7.2	12	17.5	24	36
test voltage kV (d.c.)	10	20	28	38	0	70

Insulation resistance test voltages:

- LV systems	:	500 V (d.c.)
- HV systems up to 4.6 kV	:	2500 V (d.c.)
- HV systems above 4.6 kV	:	5000 V (d.c.)

7.4 CAPACITORS

High Voltage tests on capacitor units should **exclude** the cable and be carried out between each phase to earth with the remaining phases connected to earth.
High Voltage tests: (commissioning only)
Duration 10 s, after stabilisation of charge current.

highest system voltage kV (a.c.)	1.0	3.6	7.2	12	17.5
test voltage kV (d.c.)	1.0*	10	20	28	38

7.5 RECOMMENDED INSULATION VALUES FOR EQUIPMENT

	Minimum maintenance value (NOTE 1)	Minimum for acceptance at commissioning
SWITCHGEAR Insulation resistance: HV bus LV bus LV wiring (NOTE 2)	100MΩ 10MΩ 0.5MΩ	200MΩ 20MΩ 5MΩ
CABLES Insulation resistance: HV and LV minimum length 100 m (NOTE 3)	$\dfrac{kV}{km}$ MΩ	$\dfrac{10 \times kV}{km}$ MΩ
MOTORS and GENERATORS Polarisation Index: (NOTES 6 and 8) LV and HV machines Class B and F Insulation resistance(at 25 °C) LV and HV machines: (NOTES 4, 5, and 7)	1.5 2(kV+1)MΩ	2.0 (6) 10(kV+1)MΩ
POWER TRANSFORMERS (max. 36 kV) OIL IMMERSED Insulation resistance : DRY TYPE Insulation resistance: HV side: LV side:	30MΩ 25MΩ 2MΩ	75MΩ 100MΩ 10MΩ
EQUIPMENT and COMPONENTS FIXED INSTALLATIONS Insulation resistance : MOVABLE EQUIPMENT. Hand Tools. Insulation resistance Class I: Class II: Class III: Distribution Equipment (cables, distribution boards, transformers) Insulation resistance :	1kΩ/volt 2MΩ 7MΩ 2MΩ 1MΩ	5kΩ/volt 2MΩ 7MΩ 2MΩ 5MΩ

NOTES:
1. These values are considered the lowest acceptable to allow energisation of existing equipment. Corrective action must be taken where lower values are found.
2. The above figures are to be used unless local regulations are more stringent.
3. IR to be measured with load disconnected.
   Example: Required value for maintenance (M) is kV rating/length in km.
4. Minimum insulation resistance values are given for 25°C equipment temperature; apply corrections for differing temperatures.
5. For machines < 10 MVA energisation is possible if IR or PI is above the minimum given.
6. PI values below those given can be accepted if IR is > 100(kV+1).
7. For insulation values during a machine's lifetime.
8. PI measurements on insulation class 'F' machines with IR in the Γ … range may be difficult to obtain due to meter scale compression.

7.6 GENERATOR COMMISSIONING

A detailed commissioning plan should be developed prior to commencing tests. Details of a test programme should be agreed between the Principal and the Contractor/Manufacturer. NOTE: A switching plan should form part of this procedure.

The following tests should be carried out in order to prove the satisfactory performance of the generator, its governor, automatic voltage regulator, and synchronising and protection systems.

Pre-commissioning checks (prior to running)
All electrical pre-commissioning work should have been satisfactorily completed, i.e. all possible tests should have been carried out, prior to the equipment being run and energised.

Protection relays should be tested, however some tests may only be possible with the generator running.

Pre-synchronising checks (generator running, no-load)
Open circuit tests
Verify the excitation system and generator characteristics by gradually increasing excitation and plotting generator output voltage against excitation current.

Generator phase rotation
Verify phase rotation of generator by means of an LV phase rotation meter connected to generator VTs, the VTs being fed either from the busbars (test 1) or the generator (test 2).

Synchronising circuit checks
During above test (phase rotation), verify that operation of the synchroscope, voltmeter, auto and check synchronising relays follows the 'beat frequency', i.e. the difference between generator and network frequency and voltage.

During above test (o/c tests), verify in-phase indication on all of above devices and verify synchronising relay operating parameters and close command.

As a last check, using high voltage 'phasing sticks' across the open generator breaker spouts, verify that the synchronising relay 'close' command coincides with the in-phase condition on the phasing stick.

Protection tests
In the absence of suitable 3-phase primary injection test sets, differential relay in-zone operation and out-of-zone stability can be verified by using the generator as a current source.
* Differential relay stability check:
  A 3 phase short circuit should be applied at a suitable point **outside** of the differential protection zone.
  The generator should subsequently be run up to speed and excited.
  The generator excitation should be adjusted from zero to a low figure (ref. manufacturer's test data) such that no more than full load current may flow.
* Differential relay sensitivity check:
  A 3 phase short circuit is applied at a suitable point **inside** the differential protection zone. Proceed further as for the stability check.
  Other items of protection and indication equipment may also be verified at this time if not already done by primary current injection, e.g. overcurrent, neutral displacement, negative sequence relays.

- Overspeed check:
  Operation of mechanical and electronic overspeed trip devices should be verified.

Post synchronising checks (generator running, loaded)
Synchronise to Grid
The network to which the generator will be initially synchronised should be configured such that the risk of a disturbance to normal plant operations is minimised.

Function tests
The following should be confirmed:
- Auto and Manual synchronising.
- Control of power factor over full range of power generation.
- Control of load.
- 'Bumpless' change-over from Auto AVR to Manual and vice versa.
- The change over of AVR from power factor to voltage control
  (when changing from coupled to 'island operation')

Protection tests
The operation of the Reverse Power relay should be verified by decreasing the governor setting.

The operation of the Field Failure relay should be verified by decreasing the excitation at minimum load.

Dynamic tests
Dynamic tests should comprise:
- Active load rejection, reactive load acceptance, active load sharing and reactive load sharing tests as well as 'island' proving tests if applicable.

The following parameters should be measured on a suitable recorder:
> Generator voltage/time,
> Current/time
> Speed (or frequency)/time.

The test acceptance criteria should be:
> no tripping of any protection device,
> no parameter should exceed 80% of the difference between nominal value and trip set points,
> transient response should be within the design parameters.

- Governor Tests
  a. Load acceptance and rejection: transient response should be measured by the switching IN and OUT of active load blocks.
  b. Droop settings in "island condition": speed change between zero and full load should be measured.
  c. Active load sharing: it should be verified that the load is shared equally (or for different ratings, pro rata) between the generator being commissioned and all other combinations of generators. It should be verified that the load continues to be shared during changes in total load.
  d. It should be verified that electrical load variations are within agreed limits during the change-over of fuels.

- AVR Tests
  a. Transient response: the recovery of generator voltage should be measured, during the reactive load acceptance and the load rejection tests to verify that it is within the limits of time and terminal voltage.
  b. Voltage droop*: it should be verified that voltage droop is proportional to reactive load or is according to design.

    c.   Reactive load sharing: the ability of the generator being commissioned to run at the same power factor as parallel connected generators should be verified. It should also be verified that the power factor remains equal during changes in reactive load.

    d.   Current boost*: (for AVRs whose supply is derived from own terminal voltage), it should be verified from design data that generator terminal voltage is sufficiently maintained to allow operation of protective devices in the event of a short circuit.

NOTES:
1.   For items marked *, factory test results are acceptable.
2.   'block load' values should be such as to demonstrate the equipment's compliance with its design.

Load trials

Load trials should be carried out as agreed between Contractor and Principal and will normally follow satisfactory completion of all the above tests and the completion of construction and testing of the various systems and sub-systems peripheral to the generator and prime mover.

These final tests should be a series of load tests culminating in a protracted run under design conditions at designed maximum power output

These final tests will demonstrate that the generator, prime mover and all ancillary equipment meet the designed performance levels while running continuously at full rated load.

Proof of the unit's performance during tests should be in the form of a running log.

The log sheets should be formally drawn up and the parameters to be recorded agreed between Principal and Contractor/Vendor.

REFERENCES

Advanced Electrical Technology
H. Cotton, DSc (Pitman)

Alternating Current Electrical Engineering (10th Edition)
Philip Kemp, MSc, CEng, MIEE (Macmillan)

An Introduction to Power Electronics
B. M. Bird & K. G. King (John Wiley)

Applied Electricity (6th Edition)
H. Cotton, DSc (Macmillan)

BS 2769
'Portable Electric Motor-operated Tools'
British Standards Institution

BS 3535
'Safety Isolating Transformers'
British Standards Institution

BS 7430
'Earthing'
British Standards Institution

BS 5345:

- Selection, Installation and Maintenance of Electrical Apparatus for use in Potentially Explosive Atmospheres

- Installation ... of Type of Protection '*d*' (Flameproof Enclosures)

- Installation.., of Type of Protection '*i*' (Intrinsically Safe Apparatus and System)

- Installation ... of Type of Protection '*e*' (Increased Safety)

- Installation ... of Type of Protection '*n*' (N-Protection)

- Installation ... of Type of Protection '*s*' (Special)

Cathodic Protection
J.H. Morgan (Leonard Hill)

Electrical Control Engineering, Vols 1 & 2
Poole & Jackson (London Iliffe Books)

Electrical Engineers' Reference Book (13th Edition)
M. G. Say (Butterworth)

Electrical Machines (2nd Edition)
A. Draper, BSc Eng, C Eng, FIFE (Longman)

Electrical Measurements and Measuring Instruments
E. W. Golding and F. C. Widdis (Pitman)

Electrical Technology (5th Edition)
E. Hughes, DSc, PhD, C Eng, MIEE (Longman)

Electric Motor Handbook
E.H. Werninck (McGraw Hill)

Electric Power Systems
B.M. Weedy, PhD (John Wiley)

Electronic Devices and Circuits (2nd Edition)
D.A. Bell (Reston Publishing Co mc)

Electronics: Circuits and Devices
R.J. Smith (John Wiley)

Handbuch fur Explosionsschutz (2nd Edition, 1983)
(Explosion Protection Manual)
Brown Boveri & Cie A/G

Physics and Technology of Semiconductor Devices
Grove (John Wiley)

Principles of Electrical Technology
H. Cotton, DSc (Pitman)

Principles of Inverter Circuits
Bedford & Hoft (John Wiley)

Protective Relays Application Guide
General Electric Co Measurements (GEC)

Standard Handbook for Electrical Engineers (11th Edition)
Fink & Beaty (McGraw Hill)

Storage Batteries (3rd Edition)
G. Smith, C Eng, MIEE (Pitman)

The Power Thyristor and its Applications
Finney (McGraw Hill)

The Lighting of Buildings (2nd Edition)
R.G. Hopkinson, PhD, C Eng & J.D. Kay, AA Dipl, RIBA (Faber & Faber)

Thermocouple and Resistance Thermometry Data
T.C. Limited

Thyristor Physics
Blicher (Springer)

INDEX